Adaptations of Desert Organisms

Edited by J. L. Cloudsley-Thompson

Springer
Berlin
Heidelberg
New York
Barcelona
Budapest
Hong Kong
London
Milan
Paris
Santa Clara
Singapore
Tokyo

Volumes puplished in this series

Ecophysiology of the Camelidae and Desert Ruminants
By R.T. Wilson (1989)

Ecophysiology of Desert Arthropods and Reptils
By J.L. Cloudsley-Thompson (1991)

Plant Nutrients in Desert Environments
By A. Day and K. Ludeke (1993)

Seed Germination in Desert Plants
By Y. Gutterman (1993)

Behavioural Adaptions of Desert Animals
By G. Costa (1995)

Invertebrates in Hot and Cold Arid Environments
By L. Sømme (1995)

Energetics of Desert Invertebrates
By H. Heatwole (1996)

Ecophysiology of Desert Birds
By G.L. Maclean (1996)

Plants of Desert Dunes
By A. Danin (1996)

Biotic Interactions in Arid Lands
By J.L. Cloudsley-Thompson (1996)

Structure-Function Relations of Warm Desert Plants
By A.C. Gibson (1996)

Physiological Ecology of North American Deserts Plants
By S.D. Smith, R.K. Monson, and J.E. Anderson (1997)

Ecophysiology of Small Desert Mammals
By A.A. Degen (1997)

Homeostasis in Desert Reptiles
By S.D. Bradshaw (1997)

Ecophysiology of Amphibians Inhabiting Xeric Environments
By M.R. Warburg (1997)

Avian Desert Predators
By W.E. Cook (1997)

Carnivore Ecology in Arid Lands
By J. du P. Bothma (9918)

Ecophysiology of Economic Plants in Arid and Semi-Arid Lands
By G.E. Wickens (1998)

In preparation
Arthropods of Mediterranean-Type Ecosystems
By G.P. Stamou (1998)

Gerald E. Wickens

Ecophysiology of Economic Plants in Arid and Semi-Arid Lands

With 38 Figures and 12 Tables

Springer

Dr. Gerald E. Wickens
The Triangle
Buxton Road
Aylsham
Norfolk, NR 11 6JD
United Kingdom

Front cover illustration: Cactus E*chinopsis atacamensis.*
Photograph taken by G.E. Wickens

ISBN 978-3-642-08089-0
ISSN 1430-9432

Library of Congress Cataloging-in-Publication Data.
Wickens, G.E. Ecophysiology of economic plants in arid and semi-arid lands / Gerald E. Wickens. p. cm. – (Adaptations of desert organisms). Includes bibliographical references and index.

1. Arid regions plants–Ecophysiology. 2. Plants, Useful–Ecophysiology. 3. Arid regions plants–Adaptation. 4. Plants, Useful–Adaptation. I. Title. II. Series
SB317.A69W535 1998 581.6'3'06154-dc21 97-42792

Printed in Germany

Cover design: Design & Production GmbH, Heidelberg

Camera ready by RTS, Wiesenbach:

Preface

Economic plants range from those that are commercially utilised by the industrial nations of the world to those that safeguard the survival of hunter-gatherer aboriginal communities. In a world of increasing human and livestock populations, land and environment degradation, species loss and climatic change, it is becoming increasingly important to work for the sustainable management of the world's natural resources. This is especially important for the arid and semi-arid regions where any abuse of the harsh environment is unforgiving. Plant resources are especially important since they protect the environment, provide habitats for both flora and fauna, and directly or indirectly provide the resources for man's survival.

An understanding of how plants are adapted to survive is an essential tool for the better management of the environment. Since there are over 20000 useful plants in the arid and semi-arid regions of the world, of which only a relative few have been even partially investigated, the writer makes no apologies for which plants have been included or ignored. All the plants mentioned in the introductory chapters marked with an asterisk are discussed more fully in Chapter 11.

This book is written for the generalist as a broad, multidisciplined survey of some of the many factors that interact between the environment, ecophysiology and economic plants of the drier regions of the world. Thus, it is hoped that it will stimulate readers in having an interest in the diversity of adaptations by which plants are able to survive and, hopefully, encourage such plants to be investigated more fully. Other books in this series, Gutterman (1993), Danin (1996), Gibson (1996) and Smith et al. (1997), should also be consulted since they provide more detailed information regarding some of the plants discussed in this present volume.

Aylsham, Norfolk, UK
April 1998

G. E. Wickens

Acknowledgements

I am grateful for the library facilities offered by the Royal Botanic Gardens, Kew, the British Museum (Natural History) and the Linnean Society of London, without which the writing of this book would not have been possible. The artistic skills of Sally Dawson in producing the maps and of my friend and colleague Dr. Roger Polhill in providing the line drawings are very much appreciated. Ing. Agr. J.C. Guevara, Professor Y. Gutterman and Dr. H.N. Le Houérou have earned my gratitude by their help with some of the literature references, I am also grateful for the forbearance with which the editor has tolerated the long delay in writing this book.

Contents

CHAPTER 1

Economic Plants, and Arid and Semi-Arid Lands of the World

1.1 Economic Botany

Economic botany is the study of plants that are directly or indirectly of benefit of man, his livestock and the maintenance of the environment. The latter includes species used for soil stabilisation, shade, etc. However, in order to avoid embracing the entire plant kingdom, some form of conscious or subconscious management must be involved. The benefits may be domestic, commercial, environmental or purely aesthetic; their usefulness may belong to the past, the present or the future. In this context, 'economic' is used in the sense of utilitarian rather than for monetary gain and includes plants that are commercially cultivated and marketed (agricultural, horticultural and forestry crops) as well as wild plants utilised in the domestic economy.

The decision as to what constitutes an economic plant is purely relative and is determined largely by availability, suitability and economic status of the plant and its products. What may be essential and highly valued in the life style of aboriginal hunter-gatherers might be considered worthless by civilised communities (and *vice-versa*). Nevertheless, it must be remembered that many of our food plants and pharmaceutical products have been and are still being developed from such 'primitive' sources.

It is not only the higher plants (angiosperms and gymnosperms) that interest the economic botanist; the lower plants (bacteria, algae, fungi, lichens, bryophytes and pteridophytes) are also included. The lower plants are probably best known for their contribution to antibiotics, alginates, nitrogen fixation and plant pathology. Economic botany is a multidisciplinary study that involves not only the purely botanical disciplines of taxonomy, ecology, physiology, cytology, biochemistry and pathology, but also, to some extent, those aspects of agriculture, horticulture and forestry concerned with breeding, propagation, cultivation, harvesting, manufacture and the economics of production and marketing (Wickens 1990). Other disciplines where plant life impinges on man's survival and well-being include archaeology and palaeoarcheology, anthropology, sociology, economic history and economic geography.

Taxonomy is the foundation stone on which economic botany is based. Accurate identification and correct naming are absolutely vital. The use of a toxic plant for food or medicine in mistaken identity for a very similar non-toxic species could have obvious drastic consequences. Economic botany charts the progress of the use of a plant in the wild, from its casual acceptance to its eventual role in the domestic, national and international economies. For the majority of species this implies domestication. However, there are exceptions, for example, *Bertholletia excelsa* (Brazil nut), *Crithmum maritimum* (sea samphire), algae such as *Laminaria* spp. (kelp) and rangeland fodder plants are harvested from the wild, although some form of sustainable management is implied for their sustainable production. Some ancient cultivated food plants are known to be of hybrid origin, such as *Zea mays* (maize), *Phoenix dactylifera** (date palm), *Psophorocarpus tetragonolobus* (winged bean) and some *Dioscorea* spp. (yams), and are unknown in the wild; it is only since the development of genetics that plant breeders have been able to discover their progenitors. Of course, not all plants make the grade to the international markets; *Aspalanthus linearis* (rooibos tea) and *Simmondsia chinensis** (jojoba) are among the rare examples of the development during this century from a wild species to a cultivar and the international health markets, while for the next century the legumes *Cordeauxia edulis** (yeheb) and *Lemuropisum edule** (tara nut) are potential candidates for the dessert nut market. Also, a number of plants have interesting biochemical constituents, such as *Parthenium argentatum** (guayule rubber), while others, such as the composites *Grindelia camporum** (biocrude and resin) and *Vernonia galamensis* (seed oil), could well make the grade early in the next century.

1.1.1 Ethnobotany

The term 'ethnobotany' was first used in an address to the University Archaeological Association by Dr. John Harshberger of the University of Pennsylvania to describe the study "of plants used by primitive and aboriginal people" (Anonymous 1895). This often cited newspaper reference has, in a number of publications, been wrongly attributed to Harshberger and not to the anonymous reporter (Wickens 1990). The first paragraph of this important and not easily available reference is reproduced below:

"Some new and interesting ideas were presented by Dr. John Harshberger in his last lecture on "Ethno-Botany" before the University Archaeological Association at the University. Ethno-botany deals with the plants used by primitive and aboriginal people, such as the Indians of America, and has for its object the study of the vegetable products employed for various purposes, as fibre and food. The main difficulty presented is that of the identification of the materials.

Ethno-botany is of service in elucidating the culture position of primitive people. We can divide the races of mankind into pastoral, agricultural, semi-civilised, and civilised. The American Indian never occupied the pastoral stage, never had herds of domesticated animals, but emerged directly from the hunting into the agricultural stage." (Anonymous 1895).

The full text of the address is given in Harshberger (1896a,b), where he defined the purposes of ethnobotany as:

1. An aid to elucidating the cultural position of 'tribes' according to their use of plants and plant products
2. Evidence of the past distribution of useful plants
3. Evidence of the ancient trade routes followed in the exchange of plant products.
4. Of use in suggesting that "the valuable properties they have utilised in their wild life may fill some vacant niche in our own."

The term 'ethnobotany' was immediately accepted and widely used, replacing such terms as 'aboriginal botany' and 'ethnographie botanique'. Its use has now been extended, beyond the confines of aboriginal and primitive people, to include the use of plants by both tribal and atribal communities without any implication of primitive or developed societies (Wickens 1990). Heiser (1985) follows Ford (1978) in defining ethnobotany as the study of plants in relation to people, while economic botany concerns plants and modern human society. He further speculates that since ethnobotany embraces both fields, economic botany should perhaps be considered as a subdivision of ethnobotany with monetary implications! Cotton (1996) has provided what the author would consider an acceptable definition with "... ethnobotany is considered to encompass all studies which concern the mutual relationships between plants and *traditional peoples*".

Ethnobotany, economic botany and the allied disciplines of agriculture, forestry and horticulture should be considered as a continuum, with no clear-cut distinctions between them. The involvement of the economic botanist per se from the first recognition of a plant's ethnobotanical utilisation to the development of a commercial crop will naturally decrease as the agronomist, for example, becomes increasingly involved with the technicalities of crop husbandry. Nevertheless, the economic botanist will need to retain a continuing albeit decreasing interest in the principles of husbandry through to production and marketing, since it is he or she who has the basic background knowledge of the plant and may therefore be able to influence every stage along the production chain.

CHAPTER 2

Arid and Semi-arid Environments of the World

2.1 Environment

More than one third of the earth's land surface (ca. 49 million km^2) is arid or semi-arid, receiving less than 400 mm annual precipitation, supporting more than 1.2 billion people, or 20% of the world's population, a population that is now expanding at a far greater rate than the environment can support. The pressure on the land is such that 50 000-70 000 km^2 of useful productive land is being lost each year through desertification. Furthermore, an estimated 3.8 million km^2 of the arid and semi-arid lands is also saline; in the Indian subcontinent alone some 400 km^2 are lost annually to agriculture due to increasing salinity, while in the USA, with its far more sophisticated agriculture, the area lost is 800-1200 km^2 (Bell 1985; Wickens 1992). The world's arid lands, the true deserts and the desert margins, are not only hostile to life, but also endangered.

The concept as to what constitutes a desert varies widely, from expanses of endless sand dunes and distant, palm-fringed oases to stony, cacti-littered scrub; at either end of the scale a hot dry climate and a non-montane topography are implied; the cold and often mountainous arid lands of, for example, central Asia and South America tend to be forgotten. These ideas are probably best summarised by the North American ecologist, Dr. Forest Shreve (1925), and cited by McGinnies (1981).

"*It is impossible to define desert in terms of a single characteristic, just as truly as it is impossible to differentiate species by such a procedure. A low and irregularly distributed rainfall is a fundamental feature but is not to be taken as the sole criterion of the environmental conditions. With low rainfall are associated low moisture content of the soil and low relative atmospheric humidity, often fluctuating only by reason of the daily changes of temperature. The diurnal temperatures are high, but the daily range is great, owing to the active nocturnal radiation from a bare and dry soil. The air movements are greater than in a humid climate, being manifested both by wind and small cyclonic disturbances, or tornillos. The evaporating power of the air is great by reason of the combined effects of low humidity, high temperature and great air movement... Low precipitation and torrential rains are responsible for poorly developed drainage, the occurrence of dry lake beds, the existence of steep outwash slopes, stretches of sand constantly shifted by the wind, soil highly impregnated with salts, surface mulches of small stones covering a fine soil, and many other minor features which are of great importance in determining the local distribution of plants and animals."*

2.1.1 Delimitation of Arid and Semi-arid Environments and Ecosystems

The object of environmental mapping is to delineate climatic differences and relate any difference to the natural vegetation and agronomy of a region. This is in order not only to obtain a better understanding of the ecophysiology of the plants, but also, with certain reservations regarding humidity, vectors for pollination, day length, etc., evaluate the possibility of successfully introducing plants to other ecosystems.

There are a number of ecological and climatic classifications regarding what constitutes an arid or semi-arid environment. Attempts have been made by Köppen (1936), Thornthwaite (1948), Meigs (1953), Gaussen (1954), Emberger (1955), de Martonne (1962), UNESCO (1979) and many others to define arid and semi-arid climates by means of aridity indices involving formulae based on precipitation in relation to temperature and/or evapotranspiration.

In practice, the high variability from the long-term calculated indices, especially in arid regions and to a lesser extent in semi-arid regions, is mainly due to the erratic annual rainfall. Such variability in rainfall increases with aridity and is thus inversely related to the long-term mean rainfall. Furthermore, aridity indices give an incomplete picture of the actual conditions, since the redistribution of rainwater within the ecosystem due to variations in landform, soils and vegetation creates local variations in edaphic aridity – not to be confused with climatic aridity. Neither do climatic indices per se give any information about seasonal variations, such as the distribution of rainfall, maximum and minimum temperatures and length of the dry season, all factors which affect plant growth (Walter 1963a; Wickens 1992).

The relative humidity (RH) and saturation deficit (SD) are likewise often not considered. In the Sahel the dry season afternoon air humidity from November to January and from February to May falls to below 20 and 15%, respectively. The zonal annual RH ranges from 30 to 51% with a mean of 38%, while the mean SD is 24 hPa. Based solely on mean annual rainfall *Opuntia* (cacti), *Atriplex* spp. (salt bushes) and species of *Acacia* subgenus *Heterophyllum* (wattles) can grow and are grown in Africa. However, because of the low RH and high SD they cannot be grown in the Sahel apart from the area

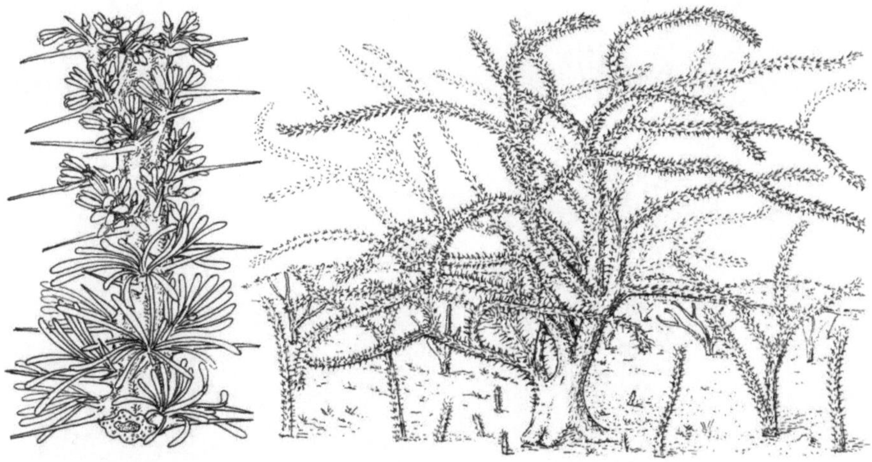

Fig. 1. *Didierea madagascariensis* (syn. *D. mirabilis*) from south-west Malagasy. *Left* Part of a flowering branch showing the spiny short and rosettes of soon deciduous leaves; *right* habit showing the spinose branches. (Adapted from Baillon 1894, Plates 261, 262)

bordering the Atlantic, where the RH is above 66% and the SD below 12 hPa, figures which approximate to those found in other arid and semi-arid parts of Africa where these introduced plants are growing (Le Houérou et al. 1993).

An ecosystem is defined as a community of interrelated living organisms (including producers, consumers and decomposers) and their environment; it may be large or small, simple or complex in structure. An ecosystem is determined primarily by the two abiotic factors climate and soil, and secondarily by the activities of the two biotic factors flora and fauna in nutrient cycling and energy flow. The organic compounds formed by the plants (the producers) during photosynthesis either directly or indirectly provide food for the animals (the consumers). The plant and animal remains are ultimately broken down by micro-organisms and small invertebrates (the decomposers), and the mineralised nutrients thus released are again taken up by the plants, thus completing the cycle. There is also a parallel energy flow where the radiant energy from the sunlight used in the photosynthetic processes is converted into chemical energy, which in turn is utilised by both the consumers and decomposers.

The movements of carnivores, ungulates and migratory birds from one ecosystem to another mean that faunal distributions are the least suited to defining the boundaries of an ecosystem. Plants are static and consequently the differentiation of boundaries between the various ecosystems can be recognised, so plant formations present fewer problems than animals. However, the life-forms represented within a particular formation are not neces-

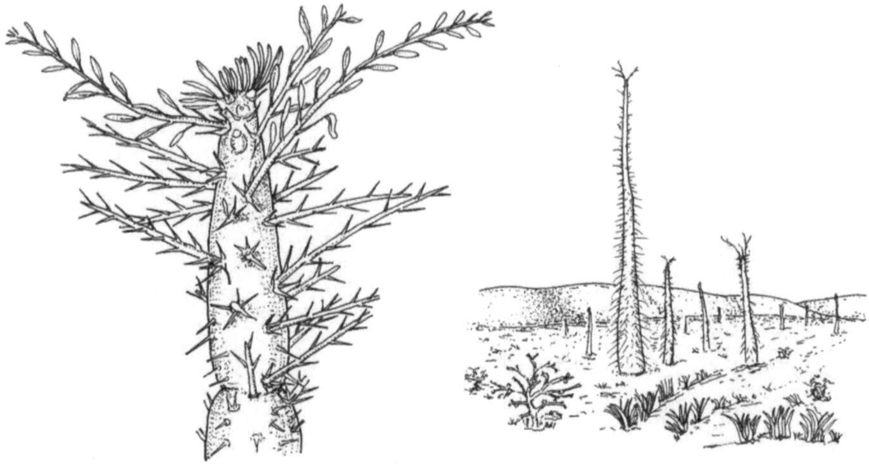

Fig. 2. *Fouquiera columnaris* (boojam). *Left* Apex of a mature plant showing spinose branches bearing primary leaves in the upper part and spinose petioles in the lower part; *right* trees growing at Calmalli, Baja California showing the tapering, columniform trunk. (*Left* Adapted from Humphrey (1935, Fig. 4), *right* from photograph in Goldman 1916, Plate 122)

sarily based primarily on taxo-ecological considerations, as shown by the convergent evolution between the taxonomically unrelated Old World Euphorbiaceae and New World Cactaceae and the even more bizarre serpentine columns of members of the CAM Didiereaceae (Fig. 1) from Madagascar and the C_3 Fouquieriaceae (Fig. 2) from south-western North America, both with wand-like branches and ephemeral leaves, the former having a greater development of water-storing tissues in the pith and cortex (Gibson 1996).

Such considerations led Walter and Breckle (1985) to consider the concept of plant formations as being artificial and oversimplified, and consequently of only secondary importance in the defining of an ecosystem. Furthermore, they consider the use of soil types as even less satisfactory because plant profiles cannot be seen, the random nature of the sampling and because soil analyses are based on agronomic rather than ecological considerations. Also, soils may be transported or sedentary, influenced by the underlying geology and weathering processes of past climates. Moreover, the soil and vegetation form an inseparable whole, where the soil supplies the water, nutrients and matrix to support the plant as well as a rich soil fauna and flora.

The climate alone is free from historical influences, for it is determined by global circulation, which in turn is controlled by the changes in the distribution of the continents and oceans. Although there have been past climatic changes, there can be no fossil climates. Climate is the sole primary factor that influences the soil, vegetation and, to a lesser extent, the fauna; it is affected by the soils and vegetation only at the microclimatic level by desertification and deforestation. The macroclimate is consequently particularly suitable for the subdivision of the geobiosphere and the delimitation of ecosystems.

By the use of ecological climatic diagrams, such as those of Walter & Lieth (1960–1967), it is possible to obtain a graphic representation of the climate as a whole, showing its seasonal course. They show:

1. The annual pattern of temperature and rainfall.
2. The wet and dry seasons characteristic of any area, as well as their intensity since the rate of evaporation is almost directly related to temperature.
3. The occurrence or otherwise of a cold season; also the months in which early and late frosts are recorded.

A 'normal climate', based on mean values does not exist since every year is different. For plants, the climatic extremes are often more important than the average conditions, and their frequency is especially important for agriculture and forestry. Thus, by constructing the climatic diagrams using values for single years (over a period of at least 20 years) instead of mean values, it is possible to show how often such extremes occur (Walter 1963).

Using such climatic diagrams, Walter and Breckle (1985) recognise nine major low-altitude climatic zones and nine major ecological zones of the world. Their extremely close correspondence led to the recognition of nine zonobiomes. These are:

ZB I	Equatorial zonobiome, with a diurnal climate (perhumid)
ZB II	Tropical zonobiome with summer rainfall (humid-arid)
ZB III	Subtropical arid biome (desert climate)
ZB IV	Zonobiome with winter rain and summer drought (Mediterranean, arid-humid)
ZB V	Warm-temperate (oceanic) zonobiome
ZB VI	Typical temperate zonobiome with short frost period (nemoral)
ZB VII	Arid-temperate zonobiome with cold winters (continental)
ZB VIII	Cold-temperate zonobiome with cool summers (boreal)
ZB IX	Arctic and Antarctic zonobiome

Intermediate zones between zonobiomes with a smooth transition between the two are termed 'zonoecotones' and given the Roman numerals pertaining to the neighbouring zonobiomes, e.g. ZE III-IV, thereby avoiding the drawing of artificially sharp boundaries. Zonobiomes III and VII are clearly relevant to this present discussion, as well as the drier parts of IV. Also recognised as a special case are the climates of the higher mountains. These seldom corre-

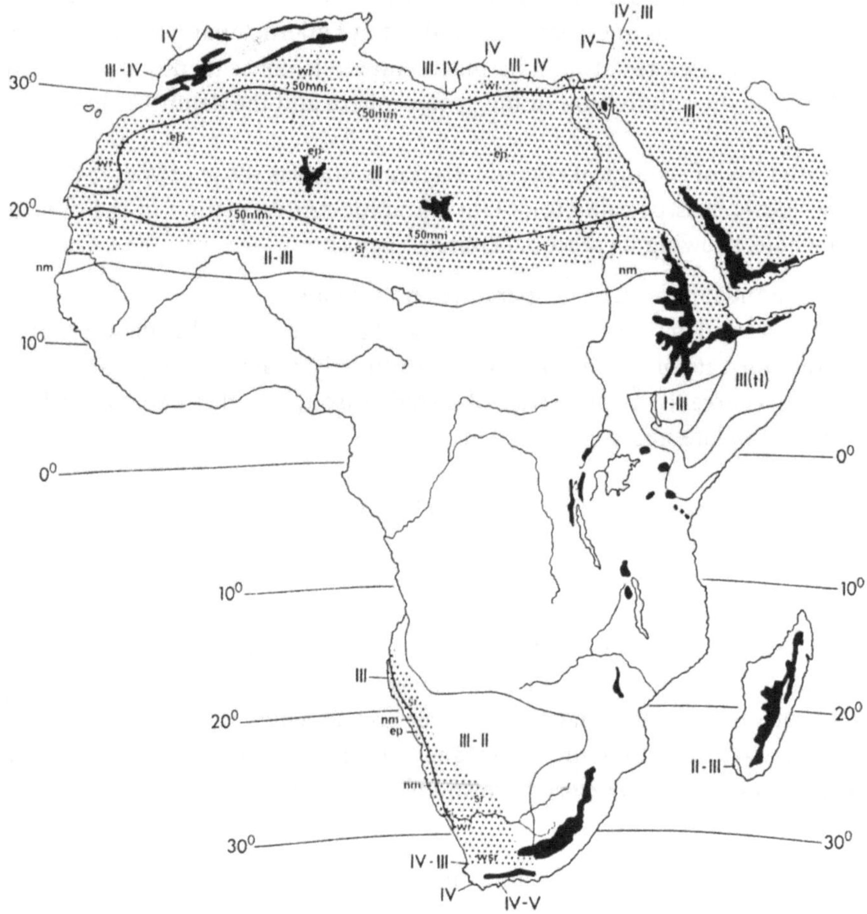

Fig. 3. Arid and semi-arid zonobiomes and zonoecotones of Africa. See text for identification of Roman numerals, etc. (Walter and Breckle 1985)

spond to any one of the zonal climatic types, although the mountain climates within any particular climatic zone have similar characteristics. The arid and semi-arid zonobiomes and zonoecotones recognised by Walter and Breckle (1985) are shown in Figs. 3-8.

While the above climatic divisions of Walter and Breckle (1985) provide a framework for the arid and semi-arid ecosystems under consideration, it should be borne in mind that the climate is no longer static. There is a major climatic change taking place (Parry 1990; Le Houérou 1996), with a marked decrease in rainfall in the dry tropics. For example, following the prolonged drought of the late 1960s, the isohyets for the Sudan have moved some 100 km

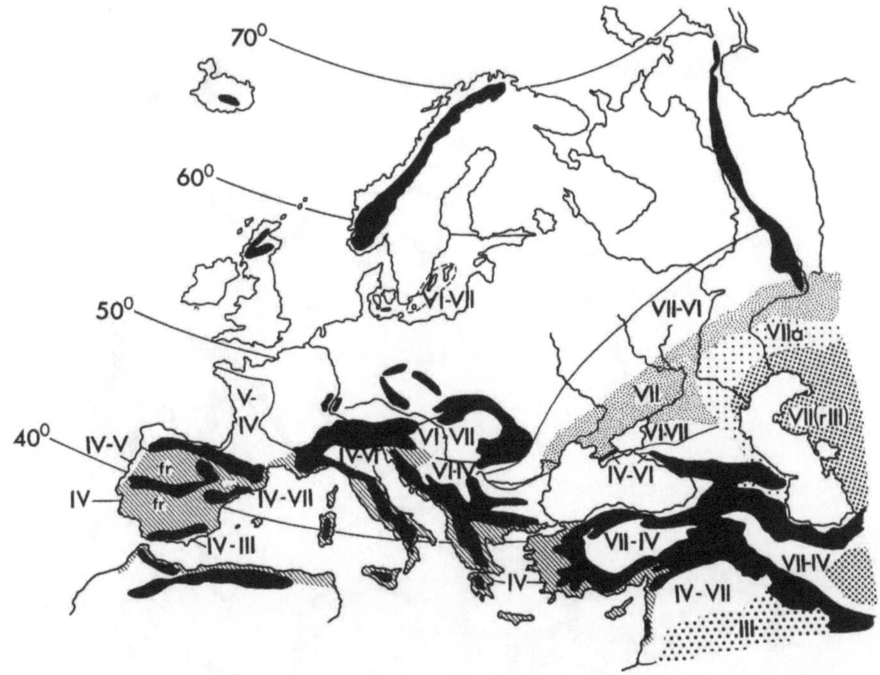

Fig. 4. Arid and semi-arid zonobiomes and zonoecotones of Europe. See text for identification of Roman numerals, etc. (Walter and Breckle 1985)

to the south of their 1967 position (Wickens 1982a). However, it may well be that it will not be until the middle of the twentieth century before any significant global change in the ecosystem boundaries will become apparent.

An alternative and widely used delimitation of what constitutes an arid or semi-arid environment is based on the classification developed by the United Nations Economic, Scientific and Cultural Organisation (UNESCO 1979) from a world map by Meigs (1953) using Thornthwaite's index. This UNESCO classification is based partly on aridity indices and partly on a consideration of all the available data on soil, topography and vegetation. This too will be subject to future reassessment as a consequence of climatic change.

Although deserts by definition have an arid or semi-arid climate, not all areas with such climates are necessarily deserts sensu stricto; the high altitude punas of the Andes and mountain steppes of central Asia and China also have an arid or semi-arid climate. The degree of bioclimatic aridity is dependent upon the relative amounts of water gained from precipitation and lost through evaporation and transpiration. The degree of aridity clearly rises as precipitation decreases and evaporation increases. The aridity index is based

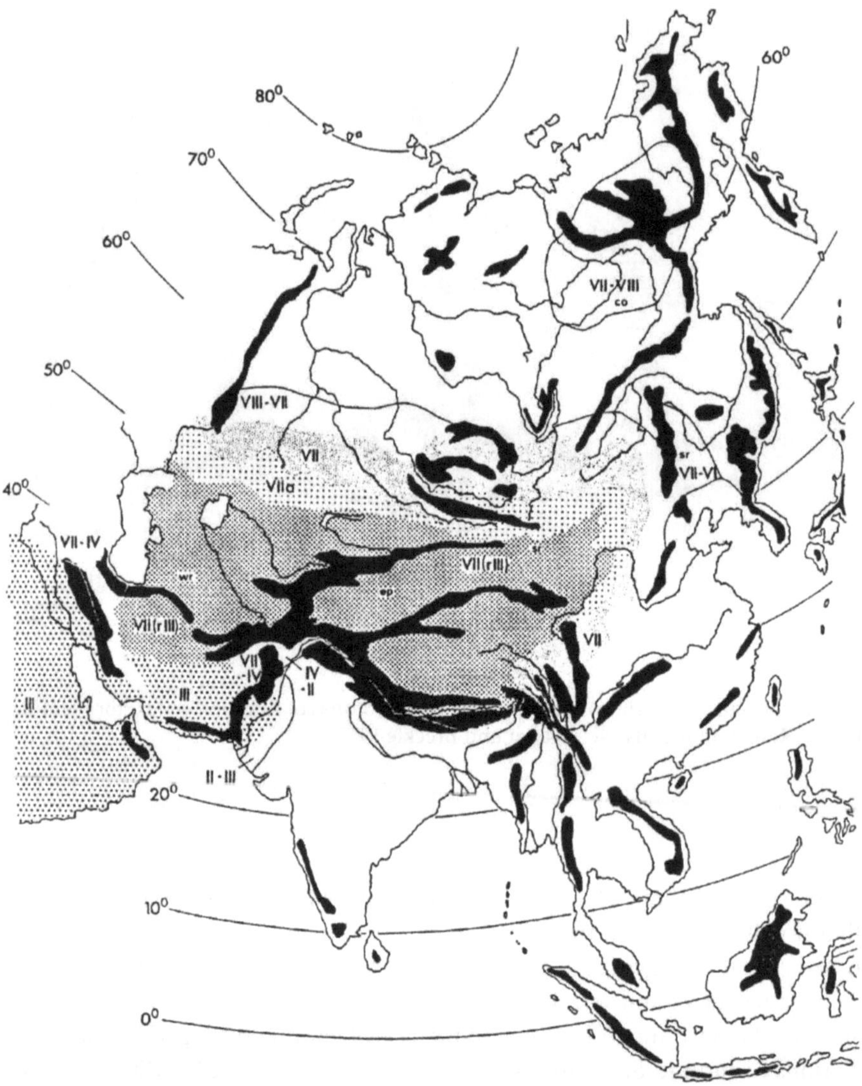

Fig. 5. Arid and semi-arid zonobiomes and zonoecotones of Asia. See text for identification of Roman numerals, etc. (Walter and Breckle 1985)

on the formula P/ETP, where P is the mean annual precipitation and ETP is the mean annual evapotranspiration. The ETP is based on the well-tried Penman formula, which takes into account solar radiation, atmospheric humidity and wind, the latter being very important in the arid and semi-arid regions on account of its drying power on the atmosphere.

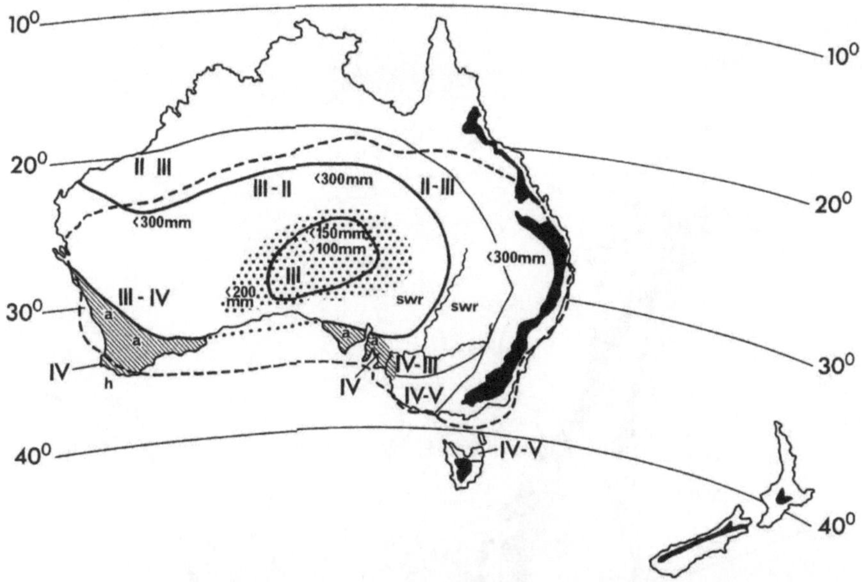

Fig. 6. Arid and semi-arid zonobiomes and zonoecotones of Australia. See text for identification of Roman numerals, etc. (Walter and Breckle 1985)

Four main classes of aridity (hyperarid, arid, semi-arid and subhumid) are recognised by UNESCO (1979), corresponding to the major geographical categories generally used by climatologists and biologists; for the purposes of this present study the latter class will be largely ignored. UNESCO have also found it necessary to introduce four subdivisions based on temperature criteria (warm, mild, cool and cold winters), since temperature and its annual variation has, together with precipitation, an important influence on plant production. This is reflected in the subdivisions based in the first place on the mean temperature, in degrees Celsius, of the coldest month of the year. These temperature classes are in turn subdivided according to the mean temperature of the hottest month of the year, the limiting values being 10, 20 and 30 °C.

There are obvious difficulties in trying to compare different climatic systems at different scales of mapping and projections [Walter and Breckle (1985) at 1:55 million and UNESCO (1979) at 1:25 million]. Not surprisingly there are varying degrees of agreement and disagreement between the two systems. Of the two, the former is the simpler to use and bears a closer relationship with the vegetation. It is used in Chapter 11 when describing the distribution of various economic species.

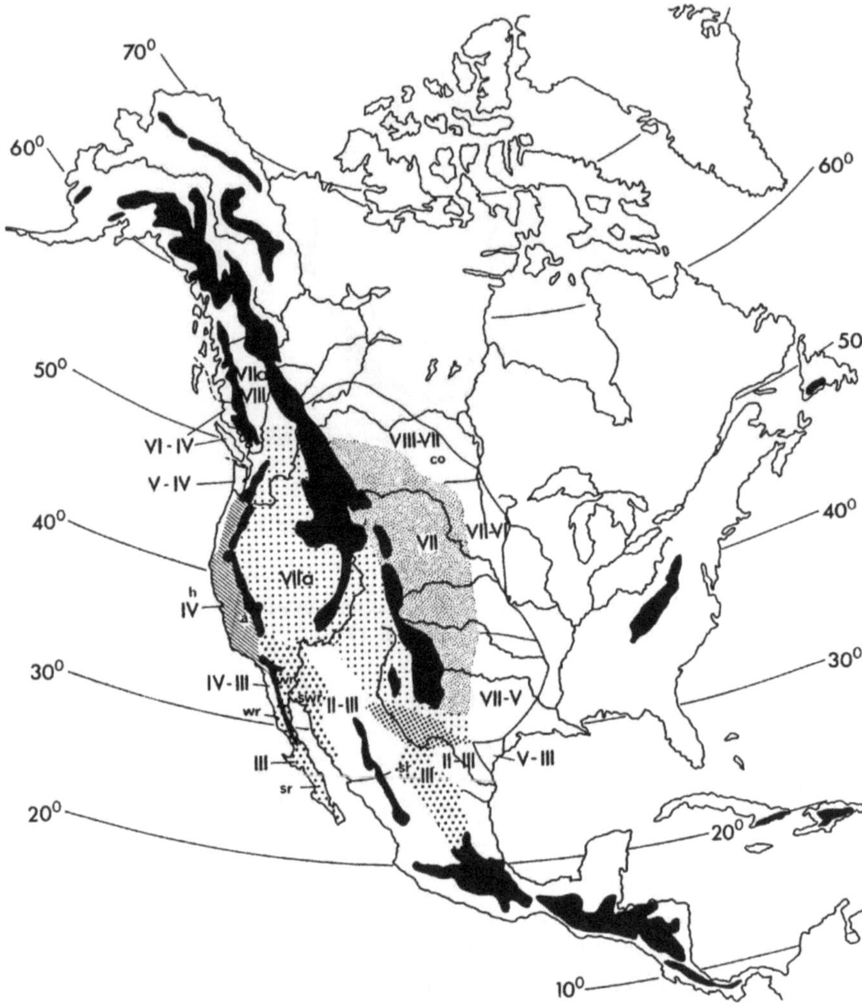

Fig. 7. Arid and semi-arid zonobiomes and zonoecotones of North America. See text for identification of Roman numerals, etc. (Walter and Breckle 1985)

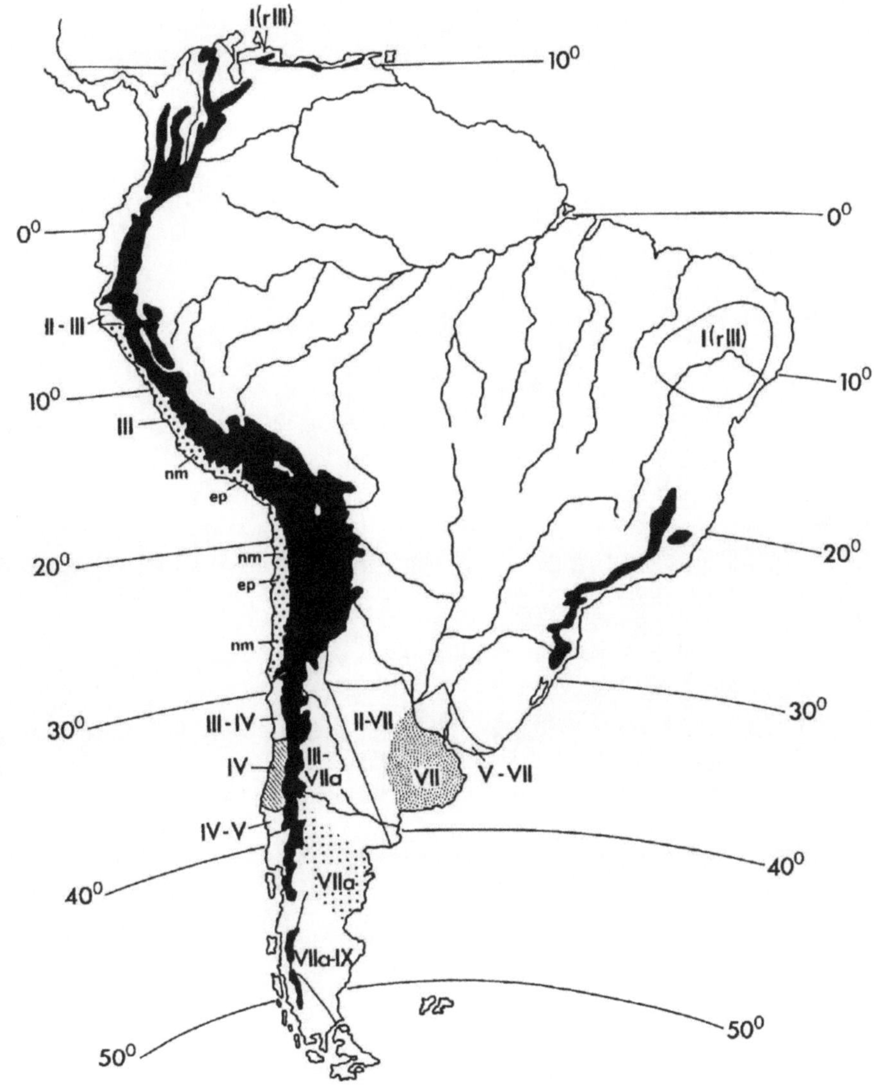

Fig. 8. Arid and semi-arid zonobiomes and zonoecotones of South America. See text for identification of Roman numerals, etc. (Walter and Breckle 1985)

CHAPTER 3

Arid and Semi-Arid Regions and Ecosystems of the World

3.1 The Saharo-Sindian Region

The Saharo-Sindian region extends westwards from the Atlantic Ocean of northern Africa to the Red Sea in the east, and extends eastwards through the Arabian Desert to the north-east of the Indian subcontinent. To the north lies the cold arid and semi-arid regions of central Asia and northern China.

3.1.1 Sahara Subregion

The vast desert of the Sahara stretches over 5000 km from east to west and over 1800 km from north to south, covering an area of 9 million km^2 (Walter and Breckle 1986). Walter and Breckle (1986) recognise three zones based on rainfall distribution: (1) a northern zone of winter rainfall of less than 200 mm annually (ZB III-IV, IV); (2) a central region with sporadic rainfall where the rain does not fall every year (ZB III); and (3) a southern zone with summer rainfall (ZB II-III). On the basis of climate and biology Le Houérou (1986) recognises four zones: (1) a northern arid zone with winter rainfall (referred to here rather loosely as the Maghreb and Mediterranean coast); (2) a northern Saharan desert with less than 100 mm annual rainfall; (3) a southern Saharan desert with less than ca. 100 mm annual rainfall; and (4) an arid zone known as the Sahel with ca. 100-400 mm summer rainfall. In the lower rainfall areas of the two desert zones the trend towards summer or winter rainfall becomes obscure and thus corresponds to the sporadic rainfall of Walter and Breckle (1986).

The accepted northern and southern boundaries of the Sahara desert roughly correspond to the borderline between contracted and scattered vegetation types. Contracted refers to perennial vegetation that is concentrated in water-receiving sites (depressions and along the hydrological network), while scattered refers to a more or less uniform, albeit sparse, vegetation on the interfluves and pediments. The limit between contracted and scattered vegetation types are affected by soil texture and substratum, since on both sides of the Sahara the vegetation becomes contracted on sands and sandy soils with ca. 50 mm annual rainfall, while on soils with a shallow substrate, pebble plains and fine textured soils, 100-150 mm rainfall is required. Thus, the northern and southern boundaries can be realistically defined as the 100-150 mm isohyet (Le Houérou 1989).

In the northern Sahara the 100 mm isohyet more or less corresponds with the northern limit for the cultivation of *Phoenix dactylifera** (date palm) and the southern limit of *Stipa tenacissima** (alfa), a characteristic species of the Maghreb and Mediterranean coast zone (White 1983; Le Houérou 1989).

In the southern Sahara the 150 mm isohyet more or less corresponds to the southern limits for the distribution of *Cornulaca monacantha**, and *Panicum turgidum** and *Stipagrostis pungens** and the northern limits of such Sahelian species as the shrubs *Boscia senegalensis** and *Commiphora africana* and the grass *Cenchrus biflorus* (cram-cram) (White 1983).

Instead of following Walter and Breckle (1986), a condensed version of the vegetation units of White (1983) is followed here because his units are a part of description for the entire African continent (Fig. 9) and provide continuity for describing the remaining arid areas of Africa which are not adequately dealt with by Walter and Breckle (1986)

3.1.1.1
Maghreb and Mediterranean Coast

The Maghreb and Mediterranean coast is the region to the north of the Sahara, between the Atlantic coast of Morocco and Egypt, where White (1983) recognises a Mediterranean regional centre of endemism and a Mediterranean/Sahara regional transition zone. The name Maghreb is derived from the Arabic 'Djezira et Maghreb', meaning 'the western islands' - the island of land between the Mediterranean Sea and the Sahara - and refers to the mountains and coastal plain of Morocco, Algeria and Tunisia. A critical bioclimatic and phytogeographical review of the flora and fauna of North Africa is given by Le Houérou (1995), who recognises the addition of an Ibero-Maghrebian subelement which, as its name implies, extends from the Maghreb into southeastern Spain. While readers should be aware of this important work, it is not followed here due to problems in correlating the work with that of White (1983) for the remainder of Africa.

Most of the rainfall in the Mediterranean regional centre of endemism falls in winter and is mainly between 250 and 1000 mm yr^{-1}; the summers are hot and dry. Along the Moroccan coast the rainfall regime is characterised by a single maximum rainfall in winter and almost total summer drought. Along the Atlantic coast there is a sharp north to south decrease in rainfall: Tangier 820 mm; Rabat 500 mm; Casablanca 400 mm and Essaouira 330 mm. On the

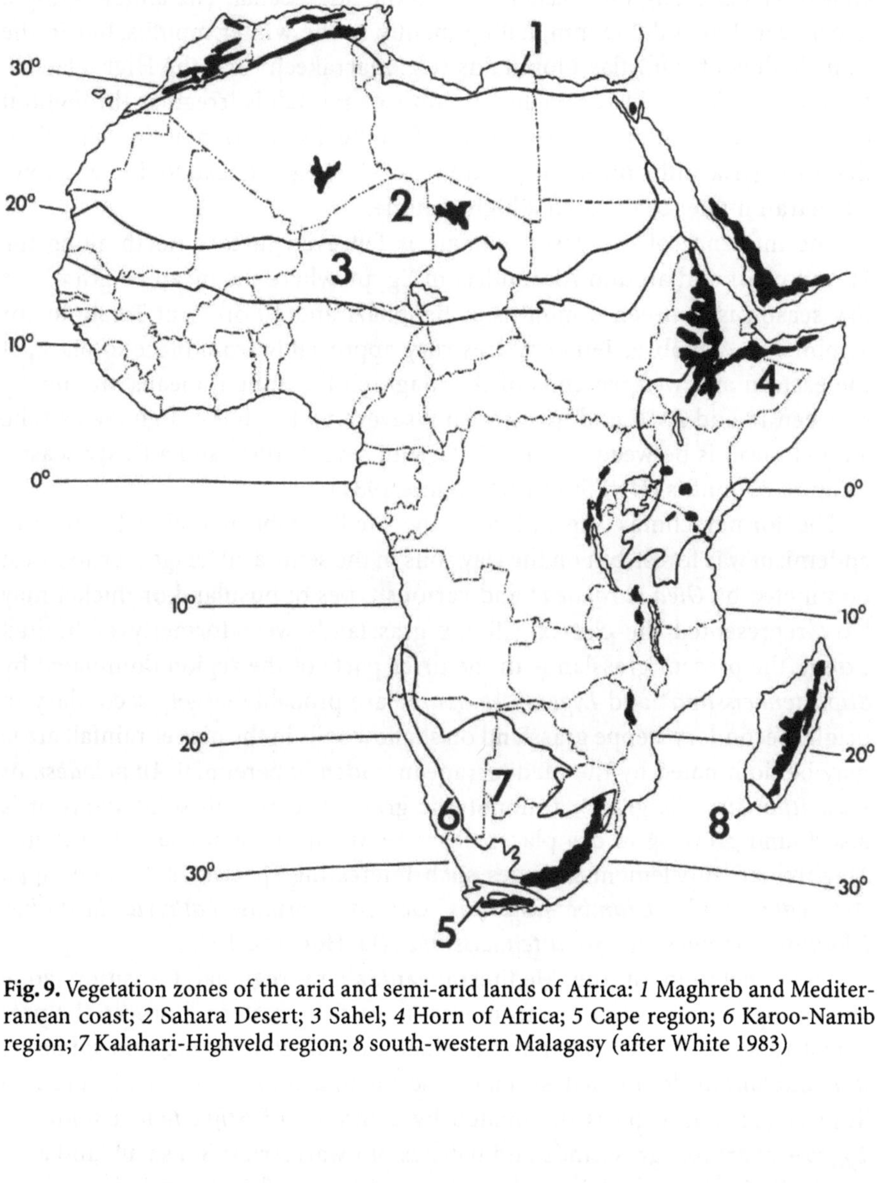

Fig. 9. Vegetation zones of the arid and semi-arid lands of Africa: *1* Maghreb and Mediterranean coast; *2* Sahara Desert; *3* Sahel; *4* Horn of Africa; *5* Cape region; *6* Karoo-Namib region; *7* Kalahari-Highveld region; *8* south-western Malagasy (after White 1983)

Mediterranean coast Algiers receives 650 mm and Tunis 420 mm. Frost is widespread, although some parts of the coastal plain are frost-free, while parts of the interior may experience up to 7 months of frost. In the high mountains of the Maghreb snow frequently lies for long periods (UNESCO 1979; White 1983).

In the west the Mediterranean/Sahara regional transition zone lies between the Mediterranean regional centre of endemism and the Sahara, extending from the Atlantic coast to the High Plateau of Tunisia, and hence from the

Gulf of Gabes along the coastal plain to the Suez Canal. The annual rainfall is between 100 and 250 mm, falling mainly in the winter months, but in the rain shadow of the Atlas Mountains (e.g. Marrakech) and the High Plateau, the main peaks are in spring and autumn, or rainfall is irregular throughout the year. The mountains considerably affect the distribution of the rainfall, so that in Algeria and Tunisia the proximity of the High Plateau to the sea allows a Saharan influence to reach a high altitude.

The influence of the desert climate is felt even further north along the Libyan coast and around Alexandria in Egypt, where the mean length of the dry season is between 8 months at Benghazi and 10 or 11 at Tarabulus or Tripoli and Ajedabya. Temperatures vary appreciably from place to place; at the eastern and western ends of the Maghreb the annual means are mostly between 18 and 20 °C and frosts are not severe, while on the High Plateau the annual mean is between 13 and 17 °C, with severe frosts and a frosty season of up to 8 months (UNESCO 1979; White 1983).

The former climax vegetation of the Mediterranean regional centre of endemism was forest, but on the clay soils in the semi-arid 'étage', scrub forest dominated by *Olea europaea** and various types of bushland or thicket may have represented the climax. Climax grasslands were formerly of limited extent; the present grasslands in the drier parts of the region dominated by *Stipa tenacissima** and *Lygeum spartum** are probably largely secondary in origin. Secondary steppe grassland on shallow soils in the higher rainfall areas may be dominated by the Mediterranean endemic perennial *Ampelodesmos mauritianicus* (dis grass), a monotypic grass species with solid stems; it is also found growing in dry places near the Mediterranean coast. The Ibero-Maghrebian subelement includes such interesting species as *Aegilops* spp., *Astragalus edulis, Crambe filiformis, Genista scorpius, Lathyrus latifolius, Rhamnus oleoides* and *Stipa tenacissima** (Le Houérou 1995).

The vegetation of the Mediterranean/Sahara regional transition zone shows appreciable variations from west to east. In western Morocco the prevalent types are scrub forest and bushland dominated by *Argania spinosa**, *and Euphorbia*-dominated succulent scrubland. From eastern Morocco to Tunisia the landscape is dominated by a mosaic of *Stipa tenacissima** or *Lygeum spartum** grasslands and patches of dwarf *Artemisia* shrubland and, especially in the mountains, relic patches of former forest consisting principally of *Juniperus phoenicea, Pinus halepensis** and *Quercus ilex*. Eastwards through Libya and Egypt the vegetation has been very severely degraded by human occupation.

White (1983) recognises the following vegetation units:

1. Mediterranean Forest. Only small relicts of the former very extensive Mediterranean forest now survive. The sclerophyllous forest is dominated by

Quercus coccifera, Q. ilex and *Q. suber*, while *Q. afares, Q. faginea* and *Q. pyrenaica* dominate the deciduous oak forest.

The dominants of the coniferous forest are *Abies* spp., *Cedrus libani* var. *atlantica* (syn. *C. atlantica*), *Cupressus sempervirins, Juniperus phoenicea, J. thurifera, Pinus halepensis*, P. pinaster* and *Tetraclinis articulata.* Among other Mediterranean forest species are *Arbutus unedo, Celtis australis, Ceratonia siliqua, Chamaerops humilis, Laurus noblis*, Olea europaea*, Pistacia atlantica, P. lentiscus** and *P. terebinthus**.

2. *Sub-Mediterranean Forest.* The Saharan Atlas enjoys a more humid climate than the High Plateau and has a vegetation similar to that of the mountains to the north and east. The surviving forests are dominated by *Quercus ilex* and *Pinus halepensis**, the latter being more resistant to forest fire as well as being less preferred for fuelwood. Associated tree species include *Juniperus oxycedrus, J. phoenicea, Olea europaea** and *Pistacia atlantica*; the ground flora include the grass *Stipa tenacissima**, which becomes dominant when the forest is destroyed.

The relic forest flora of the severely degraded Gebel Akhdar in Cyrenaica, which receives an annual rainfall of 300–600 mm, include *Arbutus pavarii, Ceratonia siliqua, Cupressus sempervirens, Juniperus oxycedrus, J. phoenicea, Laurus noblis*, Olea europaea*, Phillyrea angustifolia, Pinus halepensis*, Pistacea lentiscus*, Quercus coccifera* and *Q. ilex.* A few relic stands of *Ceratonia, Olea, Pistacia atlantica* and *P. lentiscus** occur in Tripolitania, while relic *Olea* and *Ceratonia* have also been recorded from the coastal plain of Egypt (Wickens 1977a). Few trees now remain in the coastal plain of Libya and it appears unlikely that the forest ever extended into Egypt during the present climatic era. The plains have been intensively cultivated since the Roman period, if not earlier, and, where uncultivated, have been severely degraded by overgrazing, so that only unpalatable species, such as *Thymelaea hirsuta**, remain.

3. Mediterranean Bushland and Thicket. The climax vegetation on clay soils was probably bushland or thicket dominated by *Olea europaea** and *Pistacia lentiscus** and, where not degraded, by *Pistacia atlantica* and *Chamaerops humilis.* Among the other woody associates are *Ceratonia siliqua, Pistacia atlantica, Rhamnus oleoides* and *Ziziphus lotus**, while among the more conspicuous shrubs are *Anagyris foetida, Calicotome villosa* and *Daphne gnidium.*

4. *Mediterranean Shrubland.* This is mainly secondary in origin, true Mediterranean shrubland being largely confined to the high mountains above the timber line. The altimontane Mediterranean shrubland is dominated by dense, cushion-shaped, very spinous shrubs. It nearly always occurs on skeletal soils which frequently exhibit polygonal structure due to frost action. The climate

is Mediterranean, with winter rains and very dry summers. Snow lies on the ground for several months during the winter. The secondary Mediterranean shrubland (maquis and garrigue) has usually been derived from Mediterranean forest, scrub forest or bushland. There is considerable variation in structure and composition. Classical maquis occurs on siliceous soils. It is dense, almost impenetrable shrubland up to 4 m high; it is dominated by *Erica arborea* and *Arbutus unedo*, and degrades to grassland dominated by *Ampelodesmos mauritanicus*. The term 'garrigue' is derived from the Catalan 'garric' for *Quercus coccifera*.

5. *Argania spinosa Scrub Forest and Bushland.* Restricted to south-west Morocco, the endemic *Argania spinosa** is the dominant tree. It grows on all types of soil except mobile sand. With increasing humidity it gives way to *Olea-Pistacia* bushland or *Tetraclinis* forest. Much of its distribution area has been degraded through cultivation or browsing goats. Forest associates in the semi-arid littoral include *Acacia gummifera, Ephedra altissima, Euphorbia beaumierana, Lavandula* spp., *Olea europaea**, *Periploca laevigata, Pistacia lentiscus** and *Rhamnus oleoides.*

6. *Acacia gummifera-Ziziphus lotus Bushland.* The climax vegetation of the arid Haouz-Tadla of western Morocco was bushland or thicket dominated by *Acacia gummifera, Pistacia atlantica* and *Ziziphus lotus**. The associates of the degraded remnants of this forest include *Asparagus stipularis**, *Ephedra altissima, Peganum harmala* and *Stipa capensis*. On saline soils the dominant halophytes are *Atriplex halimus** and *Lycium intricatum* accompanied by *Salsola vermiculata, Sphenopus divaricatus* and *Suaeda vermiculata* (syn. *S. fruticosa*).

7. *Succulent Sub-Mediterranean Shrubland.* The cactoid shrubs *Euphorbia beaumierana, E. echinus, E. regis-jubae* and *E. resinifera* occur in degraded forest and shrubland, especially *Argania spinosa** forest and in places may form secondary *Euphorbia* shrubland. However, on shallow rocky soils and in wind-swept places near the sea, these species may represent local edaphic or climatic climax communities; the latter certainly occurs on suitable soils to the south of the *Argania* zone. *Euphorbia echinus* succulent shrubland is characteristic of the rocky hillsides and pebble and gravel lower scree slopes in the vicinity of Goulimine, forming a mosaic with dwarf *Haloxylon scoparium** shrubland on the impermeable soils. To the south the tree and shrub associates include *Acacia gummifera, Argania spinosa** (restricted to the north), *Launaea arborescens* and *Lycium intricatum*, as well as the perennials *Anabasis aphylla, Frankenia corymbosa, Haloxylon scoparium**, *Limonium fallax, Salsola longifolia* (syn. *S. sieberi*), *S. tetragona, S. vermiculata, Seriphi-*

*dium herba-alba**, *Suaeda ifniensis, S. vermiculata* (syn. *S. mollis*), together with the annuals *Aizoon canariense**, *Fagonia cretica*, etc.

8. *Sub-Mediterranean Grassland*. The two dominant grasses are *Stipa tenacissima** and *Lygeum spartum**. They occur in a mosaic with communities dominated by the two dwarf shrubby composites *Artemisia campestris* and *Seriphidium herba-alba**, and saline areas dominated by *Atriplex halimus** and *Salsola vermiculata*.

9. *Mediterranean Halophytic Vegetation*. The principal species of the arid and semi-arid 'étages' of the Maghreb are the chenopods *Anabasis aphylla, Atriplex halimus**, *Salsola longifolia, S. vermiculata* and *Suaeda* sp., together with the herbs *Aizoon* spp., *Asparagus stipularis**, *Mesembryanthemum* spp., *Peganum harmala, Seriphidium herba-alba** and the grasses *Hainardia cylindrica* (syn. *Lepturus* sp.) and *Sphenopus* spp. The more frequent species of the more permanently moist depressions are *Juncus acutus, J. maritimus**, *Plantago coronopus**, *Scirpus holoschoenus, Spergularia maritima, Statice* spp. and *Tamarix* spp.

10. *Sub-Mediterranean Halophytic Vegetation*. The halophytic vegetation at Tauorga, the largest saltpan in Tripolitania, contains the following characteristic species: the grasses *Aeluropus lagopoides** and *Sphenopus divaricatus*, the chenopods *Arthrocnemum macrostachyum**, *Atriplex mollis**, *Bassia muricata, Halocnemum strobilaceum**, *Nitraria retusa, Salicornia arabica, Salsola longifolia, S. tetragona, S. vermiculata* and *Suaeda vermiculata*, plus *Frankenia laevis, Limoniastrum monopetalum, Limonium cymuliferum, Reaumuria vermiculata* and *Zygophyllum album*.

11. *Sub-Mediterranean Anthropic Landscapes*. The landscape in western Morocco is dominated by wheat fields and sheep pastures. The older fallows support a 2-m-high shrubland of *Retama monosperma* and *Ziziphus lotus**. Hedgerows have been planted with *Opuntia*, which has become extensively naturalised in some areas. There are also plantations of the Australian *Acacia saligna, Eucalyptus* spp. and *Pinus halepensis**. Indigenous trees are virtually absent apart from a few *Acacia gummifera* and, on the highest hills, *Quercus ilex*. The original forests of the coastal plain of Tunisia have completely disappeared, being replaced by olive groves, almond and apricot orchards and wheat fields. Further to the east, from Libya to Egypt, the land, despite the low rainfall, has been intensively cultivated since Roman times, and grazed for even longer. El-Ghareeb and Bidak (1992) examined the effect of protection on three highly palatable range species, *Cutandia dichotoma**, *Helianthemum*

*lippii** and *Plantago albicans*, from non-saline depressions at Omayed. The three species provide about 27% of the average daily diet of domestic livestock.

12. *Nile Delta.* The swamp vegetation of the Nile delta has been almost entirely destroyed following the introduction of a permanent irrigation system in the early 1800s to replace the traditional basin system. The papyrus, *Cyperus papyrus*, considered extinct, was rediscovered in 1962 in a freshwater swamp within the Wadi Natroun depression (El-Hadidi 1971).

3.1.1.2
Sahara Desert

As has already been mentioned, three longitudinal climatic zones in the Sahara Desert may be recognised: northern and southern zones with winter and summer rainfall, respectively, and a central zone with episodic rainfall of less than 25 mm average annual rainfall. In addition to the north-south differences in distribution there are also east-west differences. Thus, while the mean annual rainfall in the eastern Sahara (Nile valley) is near 0 mm, that of oceanic Sahara has a markedly higher precipitation and a slightly higher air humidity; Atar in Mauritania has an annual average of 93 mm and a relative humidity of 20-40%. The temperature also varies from north to south. While the annual mean for the whole of the Sahara is generally 20-25 °C, it is not uncommon for the absolute minimum in the north to fall between 0 and -5° C, while it almost never freezes in the south. In the major Sahara mountains the winters are cool, 0-10 °C, and temperature often fall below -1 0 °C; it may even snow (UNESCO 1979).

The phytogeographical units (domains) of the Sahara recognised by Quézel (1965) and amended by Quézel (1978) are North-Western, Northern, Atlantic, Western, Central, Southern, Saharomontane, North-Eastern and Eastern Sahara. White (1983) recognises the following divisions within the Sahara:

1. *Desert Oases.* There are relatively few places where slightly saline water issues forth as springs. Swamps of *Typha latifolia, Phragmites australis* or *Scirpus holoschoenus* occur along the edges, grading into stands of *Tamarix* spp. The original vegetation of these oases consisted of *Hyphaene thebaica* and species of *Acacia, Capparis* and *Maerua, Calotropis procera* and the cucurbit *Citrullus colocynthis*. This vegetation has now been almost completely replaced by the date palm *Phoenix dactylifera** and other cultivated species.

2. *Desert Wadis.* In the Central Sahara communities dominated by *Tamarix aphylla** occur along the large sandy wadis wherever the watertable is within

7–8 m of the surface, especially in those wadis radiating from the Saharan massifs. Under favourable conditions the *Tamarix* may form completely closed riparian forest. In more open communities associated species include the shrubs *Calligonum polygonoides* subsp. *comosum** and *Leptadenia pyrotechnica**, together with the grass *Stipagrostis pungens**.

The vegetation of the Nile as it traverses the Sahara has been described by El-Haddi (1976). The lower terraces have an annual herbaceous flora. Stands of *Acacia nilotica**, *A. seyal* and *Tamarix nilotica*, with the climbers *Cardiospermum halicacabum* and *Oxystelma alpini* (syn. *O. esculentum* var. *alpini*), occur on the silty soils of the lower middle terrace. The date palm *Phoenix dactylifera* * dominates the upper levels of the middle terraces, interrupted by clumps of *Faidherbia albida** and *Hyphaene thebaica*. The upper terraces have such xerophytic species as *Calotropis procera, Leptadenia pyrotechnica** and *Tamarix aphylla**, with much bare ground between or a sparse cover of weeds of disturbed ground. In some of the wadis of northern and north-western Sahara, receiving their water from the mountains of the High Atlas or Saharan Atlas, there occur such tree species as *Nereum oleander*, Populus euphratica* and *Vitex agnus-castus* associated with *Tamarix africana* (syn. *T. gallica sensu* White 1983).

3. *Psammophilous Vegetation*. Approximately one third of the desert surface is sand, more than half of which is without perennial vegetation, the shifting dunes being barren of vegetation. On the more stable dunes of the Western Sahara the two species that occur throughout are the grass *Stipagrostis pungens** and the intricately branched cushion shrub *Cornulaca monacantha**. In the Northern Sahara the shrubs *Calligonum* spp., *Retama raetum** and *Ephedra alata** also occur, while in the Central and Southern Sahara the shrub *Leptadenia pyrotechnica** is often present.

Annuals, apart from *Eremobium aegyptiacum* (syn. *Malcolmia aegyptiaca*), rarely occur, probably due to the disturbing influence of the constantly moving surface sand. Because the perennial plants present an obstacle to the wind, sand either accumulates either on the leeward side or between their branches; consequently, the plants tend to crown small dunes which usually retain sufficient moisture within reach of the extensive root systems for the plants to survive throughout the year without any marked periodicity in their development (Walter and Breckle 1986).

4. *Hammadas*. The stone pavements are usually devoid of vegetation due to the low soil moisture and its relatively high salt content. Plants can grow only in the rock crevices and the water-receiving depressions. Among the widespread species in the hammadas of the Northern, North-Western, Western and Central Sahara are *Anastatica hierochuntica**, *Fagonia latifolia, Farsetia*

aegyptia, Forsskaolea tenacissima, Nauplius graveolens (syn. *Asteriscus graveolens*), *Reseda villosa, Salvia aegyptiaca* and the C4 grasses *Enneapogon desvauxii* and *E. scaber.* In the Northern Sahara the particularly characteristic species are *Moricandia arvensis, Fagonia microphylla, Haloxylon scoparium** and the C4 grass *Cymbopogon schoenanthus.* In the hammada and regs of the North-Western Sahara a particularly striking plant is the cushion shrub *Fredolia aretioides**, often the only perennial present. Other characteristic species include *Limmoniastrum feei* and the almost leafless umbellifer *Pituranthos denudatus* subsp. *battandieri* (syn. *Pituranthos battandieri*). The hammada flora is virtually absent in the Central Sahara, except on the rocky flanks of the high mountains, where the two perennials *Aerva javanica** and *Anabasis articulata* are the characteristic species, accompanied by such annual grasses as *Aristida* spp., *Enneapogon* spp. and *Stipagrostis* spp.

5. *Regs.* The gravel deserts are usually so compacted that roots can barely penetrate; they are consequently often devoid of vegetation, except in the drainage channels. In the Northern Sahara the characteristic species is *Haloxylon scoparium**. Where there is a sandy cover the C4 grasses *Stipagrostis ciliata, S. obtusa* and *S. plumosa* may form a dense sward within which a number of geophytes may occur. Communities of *Cornulaca monacantha** and *Randonia africana* may occur on the heavier soils. On very shallow sand of the drainage channels of the Eastern Desert of Egypt the dominant species is *Haloxylon salicornicum**, albeit with less than 5% cover. Deeper soils support communities of the tussock-forming grasses *Lasiurus scindicus** and *Panicum turgidum**, and the herb *Zilla spinosa**

6. *Saharomontane Vegetation.* This vegetation type occurs at elevations above 1800 m in the west on the massifs of Ahaggar and its satellites Tafedest and Mouydir, the Tassil n'Ajjer and the most northerly summits of Aïr, and to the east Tibesti. It is not found on the 1900-m-high Jebel 'Uweinat, which is characterised by species of Mediterranean affinity (Osborn and Krombein 1969). The Saharan olive, *Olea europaea* subsp. *laperrinei** is the most characteristic tree of the wadis of the western massifs; it is absent from Tibesti and reoccurs on the Sahelian mountains of Jebel Gurgeil and Jebel Marra. The olive is often associated with the perennial grasses *Oryzopsis caerulescens, Stipa capensis* and *S. parviflora.* Elsewhere it may occur with the bushy *Pistacia atlantica* and thickets of *Rhus tripartita. Artemisia campestris* var. *glutinosa* dominates the more gentle wadi slopes. In the moister wadis of Ahaggar and its satellites, and the western slopes of Tibesti, especially near the permanent rock pools (gueltas), *Myrtus nivellei* and *Nerium oleander** form characteristic communities.

Tamarix nilotica may occur on both Ahaggar and Tibesti where the water-table is within 1-2 m of the surface. Except for the endemic fig *Ficus teloukai*, the larger woody species of the wadis on the southern slopes of Tibesti are of Sahelian origin. The Saharomontane grasslands are dominated by the perennials *Stipagrostis obtusa* and *Aristida adscensionis* (syn. *A. coerulescens*), and locally by *Eragrostis papposa*. Among the low shrubs present are *Anabasis articulata, Fagonia flamandii* and *Zilla spinosa**.

7. *Halogypsophilous Vegetation.* Salt pans, known as sebkhas or chottss, are formed where water evaporates from drainless depressions, although they are less common than in North Africa, where their extensive development is more dependent on the presence of salt-bearing strata than climatic factors. The hyper-halogypsophilous vegetation is dominated by the shrub *Halocnemum strobilaceum**, which either occurs in the central part of sebkhas which dry out during the summer or, if the central part is bare of vegetation, forms a belt around the zone of salt encrustation.

On drier soils the characteristic halogypsophilous shrubs of the larger chotts at the northern edge of the desert are *Salsola longifolia* and *Zygophyllum cornutum*. Further south, in the Wadi Rhir, comparable vegetation is characterised by *Traganum nudatum* and *Zygophyllum album*, a vegetation type which is widespread elsewhere in the Northern Sahara, where it occupies gypsaceous areas with only small amounts of soluble salts and often with a thin cover of aeolian sand.

8. *Absolute Desert.* Where the episodic rainfall averages less than 20 mm yr^{-1}, with many consecutive years without rain, the vegetation is extremely sparse and it is possible to travel for several hundred kilometres without finding any living plants, although perennials may survive on water-receiving sites that remain moist for several years without rain. However, the surface soil may contain abundant and viable annual seeds awaiting rain.

9. *The Atlantic Coastal Desert.* A narrow strip some 40 to 50 km wide, bordering the Atlantic and extending from the southern limit of succulent shrubs at Sagui el Hamra (ca. 27°N) in the north to the northern limit of the Sahel (ca. 20°N) in the south, is characterised by a relatively high humidity, lower temperatures and the absence of frost. The frequent mists permit the growth of crustaceous and fruticose lichens on the branches of shrubs and a dense ground growth of fruticose lichens (*Ramalina* sp.). In the north there are a few scattered succulents, such as *Euphorbia echinus, E. regis-jubae* and *Senecio anteuphorbium*; other Mediterranean linking species include *Launaea arborescens*. In the south, the Sahelian species extending northwards into Mauritania include *Acacia tortilis**, *Balanites aegyptiaca** and *Salvadora persica*.

Saharan species present include the grasses *Panicum turgidum** and *Stipagrostis pungens**, and the halophytes *Anabasis articulata, Arthrocnemum macrostachyum**, *Atriplex* spp., *Cornulaca monacantha**, *Salsola* spp., *Suaeda* spp. and *Tamarix* spp.

10. *The Red Sea Coastal Desert.* White and Léonard (1991) recognise, owing to the high percentage of endemism, a Nubo-Sindian local centre of endemism spreading from the Nubian Desert of Egypt along the coasts of the Red Sea and extending along the Persian Gulf towards the Sind and the Indian desert regional subzone. The 15 to 20-km wide Red Sea coastal plain receives very little rain. The mean annual rainfall at Hurghada, Egypt is only 3 mm, with most of the years rainless, although as much as 43 mm has been recorded for a single day.

Apart from the littoral halophytic communities, the area is largely devoid of vegetation except for the wadis. The principal tree species of the plain are *Acacia tortilis** and *Maerua crassifolia*; also present is the spiny shrub *Lycium shawii* (syn. *L. persicum*), a species not found elsewhere in the Sudan. Towards the southern part of the coastal plain there is a gradual transition, with increasing rainfall, to *Acacia asak* (syn. *A. glaucophylla*) and *A. etbaica* scrub.

Groves of the mangrove *Avicennia marina* extend southwards along the Red Sea coast. The littoral salt marshes are characterised by *Arthrocnemum macrostachyum**, *Halocnemum strobilaceum**, *Nitraria retusa**, *Suaeda monoica, Zygophyllum album* and *Z. coccineum*. On the sands at the upper edge of the salt marshes are the grasses *Aeluropus lagopoides** and *Sporobolus spicatus* (Harrison and Jackson 1958). In the coastal plain, saline reaches of the wadis have a dense growth of *Juncus rigidus* (syn. *J. arabicus*), *Suaeda monoica* and *Tamarix mannifera*. Elsewhere in the wadis the characteristic species are *Acacia tortilis**, *Calligonum polygonoides* subsp. *comosum**, *Capparis decidua, Retama raetum**, *Zilla spinosa** and the tussock-forming grasses *Lasiurus scindicus** and *Panicum turgidum**. Inland a chain of rugged mountains with peaks up to 2184 m intercept cloud moisture in the form of orographic rain, which feeds the permanent springs (nakkat) and montane runnels. A floristically depauperate variant of the East African evergreen bushland occurs on Gebel Elba including *Olea europaea* subsp. *cuspidata** (White 1983).

3.1.1.3
Sahel

The Sahel occupies a relatively narrow belt, ca. 400 km wide, of mainly flat or gently undulating terrain, extending from the Atlantic to the Red Sea. Large areas are covered by aeolian sand dating from the drier phases of the Pleistocene and by clays deposited during the wetter phases of the Pleistocene. Emergent outcrops of older rocks form the massifs of Adrar des Iforas, Aïr, Ennedi, Jebel Marra and the Red Sea Hills.

The climate is typically tropical with an annual rainfall mostly between 150 and 500 mm and falling during the short three to four summer months; the dry season is long and severe (White 1983; Le Houérou 1989). UNESCO (1979) provides the following quotation from Monod (1973) regarding the variability of the rainfall:

"*It should not be forgotten that because of the range of annual rainfall variability at a given point, the isohyets which seem so precise on our maps can only represent rough approximations: the 100-mm curve in Mauritania which 'on average' passes through Nouakchott and Adrar, in 1951-1952 lay 300 km further north, and 300 km further south in 1941-1942, giving a difference on the ground of 600 km. Trees do not move according to these episodic fluctuations, but for the nomad and his animals the steppe-desert margin shifts over hundreds of kilometres in latitude.*" Except near the coast, the mean annual temperature is between 25 and 30 °C, the mean annual minimum being 18–20 °C and the mean maximum 35–38 °C. There may be occasional light frosts in the northern Sahel and on the massifs, e.g. Jebel Marra (White 1983; Le Houérou 1989). The second half of the dry season may experience a very disagreeable kind of tropical line squall associated with the north-eastern continental trade winds, known as the 'harmattan' or 'habub'. According to Grenot (1974), winds in the Sahara can reach or even exceed 100 km h^{-1}.

The climatic picture has now been further complicated by the Sahelian drought, which has continued more or less unabated since 1968. Thus, in the Sudan, Wickens (1982a) demonstrated that during 1968–1972 there was a parallel southward shift of the pre-1965 isohyets of up to 1° latitude (ca. 110 km); a shift of similar magnitude (100–150 km) has also been recorded for Niger (Sivakumar 1989). The effects of the continuing drought are further complicated by an unacceptable increase in desertification, a man-made process arising from overcultivation, overgrazing and deforestation. The combined effect upon the vegetation has been dramatic, with a noticeable southward shift in a number of tree species, especially those, such as *Terminalia brownii*, whose northern-most limits occur in the more favourable sites of the Sahel.

The following descriptions of the vegetation based on White (1983) refer to conditions before 1970, before the devastation caused by the present drought. Provided the anthropogenic pressures can be totally removed it is possible that the vegetation may eventually recover if and when the rainfall returns to 'normal', although the longer low rainfall conditions prevail, the more likely that the soil seed bank will be depleted.

1. *Sahel Semi-Desert Grassland.* On the deep sandy soils of northern Sahel where the annual rainfall is less than 250 mm the vegetation usually contains an admixture of bushes and bushy trees and extensive areas of grassland. The chief woody species are *Acacia ehrenbergiana, A. laeta, A. tortilis**, *Balanites aegyptiaca**, *Boscia senegalensis**, *Commiphora africana* and *Leptadenia pyrotechnica**. They all, apart from *Acacia ehrenbergiana*, also occur in the southern Sahel. The most extensive dominant grasses are the annuals *Aristida stipoides, Cenchrus biflorus, Schoenefeldia gracilis* and *Tragus racemosus*; these also occur in the southern Sahel. The perennial desert grasses *Panicum turgidum** and *Stipagrostis pungens** may also occur in the northern Sahel. The former may extend as far south as the 250 mm isohyet on the loose sands of the dune crests; the Saharan shrub *Calligonum polygonoides* subsp. *comosum** may also occur in association with *Panicum turgidum** as far south as the 200 mm isohyet. *Stipagrostis* replaces *Panicum turgidum** as the dominant grass of the unstable dune crests where the rainfall is less than 100 mm.

The dry season diurnal course of transpiration and stomatal conductance of *A. ehrenbergiana*, together with other Sahel-Sudanian acacias *A. erythrocalyx, A. laeta, A. nilotica* subsp. *tomentosa**, *A. tortilis* subsp. *raddiana**, *A. senegal* var. *senegal**, *A. seyal* and *Faidherbia albida** were examined by Ullmann (1985). Although no species-specific differences in behaviour were observed, a correlation was found between types of stomatal performance patterns and leaf age. Mature leaves showed a steep, post-sunrise increase in leaf diffusion conductance followed by an early morning pronounced peak followed by a continuous decrease during the remainder of the day. Typically this decease was interrupted in the afternoon by a minor peak associated with an intermittent rise in relative humidity. The diurnal course of transpiration was dome-shaped, attaining a noon maximum associated with the rapid rise in air temperature and a decrease in relative humidity. Senescent leaves that survived the dry season experienced a midday depression of stomatal conductance and leaf water loss, while newly produced leaves exhibited a high conductance from early morning to late afternoon. There was no association with daily maximum leaf diffusion conductance and leaf age; however, there was an association with predawn water potential. The phreatophytic habit of many of these *Acacia* species is clearly an advantage over the need to respond to seasonal rainfall or changes in air humidity. This is evident in the unrestricted dry season gas exchange of trees able to tap ground water resources against the response to drought in trees without access to ground water.

In the Sudan and northern Chad the transition zone between the northern Sahel and southern Sahara is subject to an ephemeral type of succulent winter grazing know as 'gizu'; west of Chad the gizu-type grazing is known as 'asheb'. The gizu occurs in approximately 2 out of 5 years, although the climatic conditions that give rise to this phenomenon are not fully understood. The gizu is sufficiently succulent for camels, and to a lesser other stock, to maintain themselves for several months without having access to water. Among the gizu species recorded are the herbs *Aerva javanica**, *Blepharis ciliaris**, *Citrullus colocynthis* (syn. *Colocynthis vulgaris*), *Cornulaca monocantha**, *Fagonia cretica*, *Farsetia aegyptia*, *F. longisiliqua*, *F. stylosa* (syn. *F. ramosissima*), *Helianthemum lippii**, *Neurada procumbens* and *Salsola vermiculata*, plus the grasses *Centropodia forskalei* (syn. *Danthonia forskalei*), *Stipagrostis plumosa* (syn. *Aristida plumosa*) and *S. uniplumis* (syn. *Aristida papposa*) (Wilson 1978).

2. *Sahel Wooded Grassland*. This is the most widespread vegetation type on the sandy soils of the southern Sahel, where the annual rainfall is between 250 and 500 mm. In West Africa the chief woody species are *Acacia laeta*, *A. tortilis**, *Adansonia digitata**, *Balanites aegyptiaca**, *Boscia senegalensis**, *Calotropis procera**, *Combretum aculeatum*, *C. glutinosum*, *Commiphora africana*, *Leptadenia pyrotechnica** and *Maerua crassifolia*; their distribution extends eastwards to the Sudan. The pure stands of the shrub *Guiera senegalensis* may dominate exhausted fallow lands on sandy soils in the southern Sahel. The predominant grasses are the annuals *Aristida stipoides*, *Cenchrus biflorus*, *Schoenefeldia gracilis* and *Tragus racemosus*. Pure stands of the perennial grass *Andropogon gayanus* may occur on the deeper sands towards the south. In heavily grazed and disturbed areas the dominant weeds include the annuals *Boerhavia coccinea*, *Senna obtusifolia* (syn. *Cassia obtusifolia*) and *Tribulus terrestris*.

In the Sudan the characteristic tree of the deep sands is the gum acacia *Acacia senegal**, often occurring in almost pure stands which are possibly of secondary origin, the result of a mature gum acacia fallow. While the dominant grass is mainly the perennial *Aristida sieberana* (syn. *A. pallida*), with *Anthephora pubescens* (syn *A. hochstetteri*), and *Eragrostis tremula* also present, in addition to those species mentioned above for West Africa.

Pure stands of *Acacia nilotica* subsp. *nilotica** and subsp. *tomentosa** occur in the larger seasonal pools and along the banks of the Nile, northward into Egypt. These riverine woodlands are now far less extensive than in the past due to widespread felling for fuel and agriculture. Other riverine species are *Faidherbia albida**, *Hyphaene thebaica*, *Salix subserrata*, *Tamarix aphylla**, *T. nilotica* and *Ziziphus spina-christi* (Boulos 1966; Obeid et al. 1982).

3. *Sahel Deciduous Bushland*. Rocky outcrops and shallow stony soils may support dense and often impenetrable bushland dominated by *Acacia mellifera* and *Commiphora africana*. Other associates include *Boscia senegalensis**, *Dichrostachys cinerea* and *Euphorbia candelabrum*. *Acacia mellifera* is often the dominant shrub of the poorly drained depressions on the leeward side of

some of the major rock outcrops (jebels), again often forming practically impenetrable thickets. In the drier parts *Acacia tortilis**, *Balanites aegyptiaca** and *Commiphora africana* are occasional constituents. The grass cover is poor, with plenty of bare ground between the stools; *Cymbopogon nervatus* dominates the margins of the depressions, *Aristida adscensionis* the more open areas and *Schoenefeldia gracilis* the less exposed areas.

4. *Sahel Montane Scrub Forest.* On Jebel Gurgeil, a northern outlier of the Jebel Marra complex, the scrub forest of the rocky sheltered slopes between 1700 and 2200 m is dominated by *Olea europaea* subsp. *laperrinei**. The principal woody associates are *Boscia senegalensis**, *Ficus cordata* subsp. *salicifolia* (syn. *F. salicifolia*), *F. populifera, Maytenus senegalensis, Rhus vulgaris* and *Vangueria venosa*. On the Jebel Marra massif the *Olea* has a more scattered distribution between 2300 and 3000 m, the communities in which it formerly occurred having been destroyed by cultivation, although in some isolated areas it may form an almost closed canopy (Quézel 1969; Wickens 1977b).

5. *Sahel Montane Secondary Grassland.* The dominant grasses of the Jebel Gurgeil plateaux form a dense 40 to 60 cm high sward dominated by the perennial grasses *Aristida adoensis, Andropogon distachyos, Cymbopogon commutatus, Heteropogon contortus, Hyparrhenia hirta, H. papillepes* and *Themedra triandra*. The infrequent shrubs present include *Dicrostachys cinerea*; a few scattered trees, such as *Acacia tortilis**, *Albizia amara* subsp. *sericocephala, Faidherbia albida** and *Terminalia brownii* also occur (Quézel 1969).

The steeper slopes and eroded lands of the Jebel Marra massif are dominated by the tussock grass *Andropogon distachyos*, frequently in association with *Themeda triandra* and *Hyparrhenia hirta*; the commonest shrubs are *Blaeria spicata* and *Lavandula pubescens*. The grasses of the better-drained rolling ash plains form 5-cm-high sward dominated by *Hyparrhenia multiplex* and *Vulpia bromoides*, together with *Aristida congestsa, Festuca abyssinica, Panicum pusillum, Pentaschistis pictigluma* and *Tripogon leptophyllus* (Wickens 1977b).

6. *Edaphic Grassland and Acacia Wooded Grassland.* In the Nile valley, the cracking Pleistocene clays of the Butana, with 100-400 mm annual rainfall, are virtually treeless, although occasional clumps of *Hyphaene thebaica* may be found on the silty soils near the Nile. The dominant grasses are *Cymbopogon nervatus, Hyparrhenia anthistirioides, Sehima ischaemoides* and *Sorghum purpureo-sericeum*. These often occur in an arceate pattern, although under heavy grazing pressure these patterns are now rapidly disappearing (Wickens and Collier 1970). The acanthaceous perennial herb *Blepharis ciliaris** may occur locally in pure stands.

Where the rainfall is between 400 and 570 mm, *Acacia mellifera* often forms extensive areas of impenetrable thicket, which alternate with open areas of bare clays or with *Schoenefeldia gracilis* where there is a veneer of sand. Further south the grassy areas are dominated by *Cymbopogon nervatus, Hyparrhenia anthistirioides, Sehima ischaemoides* and *Sorghum purpureo-sericeum.*

7. *Swamp and Aquatic Vegetation.* Large areas of swamp vegetation are found along the shores of Lake Chad and in the valley of the upper Niger south of Tombouctou. The most frequent dominant is *Cyperus papyrus*, with *Vossia cuspidata* forming a floating mat, and rooted in the mat the grasses *Echinochloa pyramidalis, E. stagnina, Oryza longistaminata* and *Paspalidium geminatum.*

3.1.1.4 Sinai Peninsula and Negev Desert

The two contiguous units – the Sinai Peninsula and the Negev Desert – are treated separately since their ecological separation coincides with the boundary between Israel and Egyptian Sinai. This is a distinction of convenience and has no political implication. The vegetation has been mapped and described in some detail by Danin (1983), who recognises twelve districts in Sinai and nine in the Negev, of which the Mediterranean sands and salt marshes and the gravely plains of central Sinai and northern Negev are common to both. The Sinai Peninsula extends from the Mediterranean and Suez Canal eastward to Rafah; in the south it is bordered by the Gulf of Suez and Gulf of Aqaba. The northern sandy coastal strip grades into the central tableland, or Tih Desert, culminating in the southern mountains. The rainfall is the lowest along the entire Mediterranean coast, with less than 100 mm between Port Said and El Arish, rising sharply north-east from 100 mm at El Arish to ca. 200 mm at Rafah, and 400 mm at Gaza and only 13 mm at El Tor in the southern part of the Gulf of Suez (Walter and Breckle 1986).

The twelve districts recognised by Danin (1983) are summarised by Walter and Breckle (1986) under three geomorphic units:

1. *Mediterranean Sands and Salt Marshes (Northern Coastal Strip).* The groundwater is relatively high and in places reaches the surface where the continuous evaporation results in salt accumulation. The dominant halophytic vegetation of these saline interdunal depressions are *Arthrocnemum macrostachyum*, Halocnemum strobilaceum*, Juncus rigidus* (syn. *J. arabicus*), *Limoniastrum monopetalum, Phragmites australis, Suaeda aegyptiaca* and *S. vermiculata.* At slightly higher levels there are *Nitraria retusa*, and *Tamarix nilotica*, with *Thymelaea hirsuta** and *Zygophyllum album* in the even higher,

less saline habitats. The underlying watertable supports the cultivation of *Phoenix dactylifera** (date palm), *Riccinus communis* (castor oil plant), which is grown as an annual crop, as well as irrigated crops of apples, pears, guavas, figs, olive and citrus.

2. *Central Tableland of the Tai Desert.* The vegetation is sparse and scattered. In the wadis the dominant vegetation includes such shrubs and subshrubs *Anabasis articulata, A. setifera, Haloxylon scoparum**, *Lycium shawii, Retama raetum**, *Salsola tetrandra, Seriphidium herba-alba**, *Zilla spinosa**, *Zygophyllum album, Z. coccineum* and *Z. dumosum*. The dominant trees in the lower parts of the wadis are *Acacia iraqensis* (syn. *A. gerrardii* subsp. *negevensis*), *A. tortilis* subsp. *raddiana**, *Tamarix aphylla** and *T. nilotica.*

3. *Southern Mountains.* These occupy the southern third of the peninsula. The damp ravines contain such tree and shrub species as *Cupressus sempervirens, Ephedra alata**, *Crataegus × sinaica, Ficus carica, F. palmata* (syn. *F. pseudosycomorus*) and *Moringa peregrina*. Above 1000 m, *Artemisia judaica* and *Zilla spinosa** still persist but are giving way to Irano-Turanian elements that are better able to tolerate the low winter temperatures. A complete floristic change appears above 1600 m with the appearance of Mediterranean species.

The Negev Desert adjoins the eastern boundary of the Sinai Peninsula, forming a triangle with its western boundary from Rafah to Elat, the northern boundary extending north-eastwards along the Mediterranean coast to link up near the Israel border with the Palestinian desert strip (Sect. 3.5.2.1), and then following the 150 mm isohyet northwards to the Dead Sea at ca. 32°N, with the eastern boundary formed by the Wadi Arava, an extension of the Jordanian- Dead Sea rift.

The Negev Desert has a desert climate with winter rains determined by the cyclones in the eastern Mediterranean, the rainfall decreasing towards the south from 162 mm at Beersheba to 20 mm at Elat. The summer months are dry, but because of the proximity of the Mediterranean, the humidity is high, so that dew falls frequently at night. In central Negev the mean annual quantity of dew falling at Avedat is 33 mm, ranging over 195 nights between 0.35 and 0.16 mm; ground frosts with temperatures down to -7 °C have also been recorded. The days are hot and dry, especially when the south-east desert wind known as the Khamsin is blowing. Temperatures then rise to 42 °C, and in the Araba Valley up to 48 °C (Walter and Breckle 1965).

The nine vegetation associations in the Negev of Danin (1983) are summarised by Walter and Breckle (1986) under four regions:

1. *Northern Negev.* This includes the coastal plain, the adjacent north-western plain and hilly country, with alluvial sands, sand dunes, and in the south stony

soils and sierozems overlying loess. Sierozems are characterised by a lack of leaching, low organic content and an accumulation at the B and C horizon interfaces. Most of the loess and sandy loess coastal plains are cultivated and bear a flora of winter and summer weeds. The summer weeds include *Amaranthus* spp., *Conyza bonariensis, Portulaca oleracea** and *Solanum eleagnifolium*. The annual *Helianthemum ledifolium**, which also occurs on the loess soils, is the host to the much valued subterranean ascomycetous truffle *Terfezia leonis*, while the subshrub *H. lippii**, another truffle host, is confined to sandy soils. *Artemisia monosperma**, *Convolvulus lanatus** and *Stipagrostis scoparia** dominate the coastal dunes during the various stages of colonisation. Inland from the coastal plain the Negev lowlands are intensively grazed by livestock and much of the woody vegetation is destroyed by cutting for fuel. In winter much of the area is dominated by the unpalatable *Asphodelus ramosus* (syn. *A. microcarpus*). In the wetter, northern hills, the remnant shrubby vegetation consists of *Noaea mucronata* and *Seriphidium herba-alba**, with *Zygophyllum dumosum* dominant in the drier, southern areas. The salt marshes bordering the Dead Sea are dominated by *Alhagi graecorum**, *Arthrocnemum machrostachyum**, *Prosopis farcta, Sarcocornia fruticosa* (syn. *Arthrocnemum fruticosum*), *Seiditzia rosmarianus, Suaeda monoica, S. vermiculata*, and *Tamarix nilotica*. The dominant wadi trees and shrubs include the Sahelian *Acacia tortilis* subspp. *raddiana** and *tortilis**, *Moringa peregrina, Salvadora persica* and *Ziziphus spina-christi**. Among the dominant associates of calciferous soils are *Atriplex halimus**, *Salsola tetrandra*, and *S. vermiculata*, and on stony soils *Anabasis articulata, Blepharis ciliaris**, *Helianthemum kahiricum, Lycium shawii, Nauplius graveolens, Ochradenus baccatus, Salvia aegyptiaca, Zygophyllum dumosum* and the grass *Cenchrus ciliaris*.

2. *Central Negev*. This is the stone desert of the northern limestone and sandstone highland plateau with scattered trees. Scattered trees of *Ceratonia siliqua, Crataegus aronia, Pistacia atlantica*, and *Prunus korschinskii* occur at the divide of the Judean Mountains. The limestone and dolomites of the northern highlands support such species as *Haloxylon scoparium**, *Sarcopoterium spinosum, Seriphidium herba-alba**, *Thymelaea hirsuta** and *Zygophyllum dumosum*. On bare sandstone the dominant species are *Anastatica hierochuntica** and *Blepharis ciliaris**; *Anabasis articulata* is dominant on loose sands, with such associates as *Asphodelus ramosus* and *Stipagrostis obtusa*. On the deeper sands *Calligonum polygonoides* subsp. *comosum** is dominant; associated species include the grasses *Stipagrostis ciliata* and *S. plumosa*. The stony soils of the hills of conglomerate rocks support *Zygophyllum dumosum* together with such associates as *Aizoon canariense**, and *Neurada procumbens*. The wadis support large shrubs of *Calligonum polygonoides* subsp. *comosum** and *Retama raetum**.

3. *Southern Negev*. This area lies north-west of Elat, with exposed volcanic or metamorphic rocks, the gravely sandy plains of which have already been described in Section 3.1.1.2 as an extension of the Red Sea coastal desert.

4. *Arava Valley*. An extension of the Jordanian-Dead Sea rift, the arid, 100 km long, 8 to 20 km wide Arava Valley, with its northern part below sea level, drains into the Dead Sea. The soils consist of sand, marl and gravel alluvium with scattered salt pans. In the lower valley *Acacia tortilis* subspp. *raddiana** and *tortilis** and *Ziziphus spina-christi** occur.

3.1.1.5. Arabian Peninsula

Separated from Africa by the Red Sea rift, the Arabian Peninsula rises steeply from a narrow coastal plain to the Hijaz highlands in the north, the ridge continuing southwards from Mecca as the 'Asir mountains. In the north the highlands reach a height of 1850 m, while in Yemen Mount Hadhur rises to 3760 m. The 40 to 140-km-wide highlands passes gradually eastwards into the high plateau of the Arabian shield. Along a north-east transect through Riyadh the highlands descend from ca. 1500 m to 600 m, passing into a step landscape of obliquely orientated rock layers and hence into the sands of the Dahana Desert (Walter and Breckle 1986). Sand deserts, apart from certain coastal areas and the mountain areas of the Yemen, Hadramaut and Hadjar (Oman), cover the greater part of the Peninsula. To the north lies the 400-km-wide An Nafud, linked by the 1000-km-long, sickle-shaped Dahana to the Rub' al Khali. The largest continuous area of sand on earth, the 650 000 km^2 Rub' al Khali covers most of the southern third of the peninsula (Walter and Breckle 1986). There are no perennial rivers, although the Peninsula is intersected by a number of ill-defined wadis of intermittent flow. Only streams at the fringe of the Peninsula reach the sea and then only occasionally (Kaul and Thalen 1979).

Walter and Breckle (1986) recognise a northern winter-rain area, a rain-free area in the Rub' al Khali and a summer-rain area. The latter includes the area to the south of Jiddah bordering the Red Sea, the high mountains in the south-west, the seaboard south of the Rub' al Khali and extending north-west along the Gulf of Oman as far as the Strait of Hormuz. This climatic zone corresponds to that of the Sahel.

In the north, the winter rains are caused partly by the winds that rise up the escarpment from the Red Sea and by cyclones from the Mediterranean area. This also applies to the coastal area bordering the Persian Gulf. The rainfall, however, is irregular and unpredictable (Walter and Breckle 1986). For example, in Qatar at Doha Airport the annual precipitation over 17 years ranged from 0.4 to 302 mm, with a mean annual rainfall of 78 mm (Batanouny

1981). Temperatures are generally high, although absolute minima a little below 0 °C were registered in the northern part of the Peninsula and in the high mountains. Winter temperatures at Aden may reach 25 °C in January, and during the hot dry summer 32 °C in June. At Doha Airport the mean monthly, absolute maximum and absolute minimum temperatures were 17.1, 30.7 and 3.8 °C, respectively, for January and 33.6, 49.0 and 21.0 °C for June. Humidity is very high at the coast; in some places in the interior it can be extremely low. Except for the high mountains where there is precipitation from rising air, the climate is extremely arid (Walter and Breckle 1986) or even hyperarid (see UNESCO 1979).

The greater part of the flora of the Arabian Peninsula belongs to the Saharo-Sindian regional zone of endemism, in which White and Léonard (1991) recognise a Saharan regional subzone, an Arabian regional subzone and, owing to its rather high percentage of endemism, a Nubo-Sindian local centre of endemism spreading from the Nubian Desert along the coasts of the Red Sea and Persian Gulf towards the Sind and the Indian Desert regional subzone. The extreme south of the Arabian Peninsula represents an extension of the Somali-Masai regional centre of endemism.

The following vegetation units are recognised by Walter and Breckle (1986):

1. *Vegetation of the Central Red Sands.* Red sands cover the entire northern part of the An Nafud and the Dahana as far as the Rub' al Khali. Characteristic species include *Artemisia monosperma**, *Calligonum polygonoides* subsp. *comosum** and *Panicum turgidum**.

2. *Rhanterium epapposum Semi-Desert.* The composite *Rhanterium eppaposum** is typical of the gravel semi-deserts that cover large areas of Saudi Arabia between 25 and 30°N, extending as

3. *Stipa capensis Ephemeral Vegetation.* The winter annual *Stipa capensis* (syn. *S. tortilis*) appears soon after the winter rains and covers wide areas of central Arabia.

4. *Shrub Vegetation of the Wadis.* At ca. 30°N shrub vegetation of the wadis includes *Ephedra alata**, *Haloxylon persicum**, *Mesembryanthemum forsskalei** and the halophytic *Seidlitzia rosmarinus.*

5. *Vegetation of the White Sands.* The region of white sands lies between the Dahana dune area and the Persian Gulf, and includes Kuwait. Forming a low-lying plain with calcareous soils, the abrupt vegetation changes are associated with the nature of the soil. Characteristic species of sandy soils include *Cyperus conglomeratus* and *Panicum turgidum**, *Rhanterium eppaposum* and, on gravel, *Haloxylon salicornicum** and *Zilla spinosa**.

6. *Coastal Vegetation.* Typical coastal vegetation consists of open *Acacia-Maerua* savanna with *Acacia tortilis**, *Maerua crassifolia* and the grasses *Lasiurus scindicus** and *Panicum turgidum**, with the halophytes *Halocnemum* and *Limonium* along the shore (Walter and Breckle 1986). The Yemen inland plain as far as the foothills supports *Acacia ehrenbergiana, Cadaba rotudifolia, Salvadora persica* and *Tamarix aphylla** (UNESCO 1979).

The characteristic inland vegetation is dense bushland thicket, 3 to 5 m tall with scattered emergent trees up to 9 m; evergreens are scattered but rare, while succulents are usually present. The woody species present include *Acacia* spp., *Adenia venenata, Boswellia sacra, Cadaba* spp., *Ceratonia oreothauma* subsp. *oreothauma, Combretum molle, Commiphora* spp., *Conocarpus lancifolius**, *Lannea* spp. and the fan palm *Livistona carinensis* (White and Léonard 1991). Among the grasses are *Andropogon distachyos, Cenchrus ciliaris, Eragrostis papposa* and *Stipagrostis obtusa* (Walter and Breckle 1986). Small enclaves of evergreen bushland, thicket and scrub forest occur at low altitudes either where the rainfall is slightly higher or where there is mist precipitation. The characteristic species include *Euclea racemosa* var. *schimperi, Ficus ingens, Olea europaea* subsp. *cuspidata** and *Osyris compressa.*

7. *Halophytic Vegetation of the Persian Gulf.* Characteristic species of the saline soils are *Halocnemum strobilaceum**, *Nitraria retusa**, *Salsola imbricata* (syn. *S. baryosma*), *Suaeda vermiculata* and the grass *Aeluropus lagopoides**.

8. *Juniperus procera Montane Vegetation.* The affinities of the southern Arabian montane flora belong to an impoverished Afromontane flora, of which *Juniperus procera* is the most abundant and where even the densest stands appear to have a fairly open canopy (White and Léonard 1991). The grass *Themeda triandra** is also present (Walter and Breckle 1986).

3.1.1.6 Iranian Arid Coastal Region

The Iranian arid coastal region occurs along the eastern coasts of the Persian Gulf and the Gulf of Oman. The winters are temperate with temperatures of 10-20 °C and the summers very hot. The subtropical desert vegetation represents an eastern extension of the Arabian Desert. The vegetation consists of a spiny shrub and grass pseudo-steppe with such Saharo-Sindian trees, shrubs and grasses as *Acacia nilotica* subsp. *indica, A. senegal**, *A. seyal, Calligonum* spp., *Calotropis procera, Capparis spinosa, Salvadora persica, Ziziphus nummularia* and grasses *Cenchrus ciliaris* and *Hyparrhenia hirta.*

3.1.1.7
Indian Desert

The Indian or Thar Desert includes the arid portions of north-western India and eastern Pakistan. The region lies beyond the full force of both the south-west monsoon and south-east monsoons, and as a result the rainfall is both scanty and erratic. The annual precipitation is 300 mm or less, most of which falls during the monsoon season, from June to September. Winter rains are rare and droughts are frequent. The winters are cool, with temperatures sometimes falling below 0 °C. The mean maximum and minimum temperatures during January, the coldest month, are 28.5–27.5 °C and 6.0–10.5 °C, respectively; nights are always much cooler and the diurnal range large. The mean maximum and minimum temperatures during the hot season, from April to June, are 39–42 °C and 25–28 °C, respectively, with May normally being the hottest month (Bhandari 1990).

The following vegetation units are recognised by Bhandari 1990):

1. *Sandy Plains.* Dominated by such trees and shrubs as *Tecomella undulata* (syn. *Tecoma undulata*), *Prosopis cineraria** and *Salvadora oleoides*, together with *Azadirachta indica, Balanites roxburghii, Calligonum polygonoides*, Capparis decidua, Crotalaria burhii, Leptadenia pyrotechnica** and *Ziziphus nummularia*. *Crotalaria burhia* is often the dominant subshrub of the sandy soils; associated species include *Aerva javanica* vars. *bovei** and *javanica*, Calotropis procera, Farsetia stylosa* (syn. *F. hamiltonii*), *Indigofera argentea, I. cordifolia, Launaea residifolia, Leptadenia pyrotechnica** and the sedge and grasses *Cyperus arenarius, Cenchrus biflorus, C. prieurii* and *Cymbopogon martinii*. The dune crests are often colonised by *Citrullus colocynthis* and the grass *Lasiurus scindicus**. In stabilised areas, favoured by an accumulation of organic matter, the characteristic species include big, tuft-forming bushes such as *Acacia jaquemontii, Calligonum polygonoides*, Capparis decidua, Crotalaria burhia, Leptadenia pyrotechnica*, Sericostoma pauciflorum* and *Ziziphus nummularia*.

2. *Gravel Plains.* With such trees and shrubs as *Acacia leucophloea, A. nilotica* subsp. *indica, Capparis decidua* and *Ziziphus* spp.

3. *Valleys and Depressions.* With *Acacia senegal** and *Euphorbia caducifolia* supporting numerous scramblers and climbers.

4. *Rocky Habitats.* The dominant trees and shrubs include *Anogeissus* spp., *Balanites roxburghii, Commiphora wighti** and *Euphorbia caducifolia; Boswellia serrata* is also present on some hill slopes. Associated scramblers and

climbers include *Asparagus racemosus, Corallocarpus epigaeus, Rhynchosia minima, Rivea hypocrateriformis and Sarcostemma viminale* (syn. *S. acidum*). Among the herbs are *Chascanum marrubifolium, Ocimum americanum, Seddera latifolia* and *Talinum portulacifolium.*

5. *Riverine Vegetation.* An open assemblage of dwarf trees, generally not exceeding 3.5 m high but dominated by shrubs, is confined to the floodplains and riverbeds (Kaul and Thalen 1979).

6. *Halophytic Vegetation.* Among the more common plants are *Cressa cretica**, *Haloxylon recurvum, H. salicornicum**, *Portulaca oleracea**, *Salsola imbricata, Sesuvium sesuvioides, Suaeda vermiculata, Tamarix indica* (syn. *T. troupii*), *Trianthema triquetra, Zaleya redimita* and *Zygophyllum simplex* (Bhandari 1990), and the grass *Leptochloa fusca**.

3.2 Horn of Africa

Lying almost entirely below 900 m, the Horn of Africa corresponds fairly closely to the Somalia-Masai regional centre of endemism of White (1983), extending from south-east Sudan and the lowlands of Eritrea and southern Ethiopia through Somalia and Kenya to the plains of northern Tanzania; the island of Socotra is also included. The climate is arid or semi-arid, with the annual rainfall generally less than 500 mm, and in places as low as 20 mm [ZB and ZE I-III, II-III, III(t)]. In most places the rainfall is bimodal and separated by periods of drought, elsewhere the rainfall is irregular. The bimodal rainfall is related to the influence of the south-west monsoon in summer and the north-east monsoon in winter, the rainfall occurring during the intervening calm periods. The temperatures are high, with the mean monthly temperatures mostly between 25 and 30 °C.

Most of the region is covered with deciduous bushland and thicket, which grade into and are replaced by semi-evergreen and evergreen bushland and thicket on the lower mountain slopes. The following descriptions of the vegetation communities follow White (1983):

1. *Acacia-Commiphora Deciduous Bushland and Thicket.* Characteristically this community is dense, mainly deciduous bushland or impenetrable thicket, 3 to 5 m tall with scattered emergent trees up to 9 m tall. In those areas of Kenya and Somalia where the rainfall is less than 250 mm, the vegetation is intermediate between bushland and shrubland, consisting of low, 2 to 3-m-tall bushes and stunted trees, principally of *Acacia reficiens* subsp. *misera,* which form a thin cover over a ground layer of mainly small shrubs.

In the south-east corner of Ethiopia and north-eastern and central Somalia, the open *Acacia-Commiphora* bushland on gypsaceous soils supports *Commiphora guidottii* (*C. sessiliflora*), often in association with *Boswellia rivae* and *Cyphostemma betiforme*. The economically important *Cordeauxia edulis** also occurs in central Somalia and the Ogaden of Ethiopia. In northern Somalia, from near sea level up to ca. 700 m, often on limestone boulders and cliffs, the endemic *Boswellia frereana* occurs in association with *Acacia* spp., *Adenium obesum**, *Boswellia bricchettii*, *Cadaba longifolia*, *Commiphora* spp., *Dobera glabra* and *Lannea obovata*. Above 700 m *Boswellia frereana* is replaced by *B. sacra*, which reaches its highest limit at 1230 m in the evergreen and semi-evergreen bushland and thicket (Thulin and Warfa 1987). Ephemeral grasses appear after the rains, but perennial grasses are insignificant.

2. *Evergreen and Semi-Evergreen Bushland and Thicket.* This type of vegetation occurs on the drier slopes of mountains and upland areas from central Tanzania to Eritrea and beyond. Although varying greatly in composition and richness, the characteristic species are *Carissa edulis*, *Dodonaea viscosa*, *Olea europaea* subsp. *cuspidata** and *Tarchonanthus camphoratus*, together with species of *Acokanthera*, *Euclea*, *Sansevieria* and *Teclea*, with succulent species of *Aloe* and *Euphorbia* also present.

In Somalia the principal dominants of the evergreen bushland are *Acokanthera schimperi*, *Buxus hildebrandtii*, *Cadia purpurea* and *Dodonaea viscosa*. In parts of Kenya *Acacia* spp. and *Combretum molle* have invaded some areas where the evergreen thicket has been severely degraded and occur in association with such evergreen species as *Carissa edulis*, *Dodonaea viscosa*, *Euclea divinorum*, *E. racemosa* subsp. *schimperi* and *Tarchonanthus camphoratus*.

3. *Semi-Desert Grassland and Shrubland.* Where the rainfall is between 100 and 200 mm the deep sands of the semi-desert grassland are dominated by *Centropodia forskalii* (syn. *Asthenatherum glaucum*), *Eragrostis hararensis* or *Panicum turgidum**. Shrubland occurs on stony soils. On the coastal plain of Somalia the principal shrubby species are *Aerva javanica**, *Farsetia longisiliqua* and *Jatropha pelargoniifolia*. Further inland, on the gyseous soils, the former grasslands dominated by *Chrysopogon plumulosus* and *Dactyloctenium robecchii* have been replaced by dwarf shrublands.

4. *Edaphic Grassland.* In Somalia treeless plains of waterlogged grassland occur within *Acacia bussei* scrub and *Acacia-Commiphora* bushland and are dominated by *Chrysopogon plumulosus*. Small, temporary pools are dominated by *Andropogon kelleri* and larger, ill-defined areas of inland drainage are covered with *Chloris roxburghiana* and *Cynodon dactylon* with scattered *Acacia tortilis**. Clay plains within *Acacia etbaica* scrub woodland are dominated

mainly by *Andropogon* spp., *Aristida adscensionis, Cenchrus ciliaris, Chrysopogon plumulosus, Eragrostis* sp. and *Panicum coloratum.*

5. *Scrub Forest.* In those areas at relatively low altitudes where the rainfall is higher than of deciduous bushland and thicket but too low to support true forest, there is 7 to 10-m-tall scrub forest dominated by species of *Commiphora* and candelabra *Euphorbia.*

6. *Riparian Forest.* Economically important trees of the northward flowing seasonal rivers of northern Somalia are *Conocarpus lancifolius** and *Mimusops angel.*

3.3 Southern Africa

3.3.1 Cape Region

The Cape region occupies the south-western and southern part of Cape Province, South Africa, between 32-35°S and 18–27°E. The landscape is dominated by a series of folded mountain ranges of 1000–1500 m above sea level, with individual peaks in excess of 2000 m. The average annual rainfall is mostly from 300 to 2500 mm, with up to 5000 mm in some localities in the mountains (ZB IV). In the western part of the region 60–80% of the rain falls during the winter months, but is more evenly distributed to the east of Swellendam. On the higher mountains the summer drought is alleviated by moisture-bearing clouds from the south-east; on the west coast sea mists are fairly frequent. Snow falls regularly on the higher mountains during the winter, especially in the west, but only persists on the southern slopes. In general the winters are mild and frost is unknown along the coast but may occur inland and in the mountains.

The predominant vegetation is shrubland (fynbos), mainly as 1 to 3-m-tall sclerophyllous, shrubland. However, there are large areas of secondary shrubland dominated by the composite *Elytropappus rhinocerotis* (rhenosterbos). Coastal fynbos occupies much of the coastal plain, with local patches of bushland and thicket dominated mainly by tropical species. Provided there is sufficiently long protection from fire, *Protea* and other tall shrubs may form dense, impenetrable, 4 to 6-m-high thickets. However, most fynbos species are unable to regenerate under these conditions and become moribund and die; it follows, therefore, that such thickets are a transient phenomenon and are not described as a separate unit. The fynbos is highly combustible, especially during the hot dry summer months, and it is now widely accepted that periodic burning is necessary for its healt hy maintenance.

The following vegetation units are recognised by White (1983):

1. *Cape Shrubland (Fynbos).* Typically fynbos contains scattered tall bushes and, less frequently, widely spaced trees. The conspicuous, monocotyledonous 'switch plants' of the Restionaceae are invariably present, and may even be dominant in some extreme habitats. The typical restionoid growth habit of the Restionaceae, and also of some Cyperaceae, consists of tufted or rhizomatous plants ranging from 20 cm to 2 m or more tall with green, tubular or wiry stems bearing reduced, non-photosynthetic scale leaves. The scale leaves persist for more than 1 year, but usually for less than 4. Geophytes are plentiful but annuals are only conspicuous in the drier fynbos types. Grasses are now uncommon but were more abundant before the advent of the Europeans.

2. *Secondary Cape Shrubland (Rhenosterbosveld).* Rhenosterbosveld occurs at below 300 m between the foot of the mountains and the coastal plain in two main areas: in the south and in the west. The annual rainfall is between 300 and 500 mm. The fertile soils derived from shales have been cultivated for centuries and are readily invaded by the rhenosterbosveld. The rhenosterbosveld is usually up to 1 m tall, rarely attaining 2 m, of which the dominant species is *Elytropappus rhinocerotis*. Although rich in species, the characteristic fynbos families Ericaceae, Proteaceae and Restionaceae are absent from the southern block. The original vegetation was probably evergreen scrub dominated by *Olea europaea* subsp. *cuspidata** and *Sideroxylon inerme*. The southern rhenosterbosveld is rich in grass species, many of which, like *Hyparrhenia hirta*, are widespread in the tropics. The western block, however, has a greater admixture of fynbos species. Bushy species are fewer but include *Olea europaea* subsp. *cuspidata** and characteristic, non-tropical annual grasses such as *Pentaschistis patula* and *Tribolium echinatum* (syn. *Lasiochloa echinata*).

3. *Coastal Bushland and Thicket.* Certain trees and shrubs which are not true fynbos species locally form thickets or scrub forest up to 10 m tall along the south coast, including *Cassine peragua, Chionanthus foveolatus, Euclea* spp. and *Olea europaea* subsp. *cuspidata**.

4. *Riparian Bushland and Thicket.* Characteristic species throughout the length of the watercourses are *Cunonia capensis, Hartogia capensis, Ilex mitis* and *Maytenus oleoides.*

5. *Transition to Karoo.* The arid fynbos representing the transition from typical Cape to typical Karoo vegetation contains such typical shrubby Karoo genera as *Chrysocoma, Eriocephalus, Euryops, Hermannia, Lightfootia, Pteronia, Selago* and *Walafrida*, together with succulents such as *Aloe ferox* and *Euphorbia mauritanica.*

3.3.2 Karoo-Namib Region

The term 'karoo' is the Nama word for 'dry area'. The Karoo-Namib region occupies the central, northern and north-western parts of Cape Province immediately to the north of the Cape region, mostly north of 33°N and between 17 and 25°E and extending northwards as an increasingly narrow band along the entire length of Namibia into south-west Angola to about 11°S, including both the coastal plain and the escarpment of the interior plateau. The topography of the interior of Cape Province is extremely even, except where broken by dolerite dikes, sills and other intrusions. The soils are clayey and tend to accumulate salts, giving rise to extensive brackish seasonal swamps. In the north-west Cape the terrain is one of subdued relief, with rugged mountains in western Namaqualand and in the gorge tract of the Orange River.

The semi-desert Karoo shrubs are generally less than, and frequently much less than, 2 m tall. The stem succulents are represented by species of Asclepiadaceae (especially *Caralluma, Hoodia, Huernia, Stapelia* and *Trichocaulon*), *Euphorbia* and *Senecio* (*Kleinia*), and by species of *Pelargonium* and *Sarcocaulon* which shed their small mesomorphic leaves during drought. Most of these are less than 1 m tall but with the largest *Euphorbia* species up to 2 m high. Leaf succulents are pre-eminently represented by the Aizoaceae.

The Namib Desert stretches for ca. 2000 km along the west coast of southern Africa, from the Olifants River in Namaqualand along the entire length of Namibia and continuing northwards for a short distance to Moçâmedes in Angola. For most of its length it forms a belt 100-km wide between the coast and the base of the escarpment of the inland plateau. The Kuiseb River divides the southern sandy coastal plain from the northern gravel plains. In the north the up to 50-km-wide, arid coastal sand field of the Moçâmedes Desert extends southwards to the lower Swakop River. Inland the sand field merges with calcrete, gypsum and gravel plains lying at 300-600 m a.s.l., with scattered rocky outcrops rising to 500-1000 m and the Brandberg Mountain rising to 2700 m. From the Kuiseb River the coastal sand sea extends southwards to the Orange River. South of the Orange River the rather narrow and undulating coastal sandy plain becomes less impressive while the gravel valleys and plains of the mountainous landscape become increasingly dominant. The sandy plains continue as a rather wide coastal belt to south of Port Nolloth and then persist as a narrow coastal belt. The boundary is arbitrarily defined as the 100-mm isohyet, although much of the area receives considerably less rainfall (White 1983; von Willert et al. 1992).

The annual rainfall in the Namib Desert is less than 100 mm; elsewhere it rarely exceeds 250 mm. To the west of a line running from Spencer Bay in Namibia through Calvinia in South Africa to Sutherland more than 60% of the rain falls during the winter months (ZB IIIwr). East from a line from Swakopmund in Namibia through Pofadder and Fraserburg in South Africa to Willomore more than 60% of the rain falls during the summer (ZB IIIsr). The rainfall varies greatly in both amount and distribution from year to year; even in the wetter parts of the summer-rainfall belt winter rain may occur once in ca. 12 years.

At Swakopmund the mean annual rainfall is only 10 mm and on average rain occurs only once every 2 years. However, there are significant amounts of night precipitation along the coast from the coastal fogs, with 94–215 days of fog each year; these fogs are just as frequent elsewhere along the coast of southern Namib. The fog rarely penetrates as far as 50 km inland. The precipitation provided by individual fog is mostly less than 0.1 mm and never more than 0.7 mm, providing a total precipitation of 40–50 mm. The rapid evaporation of the moisture means that the fog contributes little to vascular plant life except where the fog condenses on rock faces and collects in crevices.

The winter rainfall region of the southern Namib (ZB IIIwr) supports a far more luxuriant vegetation, mainly of succulents, than areas of northern Namib with equivalent amounts of rainfall. The winter rains, falling in the cooler months of the year, are ecologically more effective due to their lower evaporation losses. Also, the winter rains tend to fall as soft drizzles or light showers whereas the summer rains are much heavier. Furthermore, truly season deciduous species will not produce new leaves, neither will the seeds of annuals germinate if the rains falls in the wrong season (von Willert et al. 1992).

The coastal belt is frost-free except in the south, where occasional light frosts occur during July, as at Port Nolloth. Further inland, in southern Namibia and throughout the interior of Cape Province, the frost period lasts 5–6 months, although the mean minimum temperature is never below zero except in the extreme east.

White (1983) recognises the following vegetation types:

1. *Bushy Karoo Shrubland.* This occurs in the Great Karoo, the Little Karoo, Robertson Karoo, and on the rocky hills of Namaqualand and the lower valleys of the Orange River and northward to south-west Angola. It consists of shrubland dotted with small bushy trees and large shrubs, Succulents are usually abundant, especially in the south, while non-succulent shrubs are always present but are usually subordinate to the succulents.

2. *Succulent Karoo Shrubland,* This is largely confined to the sandy plain of Namaqualand and the heavier stony soils of the Namaqualand escarpment foothills below 610 m. The annual rainfall is less than 200 mm, falling in winter. The dominant vegetation consists of succulents, ranging from those almost entirely embedded in the soil to shrubs up to 2 m tall. Grasses are few and usually inconspicuous.

3. *Dwarf Karoo Shrubland.* Extensive shrubland occupies arid and central Karoo, including the lower slopes of the Karoo mountains. Bushes and trees are absent. The few large shrubs present are represented mainly by *Rhigozum trichotomum.* The relatively few succulents present are generally inconspicuous. Grasses are more abundant and become increasingly so towards the east. The soils are often saline, with *Salsola tuberculata* widespread and locally dominant, while *Salsola aphylla* and other halophytes occupy the extensive brackish flats of the arid Karoo

4. *Montane Grassy Karoo Shrubland.* The original vegetation was probably more grassy; now shrublets of *Chrysocoma, Elytropappus rhinocerotis, Eriocephalus, Pentzia, Ruschia,* etc. are abundant.

5. *The Outer Namib Fog Desert.* The extremely sparse vegetation consists of isolated plants of *Barleria solitaria, Ectadium virgatum, Indigofera cunenensis, Merremia multisecta, Petalidium* spp. and the grasses *Eragrostis cyperoides* and *Stipagrostis ramulosa.* To the south of Swakopmund a large belt of shifting dunes extends as far south as Luderitzbuch. The few species known to occur include *Monsonia ignorata, Trianthema hereroensis, Stipagrostis gonatostachys* and *S. sabulicola.* In the central Namib the dune vegetation consists of scattered cushions of *Psilocaulon salicornioides, Salsola aphylla* and *S. nollothensis.*

The gravel desert to the north of Swakopmund is, apart from crustose lichens, normally devoid of vegetation except when the rare occasions of adequate rainfall permit an ephemeral halophilous succulent vegetation. In the higher rainfall gravel desert of southern Namib the perennial vegetation includes the stem succulent *Euphorbia gummifera* and small shrubs of *Sarcocaulon spinosum* and *Zygophyllum retrofractum.* Among the predominant dwarf succulents inhabiting rock crevices are a number of stem succulents such as *Hoodia currorii* and *Trichocaulon* spp., stem succulents with deciduous leaves such as *Adenia pechuelii, Pelargonium otaviense, Sarcocaulon mossamedense* and *Senecio longiflorus* as well as the leaf succulents *Aloe asperifolia, Cotyledon orbiculata** and *Lithops* spp.*, etc.

6. *The Inner Namib Desert.* Succulents are represented by *Sesuvium sesuvioides* and the grasses by *Eragrostis nindensis, Kaokochloa nigrirostis, Stipagrostis ciliata* and *S. obtusa.* South of the Orange River, on Kalahari sands with less than 40 mm annual rainfall, *Stipagrostis brevifolia* is the dominant species. There may be scattered individuals of the large shrub *Parkinsonia africana* and the few Karoo dwarf shrubs represented by the genera *Hermannia* and *Lycium.*

7. *The Welwitschia bainsii Transition Zone. Welwitschia mirabilis* (syn. *W. bainsii*) occurs in the broad, shallow and barely discernible channels between the Kuiseb River just south of the Tropic of Capricorn to San Nicolau in southern Angola that receive floodwaters from the higher land. Following rain the intervening plain is dominated by such grasses as the annual *Stipagrostis subacaulis* and annual forms of the perennial *S. hochstetterana*. At its eastern limit in Namibia near the township of Welwitschia, where the rainfall is 200 mm, *Welwitschia* occurs on gravel soils of deltaic deposits, often in the shade of *Colophospermum mopane* and *Terminalia prunioides*.

8. *River-Bed Communities*. Open riparian woodland bordering the larger rivers supports *Acacia erioloba, Euclea pseudebenus, Faidherbia albida*, Salvadora persica*, etc., with *Adenolobus, Commiphora saxicola* and *Parkinsonia africana* along the larger erosion channels. On the sand dunes in the bed of the lower Kuiseb, the cucurbit *Acanthosicyos horridus* forms dense impenetrable thickets. The river estuaries and floodplains with accessible groundwater support various halophytes including *Salsola* spp., *Sarcocornia pillansiae* (syn. *Arthrocnemum dunense*), *Suaeda plumosa* and *Zygophyllum stapfii*.

9. *The Desert of Moçâmedes*. The flora and vegetation types are similar to those of the Namib proper but with the addition of more species of tropical affinity. *Welwitschia mirabilis* is absent from the coastal belt of mobile dunes to the south of Porto Alexandre but otherwise scattered throughout.

3.3.3 Kalahari-Highveld Region

The Kalahari-Highveld region, the Kalahari-Highveld regional transition zone of White (1983), separates the savanna woodlands of South Africa, Zimbabwe and Zambia in the east from the Karoo-Namib region in the west. Most of the region occurs on the great Interior Plateau of southern Africa, extending in only a few places in the east beyond the Great Escarpment. The Kalahari basin, lying between 850 and 1000 m, occupies the central part, the land rising gradually to the east and west.

The narrow strip formed by the peripheral highlands in the west reach their maximum elevation of 2484 m in the Windhoek Mountains of Namibia. To the east and south-east the land rises gradually towards the plateau rim and the Great Escarpment to more than 2000 m. The annual rainf all is mainly between 250 and 500 mm, increasing somewhat in the east towards the Drakensberg mountains (ZE III–II). The winter temperatures are low except in the narrow north eastern extension into Angola. Frost is widespread and severe.

White (1983) recognises the following vegetation types:

1. *Kalahari Thornveld.* The characteristic vegetation of the Kalahari sands is wooded grassland. In Botswana the more or less continuous grass sward is less than 1 m tall. The principal trees and shrubs include *Acacia* spp., *Boscia albitrunca, Dichrostachys cinerea* and *Terminalia sericea*; the tuberous perennials *Harpagophyton procumbens** and *Tylosema esculentum** are also present. Broad-leaved species become increasingly more abundant in the northern parts. The area of parallel dunes bordering the Namib has a vegetation which represents a transition from Kalahari to Karoo-Namib vegetation, with a mosaic of lightly wooded grassland on the dune crests and pure grassland in the interdunal depressions, and *Rhigozum trichotomum* shrubby grassland in the deeper hollows overlying calcrete. In Angola, at the edge of the Moçâmedes desert, the scattered community of *Colophospermum mopane* and *Welwitschia mirabilis* also includes a number of species from the Zambezian woodland savanna such as *Acacia mellifera, Albizia anthelmintica, Commiphora* sp., *Schinziophyton rautanenii** and *Terminalia prunioides.*

2. *The Windhoek Mountains.* Wooded grassland with *Acacia hebeclada, A. hereroensis, A. reficiens, Albizia anthelmintica, Combretum apiculatum,* etc.

3. *Highveld Grassland.* Representing the climatic climax between 1220 and 2150 m, highveld grassland occurs over large areas of the Interior Plateau of South Africa. It extends west from the Drakensberg from the extreme south of the Transvaal through Orange Free State to the eastern Cape. In most areas the dry, extremely frosty winters preclude the development of woody vegetation, which is largely confined to riparian sites, sheltered ravines and rocky hills and escarpments. Among the main grasses present are *Alloteropsis semialata**, *Andropogon shirensis, Anthephora pubescens, Aristida* spp., *Cynodon dactylon, Digitaria* spp., *Eragrostis* spp., *Eustachys paspaloides, Heteropogon contortus* and *Themeda triandra.* The genus *Hyparrhenia* is conspicously absent, while *Themeda triandra* is the fire climax species. Among the numerous forbs present are the rhizomatous, multistemmed suffrutex *Ziziphus zeyherana* and the perennial *Harpagophyton procumbens**.

3.4 South-Western Malagasy

From a narrow coastal plain in south-western Malagasy a limestone escarpment rises to ca. 400 m. The region lies in the rain shadow of the SE monsoon, providing an annual rainfall of 300–500 mm, falling mostly in the summer as local heavy showers, although some rain can fall during any month of the year. A cold current parallel to the coast causes any atmospheric moisture to condense over the sea, thereby increasing the aridity of the winds reaching the land. Even so, nightly dew during the dry season months of March to November may be a considerable source of water for plants. The dry season normally lasts at least 8 months, interspersed with droughts lasting for 12–18 months (ZE II–III); (White 1983; Rabesandratana 1984; Walter and Breckle 1986). The vegetation of deciduous thicket is dominated by members of the Malagasy endemic family Didiereaceae, especially *Didierea* and *Alluaudia*, with arborescent species of *Euphorbia* codominant with the economically important *Lemuropisum edule** also present (UNESCO 1979; White 1983; Rabesandratana 1984; Walter and Breckle 1986).

3.5 Middle East

The Middle East deserts are the cool and cold winter, arid and semi-arid areas from Israel north to Turkey and from south of the Caspian Sea eastwards to Iran and Baluchistan. The climate ranges from a Mediterranean type in the coastal deserts, with winter rainfall, cool winters and hot summers, to an arid-temperate climate in the interior deserts, with winter rainfall, cool to cold winters and hot summers; the Lut region of central Iran is classified as hyper-arid (UNESCO 1979).

3.5.1 Desert of the Central Anatolian Plain

In Turkey, the semi-arid, central Anatolian plateau is isolated from the Syrian Desert (Sect. 3.5.2.1) by the mountains of the Anti-Taurus. The plateau lies between 800 and 1000 m, although various mountains rise much higher. In the centre, the plateau falls gradually to a treeless depression containing the large salt lake Tuz Gölü. Calcareous deposits of chalk, clay and marl predominate. Several extinct volcanoes rise from the plateau floor. This is known as the Desert of the Central Anatolian Plain.

Precipitation is low, generally less than 400 mm, and occurs mainly during the autumn and spring, with a great deal of it falling as snow and covering much of the area from November to March or even April. The winters are cold, with the average temperature for January less than 0 °C. The summers are hot, with a July-August average from 20 to 25 °C; the temperatures soar during the day and drop suddenly at night. A striking feature of the climate is the very low summer humidity, much lower than in the Mediterranean region of Turkey (ZE VII-IV).

The correspondingly high saturation deficit favours a predominantly herbaceous and suffruticose flora, dominated by species of *Acantholinum* and

Astragalus, and, apart from a few conifers, precludes the growth of evergreen trees and shrubs. Occasional *Salix* spp. (willows), and *Ulmus* spp. (elms), grow by the feeble streams, together with species of *Tamarix*. It is a region of extensive sheep and cattle pastoralism, but with the cultivation of some cereals such as wheat and barley, together with tobacco and *Papaver somniferum* (for the euphoric drug opium and opium oil) being grown in parts of the region (Naval Intelligence Department 1919; Davis 1965 UNESCO 1979).

A broad girdle of *Pinus nigra* subsp. *pallasiana* (Crimean pine) forest clothes the surrounding mountains to the north, west and south of the plateau. Lower down, the forest merges with oak scrub, with *Quercus cerris, Q. infectoria* and *Q. pubescens*. In the treeless steppes the suffruticose *Seriphidium fragrans* (syn. *Artemisia fragrans*) is often the dominant species, being replaced in some areas by the closely related *S. spicigerum* (syn. *Artemisia spicigera*). Associated species include *Noaea mucronata, Peganum harmala* and spiny cushion shrubs of *Acantholinum* spp., *Astragalus brachypterus, A. microcephalus**, *A. wiedenmannianus*, etc., together with tufted, perennial grasses (Davis 1965; Zohary 1973). The halophytic vegetation around Tuz Gölü includes many members of the Chenopodiaceae, such as *Bassia prostrata* (syn. *Kochia prostrata*), *Halanthium kulpianum, Halocnemum strobilaceum**, *Kalidium foliatum**, *Salicornia prostrata, Salsola* spp. and *Suaeda* spp.

3.5.2
Middle East Deserts

The deserts of the Middle East, in the present context, include Israel, Jordan, Lebanon, Syria, Iraq and northern Iran. Two regions are recognised; the Syrian Desert and the Iranian Desert. Although northern Saudi Arabia, with its cool winter, arid and hyper-arid climate, imperceptibly merges with that of the Syrian Desert, it is considered preferable to deal with this as part of the Arabian Desert. The Iranian coastal plain bordering the Persian Gulf and the Gulf of Oman also has a subtropical desert climate and will be considered with the Arabian Desert. An excellent source of information on economic plants is the as yet unfinished *Flora of Iraq* (Townsend and Guest 1966-1980).

3.5.2.1
Syrian Desert

Two subregions of the Syrian Desert are recognised: the coastal Palestinian desert strip of Israel, Lebanon and eastern Syria and the interior deserts of Jordan, Iraq and Syria, imperceptibly divided by the transition from a Mediterranean climate (ZB IV) to an arid temperate climate with hot summers and cold winter with winter rain (ZE IV–VII).

The following divisions are recognised by Zohary (1973):

1. *Palestinian Desert Strip*. A belt of hot and dry Mediterranean climate from Gaza northwards into Lebanon and western Syria, more or less parallel to the coastline but east of the coastal range. The annual rainfall declines from ca. 1000 mm in the north to ca. 100 mm in the south in the region of the Gaza Strip. However, the mean annual temperatures increase from ca. 16 °C in the north to ca. 23 °C in the south. From east to west the annual rainfall and mean temperatures undergo similar but less regular changes due to the rain shadow caused by the Israel and Jordan mountain ranges (Zohary 1962).

The former climax vegetation was Mediterranean evergreen oak woodland with stands of *Pinus halepensis* var. *brutia**, *P. nigra, Quercus cerris, Q. coccifera* subsp. *calliprinos* and *Sarcopoterium spinosum*. These formations have now either disappeared or are severely degraded (UNESCO 1979). Other important woodland species are *Arbutus andrachne**, *Ceratonia siliqua, Laurus noblis**, *Pistacia lentiscus**, *P. terebinthus** , *Quercus* spp., *Rhamnus* spp. and *Styrax officinalis**.

The *Ceratonia siliqua* and *Pistacia lentiscus** coastal belt is well developed in southern Lebanon. Further north the *Ceratonia siliqua* becomes increasingly rare and is replaced by *Myrtus communis*, with *Pistacia terebinthus** and *Quercus coccifera* subsp. *calliprinos* higher up in the maquis. Further north still the *Pinus halepensis* var. *halepensis** forests in southern Lebanon are replaced by *P. halepensis* var. *brutia** (Zohary 1973). An enclave of relic Saharo-Sindian species thrives in the Lower Jordan Valley; the cold winters limit the species distribution further north. They include *Acacia tortilis* subsp. *raddiana**, *Balanites aegyptiaca**, *Calotropis procera**, *Capparis decidua, Hyphaene thebaica, Leptadenia pyrotechnica**, *Salvadora persica* and *Tamarix aphylla* (Zohary 1962; Danin 1983).

Bordering the Syrian Desert the Mediterranean evergreen oak woodland grades into semi-steppe dwarf shrub with *Sarcopoterium spinosum*. Associated species include *Cistus incanus* subsp. *cretica* (syn. *C. cretica*), *C. salviifolius, Salvia fruticosa**, *Satureja thymbra**, *Thymbra spicata** and *Coridothymus capitatus**, with *Thymelaea hirsuta** present on the coastal plain, often in association with *Hyparrhenia hirta* and other psammophytes (Zohary 1973). *Artemisia monosperma** is the dominant psammophyte of the coastal plain of Israel northwards of the Gaza Strip. Other shrubs include *Polygonum palaestinum* and *Retama raetum**. Associated grasses include *Ammophila arenaria* subspp. *arenaria** and *arundinacea**. The salt desert community in the vicinity of the Dead Sea includes *Atriplex* spp., *Nitraria retusa, Salsola tetrandra, Siedlitzia rosmarinus* (syn. *Salsola rosmarinus*), *Suaeda* spp. and *Tamarix* spp. (McGinnies et al. 1968).

2. *Interior Desert.* The Syrian Desert is a triangular area based on the 30th parallel, with the apex at ca. Aleppo, bounded on the east by the Euphrates and on the west by a line drawn from Aqaba in the south along the Wadi 'Araba, the Jordan Valley and the eastern slopes of Mount Herman and the Anti-Lebanon to Homs and Aleppo. The soils are mainly non-saline sierozems (sierozems are grey desert soils or brownish-grey soils overlying a calcareous horizon or lime pan; Jacks et al. 1960). The climate is more of a continental Mediterranean type, with an annual winter rainfall of 150–200(-300) mm (Zohary 1973). Much of the former vegetation of the steppe zone to the north and west has been destroyed by cultivation and grazing. *Seriphidium herba-alba** is regarded as the indicator species for this zone and where it has been destroyed it is replaced by the chenopod *Noaea mucronata* and the grass *Poa bulbosa*. The sedge *Cyperus stenophyllus* and *Poa sinaica* occur on clayey soils, while *Noaea mucronata, Salsola vermiculata* and *Seriphidium herba-alba** grow on clay soils with crushed stones. *Anabasis* spp., *Cornulaca setifera, Haloxylon salicornicum*, H. scoparium** and *Aristida* spp. are found on the sandy soils with *Salsola* spp. and *Seriphidium herba-alba** as important associates. The salt marsh communities contain *Halocnemum strobilaceum*, Tamarix* spp. and the grass *Aeluropus littoralis. Agathophora alopecuroides*, Astragalus spinosus, Ephedra alata** and *Seidlitzia rosmarinus* also occur in the desert areas, with *Achillea fragrantissima, Alhagi maurorum, Artemisia scoparia* and *Peganum harmala* near places of habitation (Kaul and Thalen 1979).

3.5.2.2
Iranian Desert

The Iranian Desert, which includes parts of Iran, Afghanistan and Baluchistan, is one of the smallest and least known deserts. Because of its size and altitudinal range there is considerable climatic and phytogeographical diversity. The climate is a Mediterranean type with winter rainfall, cool winters and hot summers (ZE VII–IV). The main soil types are grey desert soils, sierozems, lithosols, regosols and saline soils. The flora and its phytogeographical affinities have been studied by Léonard (1981–1992).

Four regions have been identified (McGinnies et al. 1968; UNESCO 1979), including the arid coastal region, which is more conveniently considered with the Arabian Desert (Sect. 3.1.1.5). The following subdivisions are recognised:

1. *The Hyper-arid Lut Region.* The mean annual rainfall of the Dasht e Lut is less than 20 mm. Plants and animals were reputed to be rare. However, detailed studies by Léonard (1992) have revealed the presence of more than 150 plant species, including *Anabasis setifera, Calligonum* sp., *Cornulaca monacantha**,

Haloxylon aphyllum, H. persicum*, H. salicornicum*, Peganum harmala, Seidlitzia rosmarinus, Seriphidium sieberi* (syn. *Artemisia sieberi*), *Tamarix* spp. and the grasses *Stipagrostis multinerva* and *S. plumosa*. Professor Léonard also recognises an area of date palm cultivation and a postcultural vegetation with *Prosopis farcta, P. koelziana* and *Ziziphus spina-christi*.

2. *The Central Iranian Plateau*. The plateau lies between 1000 and 2000 m a.s.l. The climate is generally very dry, with an annual rainfall of ca. 170–230 mm, falling from January to March or April; there is no rain from June to October-November. The Dasht e Kavir or Great Salt Desert, the largest salt desert in the world, lies to the east of Tehran. It is devoid of vegetation. Here the climate is drier; the winters are cool with temperatures of 0-5 °C, and the summers hot to very hot with July temperatures of 26–31 °C.

The characteristic plateau vegetation consists of a plateau steppe with *Seriphidium maritimum* (syn. *Artemisia maritima*) and the grass *Stipa lagascae* (syn. *S. holosericea*), with occasional trees such as *Pistacia khinjuk, P. terebinthus*, Prunus scoparia* (syn. *Amygdalus scoparia*) and *Juniperus excelsa*. The poorly drained plateau flats, as well as the Seistan Plain in southeastern Iran, the Qila Ladgasht region of Baluchistan and the Zirreh region of Afghanistan, support low halophytic communities where a clearly marked zonation related to salinity and depth of water table is often to be found. The pioneer species is *Halocnemum strobilaceum**, appearing along the margins of the salt pans, to be succeeded centrifugally with decreasing salinity by *Seidlitzia rosmarinus*, followed by *Salicornia europaea**. *Salsola* spp. and *Suaeda* spp. accompanied by the salt-secreting grasses *Aeluropus lagopoides** and *A. littoralis** (Kaul and Thalen 1979; Walter and Breckle 1989).

3. *The Iranian Mountain Region*. The Zagros Mountains stretch almost from Lake Uurmia in the north-west to Hashtom in the south-east, and the Elburz Mountains extend eastwards from the northern Zagros to the south of the Caspian Sea. In these mountains the winters are cold, with the mean temperature for December, January and February at or below 0 °C, while the summers are hot, with July-August temperatures of 25–30 °C. The annual rainfall ranges from 400 to 460 mm, with 5 dry months, from June to October.

The characteristic vegetation is of xerophilous deciduous oak woodlands.

3.6 Central Asia

The Central Asian arid and semi-arid areas of the former Soviet Union occupy over 100 million ha, from the eastern piedmonts of the Great Caucasian Range eastwards for nearly 3000 km to the Altai Mountains on Mongolia's western border. They include the following independent republics: Turkmenistan, 87% of which is covered by the Kara Kum (black sands); Uzbekistan, south-east of the Aral Sea between the Amu-Daria and the Syr-Daria Rivers, the Kyzyl Kum (red sands); Kazakhstan, dominated by a steppe vegetation of wormwood (species of *Artemisia* and *Seriphidium*) and grasses, but with areas of deserts comparable with those of Turkmenistan, especially north of the Caspian Sea, east of the Aral Sea and south of Lake Balkash, and the upland deserts of Kirghizstan and Tajikistan on the Sinkiang border (UNESCO 1979; Nechayeva 1985). The climate is continental, characterised by cold winters and hot summers; in winter the area is covered by snow.

Two zones are recognised by Nechayeva (1985): the plains and the mountains. The plains include the northern, southern and foothill desert subzones; the mountain zone, characterised by vertical relief, is divided into low, medium and high altitude mountain subzones. River valleys are regarded as interzonal. The economic plants of the area have been listed by Nechayeva (1985), but the setting of her information into an ecological framework has, in many cases, proved difficult. See also the economic notes in the Flora of the USSR (Komarov 1934-1964), the English translation of which (1960-1977) is unfortunately incomplete.

3.6.1 Central Asian Zone of Desert Plains

Two subzones of the central Asian zone of desert plants are recognised by Nechayeva (1985), corresponding to ZB VIIa plus the northern part of VII(rIII) and the southern past of VII(rIII), respectively (Fig. 5).

3.6.1.1 Northern Desert Subzone

The altitude of the northern desert subzone does not exceed 600 m a.s.l. The main soil types are the shallow, sandy, grey-brown clays, poor in humus and frequently highly alkaline. Ground water is fairly shallow. The mean annual precipitation in Gur'yev is 163 mm, and in the Aral Sea region 103 mm. The precipitation is regularly distributed throughout the year. Evaporation is high; nine to ten times that of the total annual precipitation. The mean monthly air temperature ranges from 25.4 to 26.3 °C in July, and from -10.4 to 13.5 °C in January.

Petrov (1979) comments on the difficulties of drawing boundaries between the steppe and desert formations based on geomorphology and soils. Typical steppe vegetation occurs in depressions with chernozem-like soils within the otherwise subalkaline chestnut soils of the semi-desert (chernozems are dark, well-drained grassland soils, granulated and rich in humus to some depth, with or without concentrations of clay in the B horizon, and calcareous below; chestnut soils are dark brown over lighter coloured soils overlying a calcareous horizon; Jacks et al. 1960). The steppe is dominated by grasses belonging to the genera *Agropyron, Festuca* and *Stipa*, together with various herbs, and sometimes steppe shrubs, mostly ×*Spiraea hypericifolia*, and with the addition of species of *Seriphidium* in the steppe-desert transition zone. Typical desert vegetation is dominated by the xerophytic subshrubs *Salsola arbusculaeformis, Seriphidium semiaridum* (syn. *Artemisia semiarida*) and *S. terrae-albae* (syn. *A. terrae-albae*) The vegetation cover is fairly uniform throughout, consisting of wormwood – sod-grass (*Seriphidium – Festuca – Stipa*) formations, creating monotonous ranges dominated by two or three species of semi-shrubs.

The floristic composition over the vast area of semi-desert forming the northern boundary of this subzone is, however, extremely varied, although physiognomically similar. The most prevalent species are the xerophytic and super-xerophytic grasses. A large number of ephemerals are also present. Over the enormous distance from the Black Sea coast to the troughs of Lakes Zaysan and Alakol' in the east, the species of *Seriphidium*, and to a lesser extent the composition of the grasses, can be observed to change and a number of corresponding desert steppe communities are consequently recognised: Black Sea with *S. tauricum* (syn. *A. taurica*); Caspian-Kazakhstan with *Artemisia tschernieviana* (syn. *A. arenaria*), *S. lercheanum* (syn. *Artemisia astrachanica, A. lercheana,*), *Krascheninnikovia certatoides* (syn. *Eurotia ceratoides*) and *K. ewersmanniana* (syn. *Eurotia ewersmanniana*); western Kazakhstan with *Seriphidium lessingianum* (syn. *Artemisia lessingiana*); and eastern Kazakhstan with *S. sublessingianum* (syn. *Artemisia sublessingiana*).

Petrov (1979a) also recognises the following rangeland communities within the semi-desert:

1. *White Wormwood* (*Seriphidium lercheanum*) *Pastures.* Distributed from south-west Europe to central Asia, *Seriphidium lercheana* is a low, 20 to 40 cm high, xerophytic semi-shrub providing a high-grade forage plant, palatable throughout the year, especially in autumn. *Agropyron* spp. also provide valuable forage. In spring there is a rich herbaceous cover of annual and perennial plants with a short growing season.

2. *Black Wormwood* (*Seriphidium pauciflorum*) *Pastures.* Distributed from western Asia to Mongolia, and northern China, *Seriphidium pauciflorum* (syn. *Artemisia pauciflora*) is a widespread, low, xerophytic semi-shrub with a preference for alkaline and subalkaline brown and light-chestnut soils (brown or steppe soils are brown to light brown, nearly neutral soils; Jacks et al. 1960). It is regarded as one of the best of the semi-desert forage plants and is most palatable in the autumn, especially to sheep.

3. *Grey Wormwood (Seriphidium terrae-albae) Pastures.* Distributed from central Asia to northern China, *Seriphidium terrae-albae* is a low, xerophytic semi-shrub of the more northerly part of the semi-desert. It is palatable to sheep, especially in winter, and has a high nutritive value.

4. *Krasschennikovia Pastures.* The C_3 xerophytic, up to 40 cm high subshrub *Krascheninnikovia ceratoides* has a preference for rocky debris or sandy loam soils (Winter 1981). The plant is palatable throughout the year, especially to sheep.

5. *Atriplex cana* Pastures. These pastures are widespread on solonchak-type soils. On subalkaline soils *A. cana* is associated with *Salsola* spp. and on lighter soils with various herbs.

6. *Herbaceous-Wormwood Pastures.* These pastures are widespread on light-chestnut soils throughout the semi-desert subzones. The principal grasses are *Agropyron desertorum, A. fragile* (syn. *A. sibiricum*), *Festuca alaica* (syn. *F. valesiaca* subsp. *sulcata*), *Stipa capillata, S. kirghisorum, S. lessingiana* and *S. sareptana.*

7. *Summer Cypress Pastures.* These pastures, with a predominance of the xerophytic semi-shrub *Bassia prostrata,* are widespread; the species is widely sown in the northern desert subzone for grazing and haylands.

3.6.1.2
Southern Desert Subzone

The altitude range of the southern desert subzone is 400 m and below. The soils are sandy, grey-brown or white alkalis with gypsum intercalations, heavy and takyr-like, and also typical takyrs containing considerable concentrations of chlorides and sulphates (takyrs are wide expanses of clays, inundated in spring and dry in summer). All the soils are slightly humic, structureless and, apart from the sands, saline. Both saline and fresh groundwater are found at varying depths, generally deeper than 10 m and consequently usually unavailable to plants.

The mean annual precipitation ranges from 970 to 1130 mm, with maximum precipitation in winter and early spring. Evaporation is very high, exceeding the annual precipitation some 26-fold. The summers are long, dry and hot and the winters short and changeable, with a wide range of temperatures. The mean annual air temperature ranges from 13.2 to 16.1 °C, the mean monthly temperature from 29 to 30 °C in July, the hottest month, and from -4.1 to 2.4 °C in January. Winter temperatures as low as -26 °C have been recorded at Ashkhabad on the Iranian border, while the January averages for Chelkar just to the north of the Aral Sea are -15 °C.

Shrubs are widely distributed throughout this southern subzone, and are able to continue growing throughout most of the year. Browse from these shrubs forms an important item in the diet of sheep and camels; the shrubs are also a source of fuel for man as well as assisting in sand stabilisation. Shrubs such as *Ephedra strobilacea, Haloxylon aphyllum**, *H. persicum** and *Salsola richteri* are important browse species for the autumn and winter; the latter species is also used for range improvement. Shrubs with a very rapid summer growth, such as *Astragalus paucijugus* and *Calligonum* spp., enrich the spring and early summer ranges. They may also be grown for range improvement. Among the conspicuous and widely distributed forage semi-shrubs of the gypsum and clay deserts of this subzone are the chenopods *Anabasis salsa, Gamanthus gamocarpus, Salsola* spp. and the composites *Artemisia tschernieviana* and *Seriphidium* spp., while the forage semi-shrubs of the sandy gravel plains are *Astragalus* spp. Range improvement is carried out by sowing the halophytes *Bassia prostrata, Halothamnus subaphylla* var. *subaphylla* (syn. *Aellenia subaphylla* var. *subaphylla*), *Haloxylon aphyllum**, *Krascheninnikovia ceratoides* and *Salsola rigida.*

3.6.1.3
Foothill Desert Subzone

The foothill desert subzone is hilly with valley-like depressions, lying between 400 and 1000 m a.s.l. The soils of the higher hills are thin, stony and poorly developed; those of the lower slopes are light, non-saline sierozems. Groundwater is usually deep and saline. Due to altitudinal changes the climate is more variable and generally more favourable than in the desert subzone. The mean annual precipitation is from 160 to 260 mm, falling mainly in the winter and early spring, and favours a good herb cover and thereby prevents the spread of shrubs. Evaporation exceeds the annual precipitation five- to eight-fold. The mean annual temperature is 14 °C, ranging from 26 °C in July to ca. 0 °C in January. The vegetation cover is dominated by the perennials *Carex pachystylis* and *Poa bulbosa* which, together with the annuals, form a continuous sward.

3.6.2
Central Asian Mountain Zone

The following subzones of the central Asian mountain zone are recognised by Nechayeva (1985):

1. *Low Mountain Subzone.* This subzone lies between 600 and 800 m a.s.l. The soils vary from light chestnut and chestnut to shallow brown and sometimes

stony, with the groundwater at varying depths. The climate is arid continental, with a mean annual precipitation between 400 and 600 mm, falling mainly in the winter and spring. The mean annual temperature at Dushanbe is 14.8 °C, ranging from 28 °C in July to 1.4 °C in January. The grasses *Bothriochloa ischaemum* and species of *Festuca, Poa* and *Stipa* predominate on chestnut soils, with *Artemisia, Seriphidium, Convolvulus, Krascheninnikovia* and *Salsola laricifolia* on shallow brown or stony soils.

2. *Mid-Mountain Subzone.* The area lies at an altitude between 1000 and 2000 m a.s.l. The soils are variable and include dark sierozems, brown, dark chestnuts and chernozem-like soils. The climate is moderately dry, with a mean annual precipitation of ca. 800 mm, falling mainly in the winter and spring. The winters are variable and warm, the summers dry. The dark sierozems and brown soils carry a grass cover containing species of *Bromus, Elymus* (syn. *Elytrigia*) and *Festuca*, and the dark chestnut and chernozem-like soils *Dactylis, Elymus, Festuca, Poa* and *Stipa*. These grass clearings occur in open woodland with species of *Acer, Cotoneaster, Juniperus* and *Juglans regia**, etc. The dry slopes bear a typical cover of deciduous bushes or 'shibliak', with species of *Pistacia, Prunus* (syn. *Amygdalus*), *Rhus* and *Ziziphus.* The ranges are, like those of the low mountain subzone, used for grazing during the spring to autumn.

3. *High Mountain Subzone.* This subzone extends from 2200 to 3300 m a.s.l. The mean annual precipitation is 1200 mm. The mean annual temperature ranges from 2 °C to 7 °C; the winters are long and cold, with persistent snow. The summers are also cold. Thus, despite the high rainfall, temperatures limit plant growth so that the vegetation consists of high mountain meadows, steppes and heathlands, and is used for summer grazing.

3.7 Arid Lands of China

Unlike the great arid plains of the former Soviet Union the arid lands of China, although extensive, often appear as discontinuous units separated by altitudinal differences; they include basins, plateaux and mountains. Petrov (1976) recognises the following deserts:

1. *Takla-Makan Desert.* ('Takla-Makan' means to 'enter and never return'). It is the largest sand desert in China, occupying much of the Tarim Basin between 800 and 1500 m, much of which is occupied by shifting sands. A narrow belt of stony desert (gobi) encircles the sands. The desert is surrounded by the mountain ranges of the Tien-shan to the north, the Pamirs to the west and Kunlun in the south; eastwards it gently tilts into the Lop Nor Depression

(which comprises a large part of the Gashun Gobi). The annual precipitation (see Lipu 1988), is 70–80 mm in the south-west, 50–60 mm in the north, 35–40 mm in the south and only 10–20 mm in the centre. The meagre rainfall is supplemented by run-off from the surrounding mountains, including melted glacial water, etc. The average maximum temperature for July is 32–36 °C; the annual range is 40 °C.

2. *Ala Shan Plateau.* This is a basin depression at 1200 m. Among the bachan dunes there are numerous large and small, flat depressions containing salt or freshwater lakes and surrounded by lush meadow-swamp vegetation known as tsaidams.

3. *Busham Desert.* This consists of low ranges up to 2790 m separated by narrow, intermontane depressions.

4. *Ordos Mountain Desert.* This lies between 5000 and 7000 m and consists of slightly eroded plains containing some broad, lake-filled depressions.

5. *Tsaidam Basin.* This is a flat, high mountain plain at 2600–3000 m, lying below the Alltyn Tagh (5400 m) and the Nan Shan (at least 5000 m) to the north and Kunlun (6000 m) to the south. The annual precipitation is usually less than 40 mm. The mean annual temperature is 2.5 °C, and the mean monthly temperature for January is -12 °C and for July 17.6 °C.

6. *Dzungaria Basin.* This is a sandy plain lying at 280-400 m.

7. *Gobi.* This consists of topographically varied desert and semi-desert landscapes with level or ridged ground, gravely or pebbly-sandy surfaces, coarsely skeletal, gypsiferous, greyish-brown soils and sparse vegetation. Although lacking surface water, groundwater is generally abundant. It includes the following deserts: Gasham Gobi, Trans-Altai Gobi, Eastern Gobi (Gobi Desert) and the Dzungaria Gobi.

The desert climates are severe, cold-temperate to warm- temperate with long, cold winters and short, hot summers. In the eastern deserts the precipitation is associated with the south-eastern monsoon from the Pacific and has a summer maximum. Except for the eastern steppes and piedmont plains the total annual precipitation is less than 100 mm; the mountains and deserts of Kashgaria, Tsaidam and Bei Shan receive less than 10 mm in some areas. In the west the precipitation is related to the moist Atlantic air and is comparatively evenly distributed throughout the year. Evaporation can reach 3578 mm.

The natural vegetation cover is poor, principally of dwarf shrubs and subshrubs, tough and often succulent or woody, spiny and either leafless or with small, deciduous leaves; the grass cover is generally poor. Species belonging to the Chenopodiaceae are particularly well represented, albeit sometimes by a single genus or species; members of the Compositae, Ephedraceae, Gramineae, Leguminosae, Polygonaceae, Tamaricaceae and Zygophyllaceae are also fairly frequently present. The physiognomy, life forms, floristic composition and elementary composition of the dominant plants are closely related to the local climates and soils. Hou (1985) recognises a desert and mountain subzone. Within the former, four vegetation types are recognised: the deserts proper, floodplain meadows, riverine forest and the cultivated oases; the last three are discussed in Sections 3.7.1.7.1-3.7.1.7.3.

3.7.1 Chinese Desert Subzone

Six categories of the Chinese desert subzone are recognised by Hou (1985). The following descriptions of the vegetation are largely based on Hou (1985).

3.7.1.1 Sandy Deserts (Shamo)

Sandy deserts (shamo) are particularly widespread in China, the sandy soils favouring the growth of shrubs and subshrubs, although shifting sands occupy much of the Takla-Makan and the western parts of the Ala Shan Plateau. The semi-arid eastern part of the Ala Shan and the Ordos Plateaux receive 200 to 300 mm annual precipitation. *Artemisia ordosica* is widespread on the stabilised sand dunes and, on the deeper soils with high groundwater, the leguminous shrubs *Ammopiptanthus* spp., *Caragana microphylla**, *C. korshinskii, Hedysarum scoparium* and *Oxytropus aciphylla* are to be found, plus small trees of *Salix* spp., together with the pungent, edible herb *Pugionium cornutum* and the sand-binding grass *Psammochloa mongolica* in the moist depressions between the dunes. The species used for sand fixation along the railway lines are: *Artemisia* and *Seriphidium* spp., *Caragana microphylla**, *Elaeagnus angustifolia**, *Hedysarum scoparium, Populus* spp., *Salix* spp. and the shallow-rooted North American introductions *Amorpha fruticosa* and *Robina pseudoacacia** (Kebin and Kaiguo 1989).

The Tarim Basin is occupied mainly by the Takla-Makan sandy desert, which is almost devoid of vegetation. Concentrated along the desert border, in the dune valleys with shallow water tables and saline soils, there are stands of scattered *Tamarix* spp. accompanied by *Alhagi sparsifolia, Karelinia caspica*, etc. Trees or shrubs up to 5 -6 m tall, the stems of *Tamarix* will produce adventitious roots when buried by the shifting sands, thereby maintaining the dune fixation process (Petrov 1972; Changdu and Liyun 1988).

In the Dzungaria Basin, the more evenly distributed rainfall is sufficient to permit colonisation of the non-saline sands, first by *Stipagrostis pennata* (syn. *Aristida pennata*), to be followed by *Calligonum leucocladum, Haloxylon persicum*, Seriphidium santolinum* (syn. *Artemisia santolina*) and *S. terrae-albae* together with an ephemeral vegetation. *Haloxylon ammodendron** occurs in the saline depressions in association with *Atriplex heterosperma* (syn. *A. micranthum*), *Horaninovia ulicina, Limonium gmelinii* and *Suaeda altissima.* There are extensive areas dominated by the rhizomatous perennial grass *Leymus chinense.* In the drier Tsaidam Basin, with less than 50 mm precipitation, the typical vegetation cover consists of *Ephedra przewalskii, Gymnocarpos przewalskii, Haloxylon ammodendron*, Nitraria tangutorum, Salsola abrotanoides* and *Sympegma regelii* (Jingtai and Derbyshire 1987).

3.7.1.2 Sandy-Pebble or Gravel Deserts (Gobi)

Sandy-pebble or gravel deserts (gobi) are widely distributed in the arid regions of China. In the semi-desert area of the eastern Ala Shan Plateau, where the annual precipitation is ca. 200 mm, shrubby vegetation covers 30–50% of the piedmont plains. The dominant species are *Caragana tibetica, Potaninia mongolica* and *Tetraena mongolica*, together with xerophyllous grasses such as *Kengia mutica* (syn. *Cleistogenes mutica*) and *Stipa caucasica* subsp. *glareosa.* In some places the evergreen legume *Ammopiptanthus mongolicus* is dominant, with *Caragana microphylla** frequently present.

Gypseous gravely deserts dominated by the dwarf, 10 to 25-cm-high shrub *Reaumuria soongarica* are extensively distributed in the Ala Shan Plateau and Bei Shan and Tarim Basins, and where the annual, mainly summer precipitation is 50 mm or less, while the 50-cm-high *Salsola passerina* is codominant with *Reaumuria soongarica* in the Ala Shan area, the former mainly on saline soils in the shallow depressions. In the high foothills of all the mountain systems a sparse shrubby and subshrubby vegetation grows along the channels of temporary streams. In places the vegetation cover is less than 5–10%.

3.7.1.3 Rocky or Hilly Deserts

Rocky or hilly deserts are found mainly on the hills and low mountains of the Bei Shan and Tarim Basin, from which all the finer products of weathering have been blown away, and where the exposed rocks have been severely sandblasted. The surface is covered by a stone pavement below which there may be a water-repellent, dusty soil rich in gypsum and salt that is antagonistic to plant growth. The area is cleft with deeply eroded valleys with steep, rubble-covered slopes.

A few low-growing xerohalophytes may be found in the rock crevices, providing a less than 1% vegetation cover. The chenopods *Haloxylon regelii* (syn. *Iljinia regelii*) and *Sympegma regelii* are frequent in occurrence. On the foothills occur *Ephedra equisetina** and *E. intermedia,* while *Capparis spinosa* is a frequent shrub on the stony soils of the Turfan Depression and Tarim Basin.

In the mountain gorges of the Tsaidam Basin, ca. 3000–4000 m, where the annual precipitation is 100–150 mm, the principal desert species of the alpine shrubland are *Krascheninnikovia seratoides* (syn. *Ceratoides latens), Reaumuria kaschgarica, Salsola arbuscula, Saussurea pulvinta* and the grass *Elymus nutans*. In the mounta in gorges where the annual precipitation is 200–300 mm, the semi-desert species of the alpine steppe and shrubland are *Caragana jubata**, *Potentilla fruticosa* (syn. *Dasiphora fruticosa*), the sedge *Carex moorcroftii* and the conifers *Juniperus przewalskii* (syn. *Sabina przewalskii*) and *Picea crassifolia* (Jingtai and Derbyshire 1987).

3.7.1.4
Loess or Loamy Deserts

The loess or loamy desert soils, consisting of fine, lime-rich, yellow loams, are not excessively saline and are to be found mainly in those parts of the Dzungaria Basin where the rainfall is evenly distributed throughout the year. The vegetation is predominantly of low subshrubs of *Seriphidium* spp., with a number of herbs which behave as spring ephemerals, soon dying at the onset of the hot summer.

3.7.1.5
Solonchak or Saline Deserts

The solonochak or saline deserts occur in low lying areas, wherever brackish underground water lies close to the surface, such as in old lake beds, old valley bottoms and river deltas. The structureless, saline soils, known as solonchaks, often have salts crystallising on the surface, with 10–30% sodium chloride or sodium sulphate. The dominant halophytic vegetation consists of low shrubs and subshrubs, herbs, grasses and annual succulent hydrohalophytes.

3.7.1.6
Cold High Deserts

In the cold high deserts of north-eastern Tibet, at 5000 m and above, there is a very sparse, species-poor vegetation, consisting of small, prostrate or cushion subshrubs, adapted to the windy, cold and dry environment. Li (1979) divided the semi-arid alpine steppes of Tibet into alpine scrub steppe and alpine steppe. The former occurs below 4500 m, in the middle and upper valleys of the Yalutsangpo River. Where the mean annual temperature is ca. 0–8 °C, with 70-150 frost-free days, the most common and conspicuous species of the alpine scrub steppe are *Caragana jubata**, *Ephedra gerardiana, Hippophae tibetana, Potentilla fruticosa, Sophora moorcroftiana* and the grasses *Aristida triseta* and *Stipa bungeana*. The alpine steppe is found mainly in northern Tibet, at ca. 4500–5000 m. The mean annual temperature is ca. 3–5 °C, with 10–100 frost-free days. The annual rainfall is from 200 to 350 mm.

The highest desert in Asia occurs in north-western Tibet; it is a cold desert with a mean annual temperature <0 °C and a mean July temperature of 12 °C; the variation in temperature during the year can be as much as 26 °C. The rainfall ranges from 50 to 100 mm. In the cold deserts of the high valleys of Kashmir, on the borders of Tibet, the mean January temperature at Leh (ca. 3500 m) is -7 to -10 °C, with a July-August mean of 17–19 °C. The shrubs of the high steppe include *Caragana pygmaea. Ephedra equisetina**, *Juniperus* spp., *Populus* spp. and *Salix* spp. (UNESCO 1979).

3.7.1.7
Specialised Habitats

Within the desert subzone the plants of these basic mesophytic specialised habitats are all more or less physiologically xerophytic because of the dry air and the salinity of the soils.

3.7.1.7.1
Floodplain Meadows

Saline meadows are widespread in the contemporary floodplain of the Tarim River and areas with in the desert with a high water table. The floodplain meadow vegetation contains grasses such as *Aeluropus littoralis, Cynodon dactylon, Leymus secalinus* (syn. *Aneurolepidium dasystachys*) and *Phragmites australis* and the legumes *Alhagi maurorum**, *Glycyrrhiza* spp. and *Melilotus* spp. Other herbs present include the apocynaceous *Poacynum hendersonii*, which yields a useful fibre, and the composite *Xanthium sibiricum*. A number of shrubs occur locally, including *Halimodendron halodendron**, *Lycium ruthenicum* and *Tamarix* spp. (Jingtai and Derbyshire 1987).

3.7.1.7.2 Riverine or Floodplain Forests

The most common species in the riverine or floodplain forest is the weeping willow *Populus euphratica*. It is particularly widespread on the floodplain of the Tarim River, which is the largest river in the Chinese desert regions and fed by the meltwaters from the glaciers and snows on the surrounding mountains. The soils are slightly saline, with the water table at ca. 1.5–2.5 m. A second species, *Populus pruinosa*, occurs only on the higher parts of the Tarim River floodplain, where the soils are less saline.

3.7.1.7.3 Oases

The oases, fed by snow melt from the surrounding mountains, provide fertile areas for cultivation within the desert region. The oases of the Dzungaria Basin, with its temperate climate, produce oats, millet, buckwheat and beans. Spring wheat and early maturing rice can only be raised under good irrigation practices. The economic crops are sugar beet and flax for both oil and fibre. There is only one harvest a year.

3.7.2 Chinese Mountain Subzone

The climate of the high mountains of the Chinese mountain subzone surrounding the desert subzone is either semi-arid or humid. The associated vegetation consists of steppes, forest and alpine meadows (Hou 1985), reflecting the climate and topography. The vegetation on the lower northern slopes of the Ch'i-lien Shan (6000 m a.s.l.) and Ho-lan Shan (3500 m) Mountains are dominated by *Artemisia frigida* and *Stipa bungeana* steppe. Higher up on the northern slopes the steppe first gives way to forests of *Pinus tabulaeformis* and then *Picea crassifolia*, while on the southern slopes there are coppices of *Prunus mongolica* and *Ulmus glacescens*. The subalpine meadows include the shrubs *Caragana jubata** and *Potentilla fruticosa*, the alpine meadows being dominated by *Kobresia* spp. The summit of the Ch'i-lien is permanently capped by snows and glaciers.

The lower northern parts of the Tien Shan Mountains (5000 m), above the semi-desert, are dominated by *Festuca alaica* and *Stipa capillata* grassy steppe, giving way higher up on the northern slopes to *Picea shrenkiana* forest. Above the coniferous forest lie the subalpine and alpine meadows dominated by *Kobresia* spp. The summits are permanently capped by snow and glaciers. The lower slopes in the western parts of eastern Tien Shan, where the climate is warmer and moister, contain local patches of deciduous forests (Walter and Breckle 1989). In the southern and much drier Tien Shan the desert vegetation occurs up to ca. 2000 m, or higher. Here, the typical mountain steppes are dominated by *Stipa krylovii* and the alpine steppes by *Festuca ovina* and *Stipa purpurea*. Permanent snows and glaciers occur at 3000–4300 m.

In Singkiang, the Chinese part of the Altai Mountains reaches an elevation of ca. 3000 m. The mountain steppes are dominated by *Festuca alaica* and *Stipa capillata*. Extensive mixed forests of *Larix russica* (syn. *L. sibirica*) and *Pinus siberica* occur in the narrow, moist valleys, with coniferous forests dominated by *Abies sibirica* and *Picea abies* subsp. *obovata* (syn. *P. obovata*) present only on the moister, north-western parts of the Altai. The subalpine meadows consist of *Festuca* spp., *Poa* spp. and *Trisetum* spp., together with such herbs as *Leontopodium leontopodioides* (syn. *L. sibiricum*) and *Polygonum viviparum*. The alpine meadows are dominated by species of *Kobresi a* and *Carex*. Higher up the meadows tend to resemble alpine tundra, with thickets of *Betula rotundifolia* underlain by mosses (Hou 1983). On the extremely arid northern slopes of the Kunlun Shan Mountains the desert vegetation can reach an elevation of 3000 m. The high mountain slopes are clothed in dry steppe; alpine meadows are absent.

3.8 Australia

Somewhat more than 80% of the continent of Australia is classified as arid or semi-arid, a larger proportion than for any other continent. However, there are no true deserts for nowhere is there the extreme condition of hyperaridity found in other continents. The basic causes of this lack of extreme conditions of aridity are the relative smallness of the continent, the absence of very high orographical barriers to the prevailing winds, the extensive oceans surrounding the continent and the absence of very cold offshore currents along the western subtropical coast (UNESCO 1979).

Because various authors have proposed very different delimitations of the continent based on very slight geomorphological differences, Walter and Breckle (1986) have, for the sake of simplicity, defined the arid areas as those with an annual rainfall of less than 250 mm. Towards the north, however, where the temperatures and corresponding evapotranspiration are very high, the definition has been extended to include areas with 350–400 mm rainfall. In the south this boundary more or less corresponds with the aridity limit of the eucalypt shrublands (see Sect. 3.8.5) and in the north with the tropical and subtropical woodlands (ZB II). Adjoining ZB II, the ZE II–III is gradually replaced along a north-south transect by the ZE III-IV and only in the central region is the annual rainfall less than 200 mm (Fig. 6).

The climate of Australia is unique since, unlike other continents, Australia is too small an area to ensure a sharp seasonal distinction of rainfall. Tropical cyclones commonly extend southwards beyond 30°S, and the Antarctic lows may just bring rainfall to the northern part of the arid zone. Hence, the north has mainly summer rain and the south predominantly winter rain, while in-between the rainfall is unpredictable and can occur at any time of the year. In the central area the scanty rainfall may fall in almost any month, or not at all. The ecological significance of winter and summer rainfall is that the winter rains, falling in the cooler months of the year, have lower losses through evaporation than is the case for summer rains falling during the hotter months of the year. The winter rains also tend to fall in light showers while summer rains falls as heavy showers. Thus, for equal amounts of rainfall the winter rains will be more ecologically effective than summer rains (von Willert et al. 1992). During the summer air temperatures over the entire dry zone are high, exceeding 35 °C for approximately 140 days. Mean winter temperatures in the northern part of the arid area are between 15 and 29 °C, while in the south they fall to as low as 0 °C, with occasional frosts during July.

A number of deserts are recognised, including the Simpson Desert north of Lake Eyre and to the west the even larger and relatively sand-free Gibson Desert. These deserts are characterised by the absence of surface run-off, consisting in part of wide sandy areas with parallel dunes, and partly by largely featureless stony plains. To the south, separated from the Gibson Desert by the Great Victoria Desert, lies the monotonous and mainly Tertiary limestone plateau of the Nullarbor Plain (*'nulla arbor'* means 'no tree'), effectively drained by means of sink holes and an underground drainage system. In Western Australia the Great Sandy Desert, lying to the north of the Gibson Desert, consists of a slightly undulating plain of horizontal layers of sandstone and conglomerates, their detritus forming widely spaced parallel dunes aligned in a west-north-west direction, cutting through a former river system to form salt lakes and salt pans. It is the lack of potable water and poor fodder resources that characterise these deserts, since they usually do not differ climatically from the adjoining farmland (Beard 1969; Walter and Breckle 1986).

The flora of central Australia has slightly less than 1200 species, of which less than 200 are considered useful to the pastoral industry, of which far less than 100 are of great importance as pastures (Chippendale 1965). Many species and genera are sclerophyllous, a tribute to not only the long prevailing arid climate but also the poor soils. In general, the present flora is lacking in such morphological defences against herbivory as spines, prickles, thorns and stinging hairs, and reflects the absence of indigenous herbivore mammals. The lack of such defences is evident from the fact that in saltbush shrubland the relatively low but continuous stocking density of 1.2 sheep ha^{-1} is sufficient to kill almost all of the bushes within 2.5 years (Anderson 1982).

The uses of Australian plants, including those used by the Aborigines, are very briefly dealt with by Lazarides and Hince (1993). This reference is widely cited in Chapter 11 because it contains the original references and is, therefore, a valuable source of primary information for researchers. Timber species, especially *Eucalyptus* spp., and forage plants such as *Atriplex* spp. and *Maireana* spp., and a number of horticultural plants have been brought into cultivation and are now to be found throughout the world. However, apart from the edible seeds of forest species of *Macadamia*, the indigenous food plants of the Aborigines have, until recently, been ignored. This probably reflects the social and cultural differences between the native Aborigines and the white settlers and the failure of the latter to appreciate the value of the former's food plants. The work of Cribb and Cribb (1974) and the nutritional analyses by Brand and Cherikoff (1985, 1988) and Brand et al. (1983, 1985) have done much to rectify the situation. They have adequately demonstrated that a number of aboriginal foods are nutritionally as good as or richer than many of the traditional European foods. For example, the seeds of *Acacia, Eragrostis, Fimbristylis, Panicum* and *Portulaca* species are almost twice as high in protein as the common cereals wheat, rice, etc. As a start, efforts are now being made to bring into cultivation the quandong *Santalum acuminatum**, whose fruit has a higher protein, fat and carbohydrate level than cultivated fruits, even on a moisture equivalent basis (Brand and Cherikoff 1988; Rivett et al. 1989).

Seven major and one minor vegetation formations (Fig. 10) are recognised by Williams (1982) and nine by Walter and Breckle (1986); a composite version of their communities is described below.

3.8.1 Semi-Arid Shrub Woodlands

The semi-arid scrub woodlands are generally restricted to the north-eastern, summer rainfall area, where the 5 to 10-m-tall trees provide 10–30% cover. The soils are red duplex (parent soil overlain by superficial deposits) and massive red earths. The major community is dominated by *Eucalyptus populnea*, with associated *Eucalyptus* spp., and the shrub layer contains such species as *Casuarina cristata* and *Eremophila mitchellii*. Other communities include *Eucalyptus populnea* with either *Acacia* spp. or *Callitris columellaris*, together with smaller trees of *Acacia aneura**, *A. excelsa, Eremophila desertii* (syn. *Myoporum desertii*) and *Eucalyptus* spp. (Williams 1982).

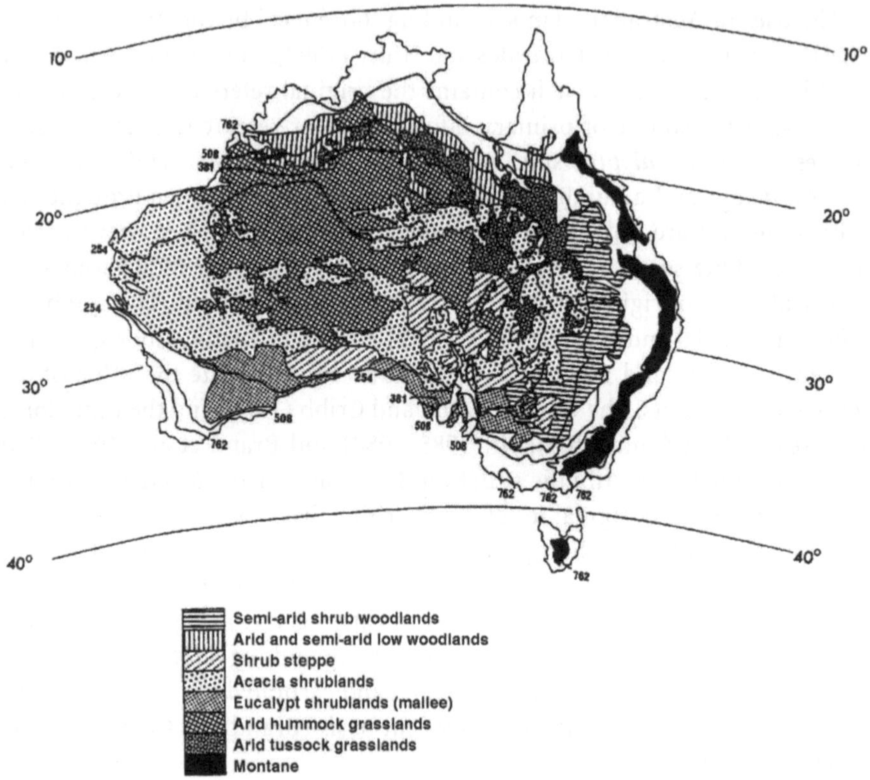

Fig. 10. Vegetation map of Australia showing isohyets; swamp, riverine and lake comunities not shown. (after Walter and Breckle 1986; Williams 1987)

3.8.2
Arid and Semi-Arid Low Woodlands

In south-eastern Australia the arid and semi-arid low woodlands consists of *Casuarina-Alectryon* (syn. *Heterodendrum*) and *Callitris* communities,while northern Australia they are represented by dwarf (4 to 8-m) *Eucalyptus* communities. Many of the species present in the semi-arid shrub woodlands (Sect. 3.8.1) also occur in the south-eastern low woodlands. However, in the northern *Eucalyptus* communities the shrubs are represented by *Acacia* spp, *Atalaya hemiglauca, Grevillea* spp., *Hakea suberea, Terminalia* spp. and *Ventilago viminalis.* Hummocks of *Triodia pungens** replace *Aristida* spp. in the drier areas (Beard 1969; Williams 1982). Between these northern and southern low woodlands are a range of communities dominated by *Acacia aneura**, *A. cambagei* and *A. georginae** (Williams 1982).

3.8.3 Shrub Steppe

The shrub steppe occupies much of the Nullarbor Plain, the Simpson Desert and areas of South Australia and New South Wales to the south and east of Lake Eyre. They occur on crusty loams, yellow-grey soils and grey, self-mulching clays, and are dominated by species of *Atriplex, Maireana* and *Sclerolaena*, with *Chenopodium* and *Rhagodia* locally dominant. Trees are uncommon and are usually confined to the more mesic sites. However, in the southern, winter rainfall region the shrubs can form the understorey to *Acacia* and *Eucalyptus* communities over extensive areas. The dominants of the low shrubland communities include the long-lived perennials *Atriplex nummularia**, *A. rhagodioides, A. vesicaria**, *Chenopodium auricomum, Maireana astrotricha, M. pyramidata** and *M. sedifolia**, with an occasional *Pittosporum phylliraeoides*, of which *Atriplex vesicaria** and *Maireana sedifolia** are the most widespread, forming both mixed and pure stands in the shallow depressions. Both species may be killed by overgrazing. In their absence, annual or short-lived perennials such as *Salsola kali* and species of *Atriplex* and *Sclerolaena* may become dominant; short-lived perennial grasses of *Enneapogon* and *Eragrostis* can behave in a similar manner (Williams 1982; Walter and Breckle 1986). The relative abundance of each of the *Maireana* species is largely determined by the depth of the underlying limestone. When close to the surface *M. sedifolia** is the predominant species, with *M. pyramidata** and then *M. astrotricha* taking precedence as the depth to the limestone increases.

The Nullarbor Plain, apart from the sinkhole depressions (dongas), is dominated by *Maireana sedifolia**, with annual grasses appearing in season, especially *Stipa trichophylla*. The centre of the Plain is treeless, becoming increasingly more and more densely wooded towards the margins. Around the extreme margins of the Plain these stands also include *Acacia aneura**, *Casuarina cristata* and *Eucalyptus oleosa* (Beard 1969).

3.8.4 *Acacia* Shrublands

Where the climate is too arid to support *Eucalyptus* as the dominant genus, it is replaced by species of *Acacia*, of which the most important is *A. aneura**. These are evergreen species of *Acacia* bearing phyllodes or rarely with transitional leaves, belonging to the subgenus *Heterophyllum* (syn. subgen. *Phyllodinae*); stipular spines are usually absent (sections *Pulchellae* and *Botrycephalae*) or often only present on juvenile plants (section *Phyllodineae*), or vestigial as in *A. victoriae** (Pedley 1981; Maslin 1988).

The varied soils of the *Acacia* shrublands consist of red siliceous sands, earthy loams, duplex massive red earths and hard pedal red duplex. In the southern regions the soils are alkaline and in the northern regions acid to neutral. The soil characteristics strongly influence the floristic composition of the communities. Among the dominant trees are *Acacia* spp., including *A. aneura**, *A. craspedocarpa** and *A. georginae**. The grasses of the *Acacia aneura* shrublands include *Chloris* spp., *Dichanthium* spp., *Digitaria* spp., *Eragrostis eriopoda**, *E. microcarpa*, *Eriachne* spp. and *Monachather paradoxus*,, together with *Rhagodia baccata** and members of the Amaranthaceae, Compositae, Cruciferae and Malvaceae (Williams1982). Colonies of *Xanthorrhoea thorntonii** (Fig. 38) occur sporadically throughout the area (Beard 1969). Wherever summer rains occur in addition to winter rains, *Acacia aneura** is accompanied by, or replaced on sandy soils by, species of *Triodia* and, less frequently, *Plectrachne* (Walter and Breckle 1986).

3.8.5
Eucalyptus Shrublands

Eucalyptus shrublands are inherently inflammable and occur extensively on grey-brown calcareous soils and red-brown earthy sands of the partly undulating topography of southern Australia. In eastern Australia the dominant species include *Eucalyptus behriana, E. dumosa* subsp. *dumosa, E. gracilis* var. *gracilis, E. oleosa, E. pileata* (syn. *E. dumosa* subsp. *pileata*), *E. polybractea, E. socialis* and *E. viridis*. In western Australia the dominants are *E. kingsmillii, E. oleosa* and *E. pyriformis*. The understorey vegetation consists of *Alectryon aleifalium* (syn. *Heterodendrum oleifolium*) and the shrubs *Cassia nemophila* and *Senna artemisioides* subsp. *sturtii* (syn. *Cassia sturtii*) with *Atriplex* spp. and *Maireana sedifolia** on the heavier textured soils and *Triodia irritans* in the sandy dunes (Williams 1982). The more stable coastal dunes of Western Australia are fixed with more or less dense mallee consisting of *Eucalyptus angulosa* and *E. foecunda* (syn. *E. leptophylla*), together with *Acacia* spp., *Eremophila* spp. and *Melaleuca* spp. (Beard 1969).

In western New South Wales, most of the common mallee species, *Eucalyptus dumosa, E. gracilis, E. foecunda, E. incrassata* and *E. socialis* may occur either as up to 10-m-tall, one to three-stemmed 'bull' mallee or as multistemmed, 1 to 3 m tall 'whipstick' mallee. The bull mallee grow in open swales where fires are very infrequent, except where the ephemeral *Stipa variabilis* provides fuel, while the whipstick mallee are associated with the highly inflammable, xerophytic grass *Triodia irritans* var. *laxispicata*. However, the *Triodia* will rarely burn before it has regenerated for at least 8 to 10 years. When bull mallee is burnt, however, it regenerates first as whip mallee and later regains its tree form (Noble 1982).

3.8.6
Arid Hummock Grasslands

The nutrient-poor siliceous sands and earthy sands of limited water-holding capacity of the arid hummock grasslands are dominated by species of *Plectrachne* and *Triodia*. These are cushion grasses with setaceous, in-rolled, mainly sclerenchymous leaves. Often referred to as spinifex, these grasslands are not to be confused with the coastal grass genus *Spinifex*. The seif dunes of the Great Sandy Desert are normally thickly vegetated on their flanks with the hummock grass *Plectrachne schinzii** and such shrubs as *Acacia salicina, A. victoriae**, *Crotalaria cunninghamii* and *Grevillea* spp. The dune crests are generally bare. *Melaleuca lasiandra* is common in the interdunal depressions. On deep sands between the dunes *Plectrachne schinzii** is dominant, but where the sands shallow or give way to lateritic gravel, it is replaced by *Triodia pungens** or, to the south of latitude 20°S, by *T. basedowii* (Beard 1969).

The sandy plains of the adjacent Balwina Desert have a ground cover of *Triodia pungens** with numerous, up to 3-m-tall shrubs which include *Acacia pachycarpa, A. victoriae**, *Cassia* spp., *Gardenia keartlandii, Grevillea* spp. and *Hakea suberosa*, together with an occasional *Eucalyptus* spp. and *Melaleuca lasiandra* in the depressions (Beard 1969). The unstable ridges of the Simpson Desert are covered with the shrubby, monotypic grass *Zygochloa paradoxa*. *Triodia basedowii* dominates the lower dune slopes and valleys and, in the north, the dune flats. A *Triodia basedowii – Grevillea juncifolia – Eucalyptus pachyphylla* community occurs on the sand flats to the north, and on the stone pavement flats the community contains *Astrebla pectinata, Atriplex vesicaria** and *Scleroleana* spp., with the Chenopodiaceae dominating the stony soils (Walter and Breckle 1986).

In the northern desert, where the rather unreliable rainfall of 250–300 mm is confined to the summer months, the dominant grass is *Triodia pungens**, forming communities in the western desert of Northern Territory with or without *Plectrachne schinzii**. In higher rainfall areas *T. pungens** is restricted to dry stony habitats. In the south *Triodia irritans* hummock grassland merges with the mallee shrublands and *Zygochloa paradoxa* occupies and stabilises dune crests in the extensive desert dunefields. Associated species may include *Allocasuarina decaisneana** and among the shrubs *Acacia kempeana, Cassia nemophila, Eremophila* spp., *Senna artemisioides* subsp. *artemisioides* (syn. *Cassia artemisioides*) and subsp. *sturtii* (Williams 1982).

3.8.7
Arid Tussock Grasslands

Arid tussock grasslands cover extensive summer rainfall areas with 250–750 mm annual rainfall. The dominant genus is the perennial *Astrebla**, which occurs on grey self-mulching clays, heavy textured brown soils and stony desert tableland soils. The two former soils may be intermittently flooded or seasonally inundated, and all three are poorly drained and dry out completely during the dry season. On the stony mesa in South Australia, with an annual rainfall of 100–150 mm, the occasional shrubs include species of *Atriplex, Maireana* and *Sclerolaena*; on the crusty red duplex soils the grasses present include *Aristida contorta, Digitaria* spp. and *Enneapogon* spp. In Northern Territory, on the grey self-mulching clays and with an annual rainfall of 250–400 mm, among the associated perennial grasses are *Astrebla elymoides* (slender, hoop or weeping Mitchell grass), *Ar istida latifolia, Panicum decompositum* and *P. laevinode* (syn. *P. whitei*), together with the ephemeral *Brachyachne convergens, Dactyloctenium radulans* and *Iseilema* spp. (Williams 1982; Walter and Breckle 1986).

3.8.8
Swamp, Riverine and Lake Communities

The larger swamp communities in south-western Australia are dominated by: *Muehlenbeckia cunninghamii* with fringing *Eucalyptus largiflorens*; swamps with *Eremophila maculata**, *Maireana pyramidata** and *Rhagodia spinescens*; and swamps with *Chenopodium nitrariaceum*, which are sometimes associated with *Eragrostis australasica* and *Maireana aphylla**, and in the northern parts with *Chenopodium auricomum*. In the arid interior there are *Atriplex nummularia** swamps associated with *Eucalyptus microtheca* (syn. *E. coolabah*) fringing the water courses, channels and rivers (Williams 1982; Walter and Breckle 1986). In South Australia, with a rainfall of 200 mm, the flood plain communities include *Eucalyptus camaldulensis** with *Acacia* sp. and *Cassia* sp. together with the grasses *Enteropogon acicularis* and *Eulalia fulva*, while the adjacent red powdery calcareous loams support *Maireana aphylla**, *M. astrotricha* and *Rhagodia parabolica* (Williams 1982). Around the salt lakes of the western desert Chippendale (1965) records samphire, *Sclerostegia tenuis* (syn. *Pachycornia tenuis*) on the edge of the lake surface, leading back to *Sclerolaena birchii* (syn. *Anabasis birchii*) and *Maireana luehmannii* (syn. *Bassia luehmannii*) with *Atriplex vesicaria** on the gypseous sandy shore, which in turn is often encircled by a thickets of *Melaleuca* spp. (ti trees) and *Acacia salicina*.

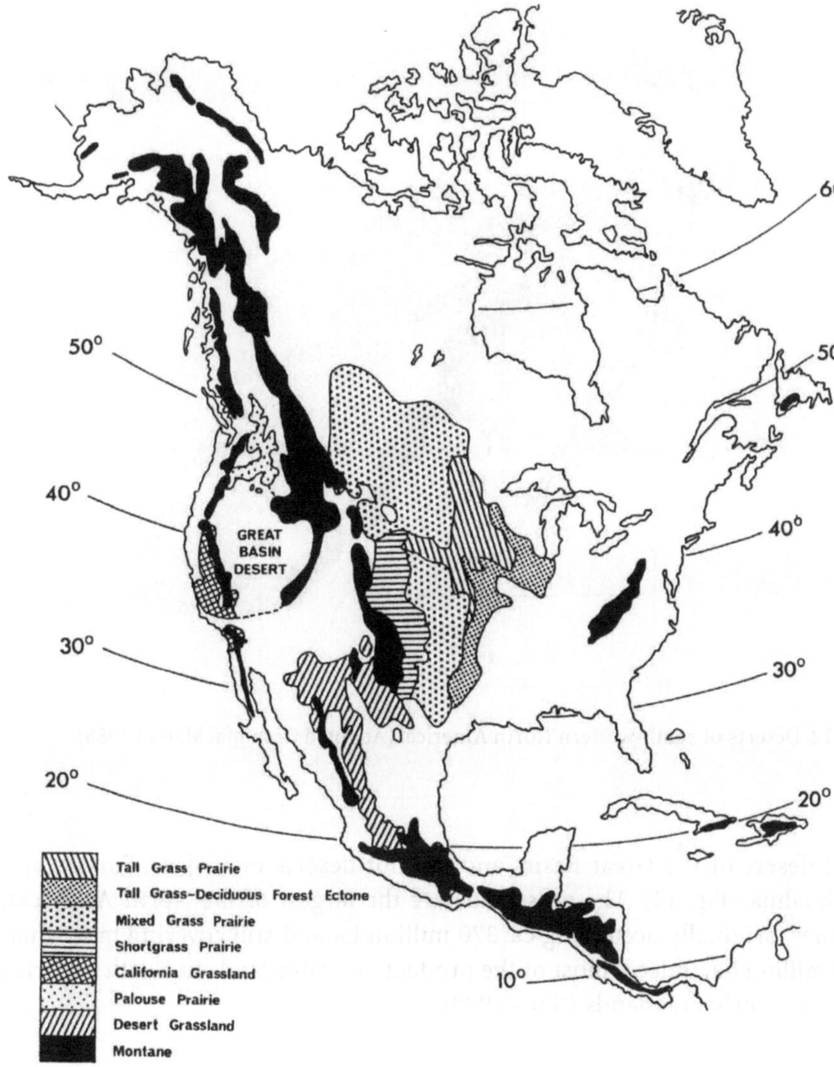

Fig. 11. Major grasslands of North America (adapted from Sims 1998)

3.9 North and Central America

The arid and semi arid ecosystems of North and Central America are represented by the grasslands and deserts. The grasslands are represented by the prairies of the Great Plains, desert grasslands of the south-west, California grassland and the Palouse prairie of the intermountain region (Fig. 11), the

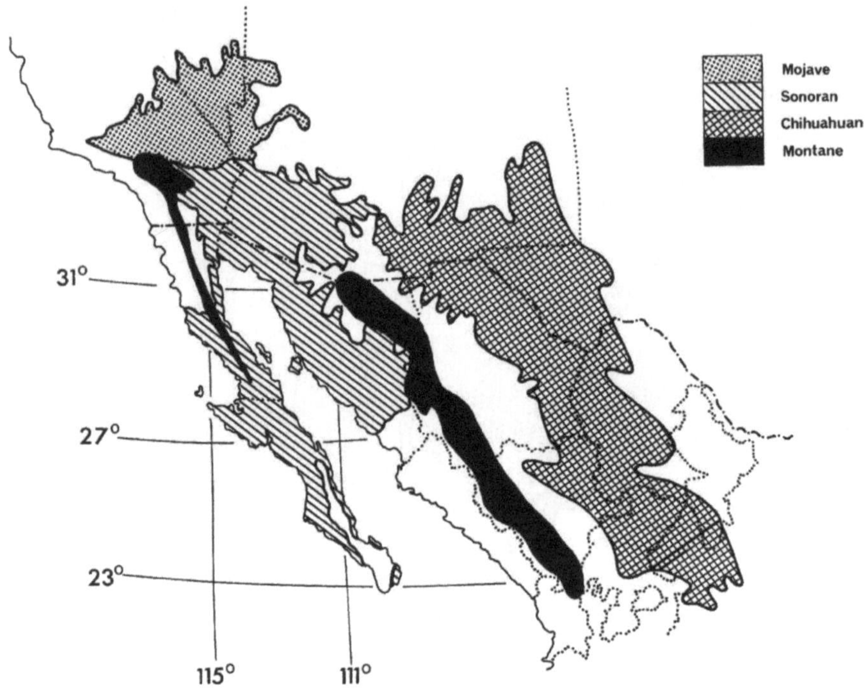

Fig. 12. Deserts of south-western North America. (Adapted from MacMahon 1988)

cold desert of the Great Basin, and the hot deserts of Mojave, Sonora and Chihuahua (Fig. 12). The grasslands are the largest of the North American biomes, originally occupying ca. 370 million ha and still covering more than 150 million ha. Indeed, most of the productive arable lands in North America were formerly grasslands (Sims 1988).

3.9.1 The Great Plains

The area of the Great Plains forms a vast, gently inclined plain, rising to ca. 1600 m at the foot of the mountains in the west and sloping eastwards at about 2 m km^{-1} to as low as 290 m km^{-1} in the east. Most of the major rivers rise in the west and eventually reach the Mississippi River, while the Red River and its tributaries flow north to the Hudson Bay. Eastwards from the eastern foothills of the Rocky Mountains the isohyets are orientated in a north to south direction; the mean annual precipitation increases eastwards from ca. 400 mm to ca. 750. Precipitation is generally heaviest during the summer, with some winter rain and snow. The annual precipitation varies considerably from year to year and periods of abnormal rainfall followed by periods of abnormal dryness are not uncommon. The summers are hot to very hot, even in the north, and punctuated by thunderstorms, while the winters tend to be cold and windy, and in most years the springs are relatively moist. The growing season is, as expected, shorter in the north and at the higher elevations in the west. Thus, a spring-flowering species blooming in March in Oklahoma may not bloom until May or June in North Dakota and, conversely, an autumn-flowering species may bloom in August in North Dakota and in October in Oklahoma.

Bordered on the west by the evergreen coniferous forests of the Rocky Mountains, the grasslands of the Great Plains extend eastwards to the deciduous hardwood forests of eastern North America. Elements of the coniferous forests extend along the hills and escarpments to midway across the Great Plains, while the deciduous forest extend westwards along the rivers and into the grasslands or prairies. The grasslands are often considered as conveniently forming three, roughly parallel, north-south zones; however, the boundaries are poorly defined. The zones are largely determined by the east-west precipitation gradient, evaporation and soils, with numerous subzones recognised by Küchler (1964) within the three zones roughly paralleled by the north-south precipitation gradient. These Canadian and US prairies are one of the world's great cereal-producing regions, especially wheat. So much so that little of the natural grassland now survives (Kaul 1986).

3.9.1.1
Tall-Grass Prairie

The most mesic of the grasslands of the Great Plains, the tall-grass prairie is dominated primarily by long-lived perennials, although much is now under cultivation or affected by grazing. Two major grassland associations are recognised:

1. *Bluestem Prairie, Also Known as 'True Prairie'.* This association extends from the southern tip of Manitoba through eastern North Dakota and western Minnesota southwards to eastern Oklahoma. The dominant species are *Andropogon gerardii*, Panicum virgatum, Schizachyrium scoparium* and *Sorghastrum nutans.*

2. *Wheatgrass, Bluesgrass and Needlegrass Prarie.* Originally extending from south-central Canada southwards through east-central North and South Dakota and Nebraska to north-central Kansas, the association is dominated by species of *Agropyron, Andropogon, Calamovilfa* and *Stipa.* with sod and bunch grasses such as *Bouteloua racemosa, Elymus smithii* (syn. *Agropyron smithii*)

and *Koeleria pyramidata* (syn. *K. cristata*), and often associated with *Andropogon nutans* and *Stipa comata**.

3.9.1.2. Mixed-Grass Prairie

As the rainfall decreases westwards there is a gradual transition from tall-grass prairie to mixed-grass prairie, the taller and more mesic grasses being replaced by shorter and more drought tolerant species. Situated in the northern Great Plains the mixed-grass prairie lies in western North and South Dakota, north-western Wyoming, eastern Montana, south-eastern Alberta, Saskatchewan and Manitoba. In Wyoming and Montana the mixed-grass prairie is similar in appearance to short-grass prairie, while south of the Nebraska Sandhills a broad belt of mixed-grass prairie extends from south-west Nebraska through western and central Kansas and Oklahoma into central Texas (Sims 1988). The four major mixed-prairie grasses are *Elymus smithii*, *Koeleria pyramidata*, *Sporobolus cryptandrus* and *Stipa comata** (Weaver and Albertson 1956).

3.9.1.3 Short-Grass Prairie

The short-grass prairie lies to the east of the Rockies, from the Nebraska panhandle and south-eastern Wyoming through eastern Colorado and western Kansas southward through the high plains of Oklahoma, Texas and New Mexico. According to Weaver and Albertson (1956) the short-grass prairie represents a grazing disclimax that has been derived from mixed-grass prairie as the result of overgrazing.

The principal range forage grasses on the heavier soils of the southern Great Plains are the valuable, turf-forming, short grasses *Bouteloua gracilis* and *Buchloe dactyloides*, while taller bunch grasses ('midgrasses') favour the lighter and often riverine soils. *Bouteloua curtipendula*, *Chondrosum hirsutum* (syn. *Bouteloua hirsuta*) and *Schizachyrium scoparium* occur on rough broken lands with shallow soils underlain or intermixed with gravel. The latter also occurs with *Andropogon gerardii** on the eastern borders of the subzone. *Stipa comata** and *Koeleria pyramidata* often accompany the short grasses on the high plains of Colorado (Savage and Costello 1948).

Less valuable midgrasses which may invade the short-grass association following drought or overgrazing. These include the decidedly unpalatable *Aristida* spp. (three-awn) on the heavier soils and *Sporobolus cryptandrus*, which is of only fair palatability, on the lighter soils. *Hilaria mutica*, a grass of secondary grazing value, is often an important associate of the short grasses in south-eastern New Mexico and the southern part of the Texas panhandle, while *Hilaria jamesii* often replaces the short grasses on the misnamed 'tobosa' flats of north-eastern New Mexico. In the drier parts *Bouteloua eriopoda* occurs sparingly in association with *B. gracilis*, or even replaces it (Savage and Costello 1948). The comparatively dense short grass turf usually supports fewer herbs and shrubs than the midgrass communities of the sandier soils. Among the few browse shrubs at higher elevations of Colorado and New Mexico are *Atriplex canescens** and *Eurotia lanata* (winterfat); these are of especial value as winter browse in Colorado.

The sandier soils, with their more open stands of bunch grasses, support many shrubs and herbs. The most valuable tall grasses of the sand hills include sand bluestem, *Andropogon hallii* (syn. *A. gerardii* var. *paucipilus*), *Eragrostis trichodes* and *Panicum virgatum*, together with *Andropogon gerardii** and *Sorghastrum nutans* (Indian grass). Little bluestem, *Schizachyrium scoparium*, dominates the oak hangers, and is frequently abundant on rocky outcrops bordering the sands and sometimes in other sites. Sand dropseed, *Sporobolus cryptandrus*, is a prominent transition grass on sandy land. The less valuable grasses include the widespread *Redfieldia flexuosa* (blowoutgrass) and *Calamovilfa longifolia* in the north, and giant reedgrass, *C. gigantea*, in the south. *Aristida purpurea** and *Triplasis purpurea* (purple sandgrass) are regarded as indicators of overgrazing (Savage and Costello 1948). On moist soils of moderate salinity the less palatable *Distichlis spicata* subsp. *stricta**, *Panicum obtusum* and *Sporobolus airoides* occur (Savage and Costello 1948).

3.9.2
Great Basin Desert

Lying within the western flanks of the Rocky Mountains, the Great Basin, also known as the Intermountain Region, is bounded by the Wasatch Range (ca. 3500 m) to the east, the Sierra Nevada (4418 m) and Cascade ranges to the west, the watersheds of the Snake and Colombia Rivers to the north and the Colorado River drainage system to the south. Embracing Nevada, it has substantial extensions into Utah and California, with smaller excursions into Oregon, Idaho and Wyoming.

It is a high elevation, mountainous, cool desert, largely above 1200 m a.s.l. and traversed by numerous, relatively parallel, predominantly north-south orientated, fault-block mountain ranges. While all the margins of the Great Basin eventually drain into the Pacific, the interior contains a complex of more than a hundred basins without any outlet, resulting in extensive playa lakes or salt flats, and with sloping bajadas (alluvial fans) occupying much of the valley areas (Irwin-Williams et al. 1990; Smith and Nowak 1990).

The varied topography has important climatic implications, resulting in considerable local variations within relatively small areas. While on the western or windward side of both the Sierra Nevada and Cascades the flanks and ridge tops receive most of the precipitation, on the leeward side there are strong local orographical precipitation gradients where the precipitation decreases with descending altitude, so much so that the valleys and lowlands are often dry. The average annual precipitation for much of the Great Basin lowlands ranges between 50 and 300 mm, and typically below 200 mm. The lowest precipitation is in the south-west (which is also the lowest in altitude) and increases to the north, east and south with increasing altitude. The annual lake evaporation ranges from 1100 mm in the north to over 2000 mm in the south (Smith and Nowak 1990).

Diurnal temperatures in the valleys can vary by more than 15-20 °C, and seasonal averages range from -40 °C in winter to above 40 °C in summer. Such low winter temperatures are considered to be the primary limiting factor differentiating between the 'cold desert' vegetation of the Great Basin and that of the Mojave 'hot desert'. In the more northern and higher altitudes freezing conditions predominate for much of the year, with the frost-free period as short as 80–90 days in the northern and central Great Basin (Hidy and Klieforth 1990; Smith and Nowak 1990).

For most of the year the average lapse rate is 6.5 °C km^{-1}, although at midday in summer with strong thermal activity the midday lapse rate on sunny mountain slopes approaches the adiabatic value of ca. 10 °C km^{-1}. On clear, calm nights, especially in winter, cold air drainage and differential radiation effects greatly reduce or reverse the temperature gradient, forming inversions in which the temperature increases with altitude in the lower hundred metres or so. Even during summer small mountain valleys may have minimum air temperatures of 10 °C or more cooler than the neighbouring slopes, a phenomenon that has important ecological implications (Hidy and Klieforth 1990).

The cold desert vegetation zone of the Great Basin lowlands is comprised of sagebrush grassland and salt desert, with the two zones intermingling over vast portions of the intermountain USA. The sagebrush grassland generally occupies the more semi-arid, higher elevation basins, ca. 1500–2200 m a.s.l., and less saline soils of the northern and eastern Great Basin, while the salt desert is largely in the arid troughs of the west. Both types are typically frost-free for less than 180 days (Holechek et al. 1989; Smith and Nowak 1990).

3.9.2.1
Sagebrush Grassland

Sagebrush grassland is a rangeland type that occurs primarily in Oregon, Idaho, Nevada, Utah, Montana and Wyoming, between the Cascade-Sierra Nevada Mountains and the Rocky Mountains. The climate is semi-arid, with the annual precipitation ranging between 200 and 500 mm, with an average of 250 mm. In the northern half 50-60% of the precipitation falls during the late autumn to early spring as snow, while the summers are quite dry. In the southern part most of the precipitation is also during the winter, but with some summer rain. The rain-shadows caused by the smaller mountains within the Great Basin account for much of the aridity in the area. Temperatures in the area are extreme, falling as low as -37 °C in winter and rising to 38 °C in summer. The growing season is approximately 100 days.

The grasslands are characterised by *Seriphidium tridentatum* (syn. *Artemisia tridentata*), with *S. novum* (syn. *Artemisia nova, A. arbuscula* subsp. *nova*), on the rocky and shallow soils in the northern and central areas, and *Seriphidium arbusculum* (syn. *Artemisia arbuscula* subsp. *arbuscula*) at higher elevations. In the central and northern part *Chrysothamnus puberulus*, occupies sandy soils with a low salt content, with *Purshia tridentata* on the sandy, rocky soils at higher elevations. *Ephedra viridis** occurs on coarse soils in the central and southern part.

Sitanion hysteryx is an important grass throughout the area. *Agropyron spicatum* (bluebunch wheatgrass) dominates the understory in the northern half and, where the grazing has not been excessive, *Elymus smithii* (western wheatgrass) dominates the understory in the southern half, with *Elymus lanceolatus* in the more mesic eastern portion. In the south-west the codominants *Stipa arida, S. speciosa* and *S. thurberiana* form a complex gradient associated with decreasing altitude; other important grasses are *Elymus cinereus, Festuca idahoensis, Oryzopsis hymenoides** and *Stipa comata**. With overgrazing the invasive exotic annual *Bromus tectorum** from Eurasia can replace some of the perennial grasses (West 1988; Smith and Nowak 1990).

Following the control of the sagebrush the perennial Asiatic grass *Agropyron desertorum* (crested wheatgrass) has been successfully seeded over large areas of the grassland, thereby providing high-quality spring forage before the native grasses are available. This allows the grazing of the native cool-season grasses to be delayed until the late spring/early summer, by which time they have completed much of their growth and can withstand grazing much better than during the spring (Holechek et al. 1989).

3.9.2.2
Salt-Desert Shrubland

The salt-desert shrubland occurs primarily in the states of Utah and Nevada, with smaller areas in Wyoming, Montana, Idaho, Oregon, Colorado and New Mexico. It is found in the lower areas of the Great Basin between 1200 and 2200 m a.s.l., where the drainage is often restricted and the water table high, forming a mosaic with the sagebrush grassland. The annual precipitation ranges from 80 to 250 mm, with an average of 120 mm, and is highly variable from year to year. Approximately half of the precipitation occurs as snow, the remainder as autumn or spring rain; the summers are generally dry.

The vegetation is characterised by a few widely spaced species of low, spiny, greyish, microphyllous, chenopod or composite shrubs of generally low palatability. The relatively unpalatable *Atriplex confertifolia* is the dominant shrub over much of the area, with the highly palatable *Eurotia lanata* as the second most common species; other associates are *Artemisia spinescens* and *Grayia spinosa*. Important understory grasses of the association include *Distichlis spicata* var. *stricta**, *Hilaria jamesii, Oryzopsis hymnenoides** and *Sitanion hysteryx*. With deterioration the understory is occupied by *Bromus tectorum*, Halogeton glomeratus* and *Salsola kali*. The associated overstory species include *Atriplex confertifolia, Chrysothamnus nauseosus*, Grayia spinosa* and *Tetradymia canescens*. The valley bottoms may be dominated by *Sarcobatus vermiculatus** in association with *Allenrolfea occidentalis, Salicornia rubra, Sarcocornia utahensis* (syn. *Salicornia utahensis*), *Suaeda* sp. and the grasses *Distichlis spicata** and *Sporobolus airoides* (Shreve 1942a; Smith and Nowak 1990).

3.9.3 Mojave Desert

The smallest of the North American deserts, the Mojave Desert is located largely in south-eastern California, with extensions into southern Nevada, south-western Utah and north-western Arizona. Lying mainly between 600 and 1200 m a.s.l., it descends in the north to 86 m below sea level in the 200-km-long Death Valley. The terrain is even more mountainous than that of the Great Basin. The drainage is largely internal, with relatively few tributaries reaching the sea by way of the Colorado River.

It is the driest of the three hot deserts, with the elevated western and southern inner margins receiving only 130 mm annual rainfall. The erratic rainfall falls largely during the winter and early spring, although some summer storms can be expected along the eastern margins. Mean annual temperatures at Las Vegas are 7 °C in January and 32 °C in July, while 150 km to the west, at Furnace Creek in Death Valley, the mean rainfall is 42 mm, 75% of which falls during November-April. The daily maximum and minimum temperatures in Death Valley are 47 and 39 °C (the highest maximum recorded reaching 56.7 °C; Kendrew 1961), while ground surface temperatures may exceed air temperatures by as much as 25–35 °C. From January to April the mean daily maximum temperature is 23–35 °C (Pearcy et al. 1974).

The Mojave Desert represents a transition zone between the 'cold' Great Basin Desert to the north and the 'hot' Sonoran Desert to the south. It has the least diversity of vegetation, and is primarily a shrubland. About one-fourth of the flora is endemic, of which nearly 80% of the ca. 250 annuals are endemic (MacMahon 1988). Its sinuous northern boundary, which extends across southern Utah, Nevada and California, is considered to represent the transition to the Great Basin Desert, where the creosote bush, *Larrea divaricata* subsp. *tridentata**, and its associates replace the *Atriplex confertifolia* and *Seriphidium tridentatum* communities. Due to the low productivity and grazing resistance of the rangeland, livestock grazing is not a practical proposition for much of the area (Shreve 1942a; Holechek et al. 1989).

In the transition zone between the 'cold' and 'hot' deserts *Bromus tectorum** intergrades with, and is eventually replaced by another Eurasian annual, *B. rubens,* which is frequently the dominant winter annual of the *Larrea divaricata-Ambrosia dumosa* and *Coleogyne ramosissimus* communities (Shreve 1942a; Smith and Nowak 1990). The *Larrea divaricata-Ambrosia dumosa* association dominates the bajadas and lower foothills of the Mojave Desert. On the finer textured soils of the valley bottoms, *Larrea* and *Ambrosia* are replaced by dense stands of *Atriplex hymenelytra** on soils derived from rocks rich in carbonates and *A. polycarpa** in less alkaline soils; *Grindelia camporum** is among the other halophytic shrubs present.

Where the bajadas become rockier, steeper or higher, a mixed shrub assemblage dominates, with *Acamptopappus sphaerocephalus* var. *hirtellus, Atriplex hymenelytra*, Ephedra nevadensis, Grayia spinosa*, widely scattered *Larrea, Yucca* spp. and various cacti, including *Opuntia basilaris**, the many-stemmed *Echinocactus polycephalus* and the barrel cactus *Ferocactus cylindraceus**. Among the plants of Death Valley are *Atriplex canescens*, A. confertifolia, A. hymenelytra*, A. lentiformis*, A. polycarpa, Suaeda torreyana* and *Tidestromia oblongifolia.*

The major washes support extensive stands of *Acacia greggii, Chilopsis linearis** and *Prosopis pubescens. Populus fremontii, Salix* spp. and *Prosopis* are dominant along the more perennial streams and springs, although many such sites have now been invaded and taken over by the exotic *Tamarix ramosissima* (syn. *T. pentandra*). Associated with the fan palm *Washingtonia filifolia* in the oases of the Mojave Desert and extending into California, Baja California and Arizona are *Bebbia juncea, Haplopappus acradenius* (alkali goldenbush), *Prosopis pubescens* and *Tamarix ramosissima.*

The calcicole *Coleogyne ramosissimus* replaces *Larrea divaricata** at higher elevations, dominating old pediment slopes and bajadas with petrocalcic (caliche) horizons; it may also occur in association with *Yucca brevifolia* (West 1988; Smith and Nowak 1990). Other plants found with *Coleogyne* are *Acacia greggii, Atriplex confertifolia, A. spinosa, Encelia frutescens, Psorothamnus spinosus* and *Tetradymia spinosa*; associated grasses include *Aristida* spp., *Bouteloua eriopoda, Hilaria jamesii, Muhlenbergia porteri, Oryzopsis hymenoides** and *Stipa* spp. (Shreve 1942a; MacMahon 1988; West 1988).

3.9.4 Sonoran Desert

Embracing much of Arizona, southern California, Baja California and Sonora, the Sonoran Desert is probably the world's most investigated and best documented desert. Its ecology has been comprehensively described by Forrest Shreve in Shreve and Wiggins (1964), and summarised by McGinnies (1981). The seven divisions recognised by Shreve in Shreve and Wiggins (1964) have been reduced to six by Brown and Lowe (1980) and Brown (1982) by excluding the Foothills of Sonora, which lie to the south of the Magdalena Plain and are now included in the more mesic subtropical thorn scrub. The revision is accepted by MacMahon (1988) and is followed here (Fig. 13). The vegetation descriptions of the divisions are compiled from the above references.

Most of the Sonoran Desert drains into the Gulf of California, with only a small fraction draining into the Pacific Ocean. All the major rivers rise outside the desert, in the mountains of Arizona, Colorado and Sonora, of which only the Colorado is perennial. There are also a number of undrained basins at low elevations such as the Salton Sea in California and the Laguna Chapala in Baja California.

The climate is relatively uniform throughout the Sonoran Desert, with regional differences due to latitude, altitude and geomorphic configuration. The climate is distinctly continental, especially in the northern parts. In the south the influence of the sea results in a slight lowering of diurnal temperatures, the intermittent effect of fog and the almost constant onshore wind. In Baja California the main effect of the ocean is a slight lowering of the sensible daytime temperature within 40 km of the coast. The north-easterly trade winds and the south-westerly winds of the prevailing westerlies have their origins within the latitudes of the Sonoran Desert. The cyclonic storms of the westerlies in winter and the descending air of the subtropical horse latitudes in spring and autumn give rise, respectively, to the expected rain and aridity. However, due to the size of the North American continent, the expected summer trade winds do not appear and are replaced by relatively moist winds from the south.

In Mexico the first showers may begin in May or even April, while in southern Arizona they usually begin towards the end of June or early July; further north the rains do not arrive until mid-July. The mean annual rainfall is usually less than 300 mm and may be as low as 75 mm. The winter temperatures vary considerably with latitude, but summer temperatures are little changed. Like the Death Valley in the Mojave Desert, parts of the Sonoran Desert have the highest sustained air temperatures in North America, where periods of 90 consecutive days with a maximum of 38 °C are not uncommon. Even in winter the February temperature can rise to 32 °C. There is a slight

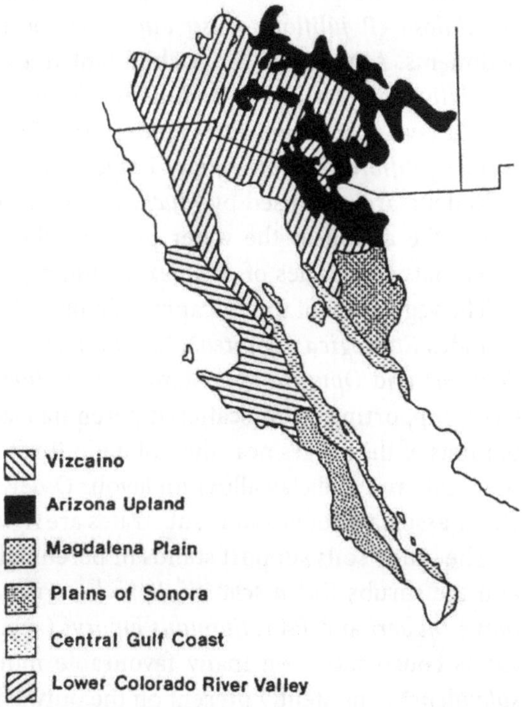

Fig. 13. Subdivisions of the Sonoran Desert. (Adapted from MacMahon 1988)

fall in temperature associated with the gradual rise in altitude from the coast to the interior. Occasional severe frosts may occur at the higher elevations. The low productivity of the rangelands makes much of the Sonoran Desert unsuitable for livestock grazing (Holechek et al. 1989). The six subdivisions within the Sonoran Desert recognised by Brown and Lowe (1980) are described below.

3.9.4.1
Lower Colorado Valley

The Lower Colorado Valley extends westwards from Phoenix and Ajo Mountain in Arizona to Indo in California, and includes the lower drainage of the Colorado and Gila rivers, the Salton Basin, the east coast of Baja California as far south as Bahia Los Angeles, and all of Sonora lying below 390 m as far south as the valley of the Rio Magdalena. It is one of the most arid regions of the Sonoran Desert.

The upper bajadas (scree slopes), level fields of the volcanic malpais (badlands), hills and mountains are dominated by *Ambrosa dumosa* and *Larrea divaricata* subsp. *tridenta**, with *Parkinsonia microphylla** and *Prosopis glandulosa* (*P. juliflora sensu auct.*) on the rocky soils of the bajadas and pediments. Among the more abundant trees along larger drainageways are *Condalia lycioides* (syn. *Condaliopsis lycioides*), *Olneya tesota**, *Parkinsonia florida* (syn. *Cercidium torreyanum*) and *Prosopis glandulosa* var. *torreyana* (syn. *P. juliflora* var. *torreyana*). Large areas of the alluvial plain of the Colorado delta are colonised by *Pluchea sericea* (arrow weed), *Prosopis* spp. and, where the action of the water prevents the establishment of larger woody perennials, by species of *Atriplex*, including *A. canescens**.

The vegetation of the volcanic malpais of the Pinacate Mountains in Sonora includes *Carnegiea gigantea**, *Encelia farinosa**, *Larrea divaricata* subsp. *tridentata** and *Opuntia* spp. Many of the older volcanic ranges are devoid of trees, supporting a few scattered perennials and grasses. Even in the canyon bottoms of the ranges near the Colorado River there are few trees except where the local sand and clay alluvium favour *Olneya tesota** or *Parkinsonia florida*. Other associates of volcanic substrates are *Agave deserti** and *Hyptis emoryi**.

The sandy soils support stands of perennial grasses, such as *Hilaria rigida*, and the shrubs *Dalea schottii, Ephedra trifurca, Erigonum deserticola, Petalonyx thuberi* and *Psorothamnus emoryi* (syn. *Dalea emoryi*). *Encelia farinosa** is conspicuous on many favourable mountain slopes, with *Fouquieria splendens** consistently present on the outwash slopes below. One of the most ubiquitous plants on the bare slopes is the grass *Heteropogon contortus*. The most frequent cacti are *Ferocactus cylindraceus** and Johnson's echinomastus, *Neolloydia johnsonii* (syn. *Echinomastus johnsonii*).

3.9.4.2
Arizona Upland

The Arizona Upland is that north-eastern part of the Sonoran Desert lying mainly in Arizona and partly in northern Sonora, and lying at elevations between 150 and 900 m. In the south and west it merges into the plains of the Lower Colorado Valley; in the north and east it is flanked by a broken series of mountains, most of which are sufficiently high to provide a more mesic environment than the semi-arid Arizona Upland. The mean annual rainfall ranges from 225 mm in the west to 325 mm at the higher elevations in the east.

The Arizona Upland claims to have more large perennials, associated smaller perennials and greater species variation per unit area than any other group of habitats in the deserts of the world. The most abundant species present throughout the area are *Celtis pallida, Condalia warnockii, C. lycioides, Fouquieria splendens**, *Larrea divaricata* subsp. *tridentata**, *Lycium andersonii, Olneya tesota**, *Opuntia engelmannii, O. versicolor, Parkinsonia microphylla** and *Prosopis velutina* (velvet mesquite). Saguago, *Carnegiea gigantea**, is far less numerous, but its height makes it a conspicuous element of the vegetation. Other conspicuous species are *Echinocactus polycephalus*, and a wide variety of prickly pears, *Opuntia* spp. Perennial grasses are locally abundant, and both winter and summer ephemerals carpet the ground in favourable seasons. The latter include *Amaranthus* spp. and *Proboscidea parviflora**.

In the broader valleys the dominant shrub is *Larrea divaricata**, with the facultatively evergreen *Ambrosia deltoidea** and *A. dumosa* (white bursage) as subdominants. The climbing cucurbit *Marah gilensis** also occurs in the thickets bordering the streams, while *Cucurbita foetidissima** and *Phaseolus acutifolius* are found on the alluvial soils in the southern uplands. The lower mountain slopes throughout the central uplands are dominated by *Acacia constricta, Carnegiea gigantea*, Encelia farinosa*, Fouquieria splendens**, *Opuntia* spp. and other cacti and *Parkinsonia microphylla**. On the upper slopes *Parkinsonia florida* and *Prosopis velutina* become increasingly abundant, while *Carnegiea gigantea** and *Fouquieria splendens** are closely restricted to warm, frost-free sites. In the rugged area drained by the Gila River *Simmondsia chinensis** dominates many of the northern slopes. Species confined to the sides of streams are *Ambrosia ambrosioides* (syn. *Franseria ambrosioides*), *Baccharis sarothroides, Lycium* sp., *Parkinsonia florida* and *Opuntia* spp. while *Atriplex canescens** may form almost pure stands on the alkaline soils of the flood plain.

3.9.4.3
Central Gulf Coast

Split into two portions on either side of the Gulf of California, the mainland area of the central Gulf coast region in Sonora consists of a narrow band of vegetation extending from 27 to 30°N. In Baja California it extends nearly as far north but southwards reaches the tip of Baja at 23°N. The two areas are occupied in part by hills and small mountains rising abruptly from the coast and in part by bajadas extending far inland. The soils are shallow and stony; sand dunes are small and infrequent, and there are also a few lagoons, marshes and mangrove swamps. The rainfall is low, ranging from 100 to 250 mm, usually falling either in the late winter or in midsummer, but sometimes absent for several successive seasons. Summer temperatures are very high, except near the coast where they are ameliorated by onshore winds. The mainland vegetation from the mouth of the Rio de la Concepción south to Guaymas has been described in some detail by Felger and Moser (1985) and their description is incorporated here with that of Shreve and Wiggins (1964) and McGinnies (1981).

The transition from the dunes and sandy plains of the Lower Colorado Valley to the hard soils of the central Gulf coast occurs near the mouth of the Rio de la Concepción, ca. 43 km north of Libertad. Here, *Ambrosia dumosa* and *Larrea divaricata** are no longer dominant, giving way to a mixed desert scrub, of which representative perennials are *Capparis atamisquea* (syn. *Atamisquea emarginata*), *Bursera microphylla**, *Fouquieria splendens**, *Jatropha* spp. and *Parkinsonia microphylla*. Although very abundant throughout central Baja California south of El Rosario, the only mainland colony of *Fouquieria columnaris** is on the coast, 10 km south of Libertad. Here, it occurs on the north slope of a low granite ridge which runs east from the beach. It does not occur on the south slope.

Among the more or less obligate species of the dunes are *Astragalus magdalenae*, *Dicoria canescens*, *Euphorbia leucophylla* and *Triteleiopsis palmeri*. Associated species include *Abronia maritima**, *A. villosa*, *Atriplex canescens**, *Croton californicus*, *Jatropha cinerea* and the grass *Aristida californica*. Characteristic trees and shrubs of riparian sites are *Acacia greggii*, *Ambrosia ambrosioides*, *Hymenoclea salsola*, *Hyptis emoryi**, *Olneya tesota**, *Parkinsonia florida*, *Prosopis glandulosa* and *Sideroxylon occidentale* (syn. *Bumelia occidentalis*).

Also scattered along the coast are conspicuous stands of columnar cacti, with *Carnegiea gigantea**, *Lophocereus schottii**, *Pachycereus pringlei*, *Stenocereus gummosus* and *S. thuberi* (syn. *Cereus thurberi*), together with *Atriplex polycarpa**, *Bursera microphylla**, *Fouquieria splendens**, *Jacquinia macrocarpa* subsp. *pungens* (syn. *J. pungens*), *Maytenus phyllanthoides*, *Olneya tesota**, *Opuntia fulgida* and *Parkinsonia florida*. Bordering much of the shore is coastal scrub dominated almost solely by *Frankenia palmeri**. Tiburón Island and the neighbouring mainland coast are the southernmost limits in Sonora for *Ambrosia deltoidea**, *Atriplex polycarpa**, *Ferrocactus cylindraceus**, *Koeberlinia spinosa*, *Opuntia arbuscula* and *Simmondsia chinensis**, as well as the northern limit for *Acacia willardiana**.

3.9.4.4
Plains of Sonora

In central Sonora there lies between the foothills and the coastal strip an area, known as the plains of Sonora, with some physical features reminiscent of the Lower Colorado Valley. The general surface is extremely regular, with an east to west inclination from 750 to 100 m. There are few mountains or hills, of which only four mountains rise above 1000 m. The mean annual rainfall ranges from 250 to 375 mm. The summer temperatures are lower than in the Lower Colorado Valley. Frosts too are fewer and less severe than further north.

Over much of the area the vegetation is of open woodland, with irregularly scattered colonies of shrubs and large but widely spaced columnar cacti. *Capparis atamisquea, Olneya tesota**, *Parkinsonia aculeata, P. florida, P. microphylla**, *Cercidium praecox* and *Prosopis juliflora* are the dominant trees, with *Bursera hindsiana** or *B. microphylla** among the codominants. *Ambrosia deltoidea**, *A. dumosa* and *Larrea divaricata** are of local occurrence and colonies of *Opuntia* spp. relatively infrequent. To the north of Carbo colonies of *Larrea divaricata** alternate with open woodlands of *Fouquieria splendens**, *Olneya tesota**, *Parkinsonia florida, P. microphylla**, *Prosopis* spp. and the cactus *Stenocereus thurberi*; the subdominants include *Lophocereus schottii** and *Opuntia* spp.

In the central region between Carbo and Hermosillo the vegetation increases in density. The colonies of *Larrea divaricata** are still frequent, and summer annuals and perennials are abundant. *Bursera microphylla**, *Fouquieria macdougalii**, *Guaiacum coulteri** and *Parkinsonia microphylla* x *Cercidium. praecox* are present, while the cactus *Stenocereus alamosensis*, although not particularly abundant, give the vegetation a distinctive appearance. In the southern region, south of the Rio Sonora, *Larrea divaricata** is no longer present, *Carnegiea gigantea** reaches its southernmost limit and *Encelia farinosa** becomes less abundant. There are a number of additional species from further south, including *Acacia willardiana**, *Ceiba acuminata**, *Cordia sonorae* (Sonoran cordia), *Forchhammeria watsonii, Pachycereus pecten-arboriginum* and *Pithecellobium sonorae* (ape's earrings).

3.9.4.5
Magdalena Plain

The Magdalena Plain forms the southernmost part of the Sonoran Desert in Baja California, with its eastern boundary formed by a continuous series of mountains. The northern half, between San Ignacio and Comondu is largely volcanic in origin, with a narrow, sandy, coastal plain. The rainfall is low, generally less than 200 mm. *Larrea divaricata* subsp. *tridentata** is dominant on many sites; *Bursera hindsiana** and *Fouquieria columnaris** are absent and *Agave* spp., *Pachycereus pringlei* and *Yucca* spp. are few and poorly developed. There are few cacti, especially the tall ones, on the volcanic mesas (plateaux), but with *Opuntia cholla* locally frequent throughout the region. *Parkinsonia microphylla** and *Cercidium praecox* occur in combination with *Bursera microphylla**, *Encelia farinosa**, *Fourquieria diguetii* (syn. *F. peninsularis*) and *Jatropha* spp.

On the plain from San Domingo to Venencio the irregular vegetation consists of *Lycium brevipes* or *Prosopis palmeri*, with emergent *Pachycereus pringlei* plus *Lophocereus schottii** and *Stenocereus thurberi*. In the southernmost portion *Fouquieria diguetii* is abundant, while *Yucca valida* is locally conspicuous. The prevailing shrubs are *Capparis atamisquea* (caper), *Encelia farinosa**, *Lycium brevipes* and *Simmondsia chinensis**.

3.9.4.6
Vicaíno Region

The Vizcaíno region occupies the Pacific drainage basin of central Baja California from the southern end of the Sierra San Pedro Martir southwards to nearly latitude 26°N. It includes the plain of the Vizcaíno Desert, the rough mountainous interior and a wedge of coastal plain that extends south of Punta Pequeño. The variable rainfall is almost entirely confined to the winter and early spring, from January to April. The remainder of the year is invariably dry, although in some years summer storms may occur to the south of the Vizcaíno plain. During spring and early summer morning fog occurs along the coast as far south as the Vizcaíno plain and extends inland over the coastal shelf for 5–6 km. The morning fog is reported to be most frequent in July and August. There is a noticeable difference in the vegetation on north and south facing slopes and, within 4–5 km of the coast, a marked difference between the leeward and windward slopes. Large perennials are generally absent on the windward slopes, with a few scanty low perennials or bare apart from the lichens covering the rocks.

3.9.5
Chihuahuan Desert

The Chihuahuan Desert includes south-western Texas adjacent to the Rio Grande, south-central New Mexico, eastern Chihuahua, western Coahuila and parts of Durango, Zacatecas, Nuevo León and San Luis Potosí. Roughly lozenge-shaped in outline, its northern boundary is formed by the high plains of south-western New Mexico and its western flank by the Sierra Madre Occidental and by the Sierra Madre Oriental in the east (Fig. 12). A small part of the area along the Rio Grande lies below 900 m a.s.l., nearly half is over 1200 m, rising to over 1800 m. Vast plains and immense undrained basins alternate with mountain ranges rising 600–1500 m above the plains, or with limestone hills. The drainage is either internal or via the Rio Grande eastwards into the Gulf of Mexico.

The mean annual rainfall ranges from 75 mm over the plains of Coahuila to as much as 300–400 mm at the higher elevations on the western and southern boundaries. From 65 to 80% of the rain falls during the summer months of June to September, with a light precipitation from October to December; January to May is very dry throughout the area. The summer daytime temperatures are from 5.5 to 11 °C lower than those of the Sonoran

Desert, except at low elevations along the Rio Grande. Moderate frosts are widespread in the plain, with severe frosts above 1700 m.

Desert plants may often be found at 2100-2500 m. Shrubs and subshrubs are the predominant life form, trees are small, few and confined to the waterways or rocky slopes. Stem succulents are abundant but not conspicuous. There are no winter herbs, no winter germination and no winter flowering, although many plants remain wholly or partly in leaf during the cold winter months. Like the Sonoran Desert, the Leguminosae and Compositae are both richly represented, although the former is better represented in the Sonoran and the latter in the Chihuahuan (Shreve 1942a; McGinnies 1981). Livestock grazing is an important rangeland use (Holechek et al. 1989). The following description of the vegetation is based on Shreve (1942a) and McGinnies (1981).

From the region of the dunes in northern Chihuahua eastwards to the Rio Conchos there is a uniform and monotonous stand of *Larrea divaricata* subsp. *tridentata**, or sometimes of the glutinous 2-m-tall shrub *Flourensia cernua* and *Acacia neovernicosa* (syn. *A. vernicosa*) (Chihuahua whitethorn), with low, buried trees of *Prosopis glandulosa* var. *torreyana* (mesquite) and the leafless, green-stemmed *Koeberlinia spinosa* (spiny allthorn) on the low ground. Along the small drainage lines of the desert Shreve (1939) observed a slight increase in the density of species of *Acacia, Celtis, Prosopis* and *Condalia warnockii,* but no significant increase in height. Along the larger arroyos (seasonal streams), in addition to larger species of *Prosopis, Acacia constricta, Berberis trifoliata, Celtis pallida, Chilopsis linearis*, Porlieria angustifolia* and *Rhus choriophylla* also occur. East of the Rio Conchos an increasing area is occupied by hills and small mountain ranges, the general elevation of the intermontane plains reaching 1500 m. Here, the desert merges irregularly with grassland or evergreen oak shrubbery, the desert boundary being in northern Coahuila. *Parthenium argentatum** is to be found on the outwash fans and slopes of the calcareous soils.

In northern Chihuahua there are extensive flat grasslands or llanos, with pure stands of the perennial *Hilaria mutica* (tobosa grass) growing in contiguous or close-set tussocks and on the deep fine soil of the closed basins (bolsons) into which there has been insufficient drainage for the development of a central alkaline dry lake. The llanos are regarded by Shreve (1942b) to be edaphic desert grasslands and not part of the climatic grassland formation of the Great Plains. Surrounding the llanos are invariably long outwash slopes covered with typical desert species such as *Acacia neovernicosa* and *Larrea divaricata**, plus numerous cacti. The rolling, hilly or irregular limestone a reas of northern Coahuila support species of *Fouquieria splendens*, Agave, Yucca* and the bromeliad *Hechtia*, together with *Euphorbia antisyphilitica* and *Jatropha spathulata.*

The plateau of southern Coahuila is dominated by *Allenrolfea, Atriplex, Suaeda, Prosopis* and other halophytic species, with *Agave striata* subsp. *falcata* (syn. *A. falcata*), *A. lechuguilla, Fouquieria* and *Larrea* on the upper, slightly stony levels. Near the southern boundary of Coahuila there is an important biological demarcation formed by a series of east-west orientated mountains, including the Sierra de Parras and Sierra Jimulco. The higher elevation, 1200–2100 m, and higher rainfall, 200–400 mm, results in a denser and better developed vegetation. The characteristic vegetation includes *Acacia, Condalia, Cordia, Larrea, Leucophyllum, Lippia, Lycium* and *Prosopis*. Large and abundant *Yucca australis* occurs in suitable sites, while *Opuntia imbricata* and *O. kleiniae* are common and ubiquitous.

In the more arid south-eastern arm of the desert of northern San Luis Potosí the rainfall is only 100 mm. The open expanse known as the Valle del Salado has no drainage outlet and as a consequence the soils are either very alkaline or with a high gypsum content. Immense tracts are covered with stunted *Larrea*, either in more or less pure stand or with *Flourensia cernua* on the moister sites. Open stands of *Prosopis* occupy favourable sites, with a heavy ground cover of *Atriplex*.

3.9.6 Grasslands and Oak Woodlands of Northern Mexico

The grasslands and oak woodlands of northern Mexico occur in the elevated plain lying between the Chihuahua Desert and the eastern base of the Sierra Madre Occidental. The plain forms a belt from 25 to 110 km wide, at elevations of 1600–2150 m, and extends through the states of Chihuahua, Durango and Zacatecas into Aguascalientes and northern Jalisco. The soils are composed of thoroughly disintegrated outwash material derived from the Sierra Madre, which are prevented from spilling onto the Chihuahua Desert by a series of barrier ridges. The annual average rainfall is approximately between ca. 400 and 500 mm, three-quarters of which falls during the summer months of June to September, while the late winter and spring are dry. Grassland and oak woodland occupy a similar altitudinal and climatic range, edaphic conditions being the determining factor; grassland prefering level, non-rocky and well-drained sites (Shreve 1939, 1942b).

3.9.6.1 Grassland

An extension of the short-grass prairie (see Sect. 3.9.1.3), in northern Chihuahua the grassland is found between 1600 and 2150 m, further south typical grassland occurs at slightly increasing altitudes, attaining 2500 m in the broad grassland valleys of Zacatecas (Shreve 1939, 1942b; Sims 1988).

3.9.6.2
Oak Woodland

Among the oak woodland of the lowest altitudinal range are *Quercus chihuahuensis* and *Q. emoryi*, and these descend locally into the upper edge of the desert-grassland transition zone. The oaks of the highest elevation are *Quercus epileuca, Q. hypoleucoides* (syn. *Q. hypoleuca*), *Q. pennivenia* and *Q. sipuraca* and are associated with the pines at 2150–2300 m and become minor constituents of the pine forest at higher elevations.

3.10
Brazilian Caatinga

Caatinga is an Amerindian word meaning 'white forest' in the Yupi language and probably alludes to the whitish appearance of the scrub during the dry season when most species are leafless (Lewis 1987). The Brazilian caatinga lies just to the south of the equator in north-eastern Brazil. It experiences the only true semi-arid climate in Brazil (Fig. 8). The average annual rainfall is ca. 260 mm in the drier interior or 'sertao' (a term denoting sparsely-inhabited backlands), rising to ca. 800 mm bordering the mainly evergreen cerrado woodlands along its western and south-western edges and 1000 mm in the east, where it meets the coastal Atlantic forests (Fig. 14). These boundaries are poorly defined and interdigitate with the more mesophytic cerrado vegetation, which ranges from closed-canopy forests and open-canopy woodlands to scrub and tall-grassland (Eiten and Goodland 1979). The anthropogenic nature of the campo cerrado vegetation due to the transporting and planting of useful wild plant species by the local Amerindians suggests the possibility that the physiognomy and floristic composition of the caatinga might too have been manipulated by man (Hayashi 1981).

The dry season lasts for 6–8(-11) months. The rainfall is somewhat erratic from year to year; sometimes the interior of the caatinga may experience little or no rain for a year or more (Eiten and Goodland 1979). Temperatures are high throughout the year; mean annual temperatures of 27.5 and 32 °C have been recorded for Quixeramobim and Paolo Alfonso, respectively (Walter and Breckle 1986). Strong winds also blow most afternoons and evenings. According to Eiten and Goodland (1979) Gaussen's xerothermic index is given as 100–250. The rainfall/evaporation quotient is less than 0.5. The annual evaporation is 1000 mm at the boundary with the Atlantic forest, 1400 mm at the boundary with the cerrado, and with higher values within the caatinga; for

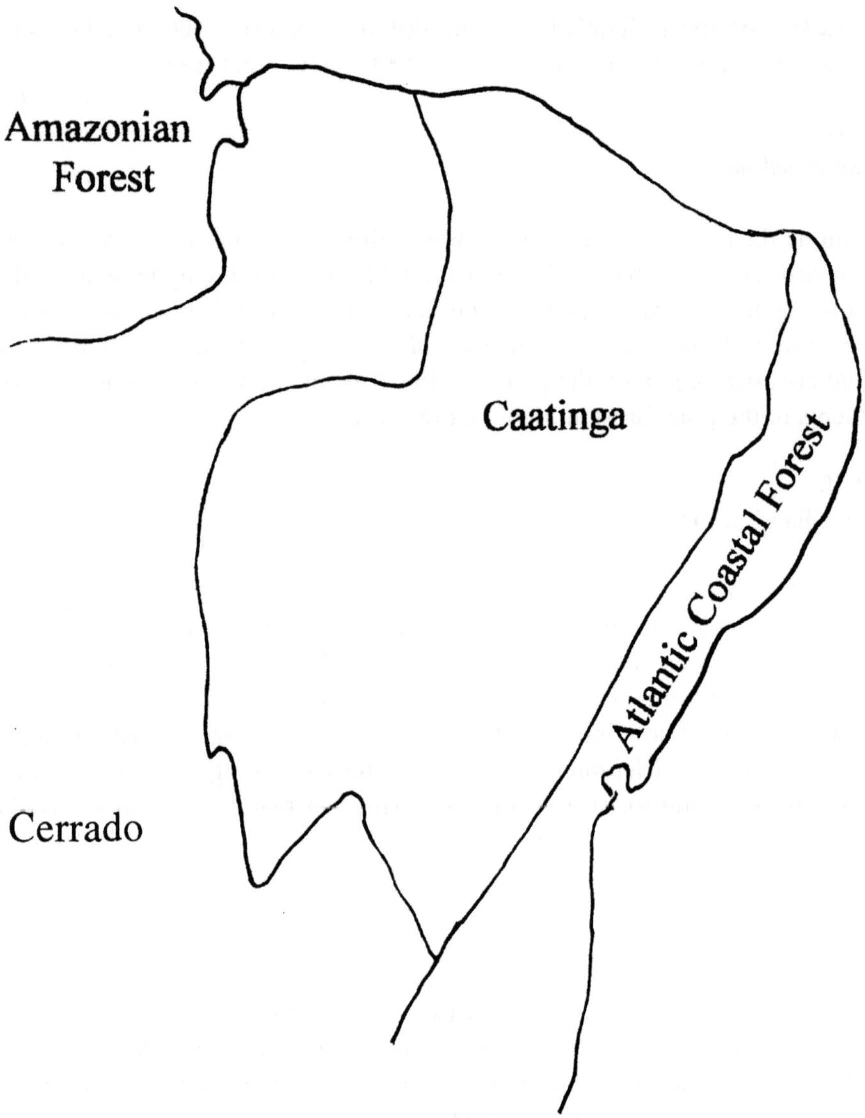

Fig. 14. The caatinga of north-eastern Brazil (After Eiten 1982)

Paulo Afonso in Bahia, Ferri (1954, 1961) gives figures of 3941 and 3310 mm for 1953 and 1954, respectively, and respective rainfall totals of 0 and 2.9 mm.

The soils are derived from a wide range of igneous, metamorphic and sedimentary rocks, with limestone rocks predominating. The soils are generally shallow, 30–50 cm in depth, rarely exceeding 1 m. Soil permeability is poor; run-off from what little rainfall occurs is heavy, with little infiltration into the soil (Ferri 1954). The available water-holding capacity is only ca. 1 mm for each centimetre of soil (Porto and Silva 1988), permitting little water to be stored by the soil for use by the plants during the dry season. However, the soils are relatively fertile, with little leaching of nutrients due to the low rainfall (Eiten and Goodland 1979). In a few areas salts may accumulate but this is not a general problem within the caatinga.

Eiten and Goodland (1979) describe the structure of the undisturbed caatinga vegetation as a mixture of communities, consisting of low to medium tall trees forming an open to closed canopy, medium tall to tall (2 to 5 m) closed scrub, often with emergent evergreen trees up to 7–8 m high, medium-tall open scrub and scrub savanna. In the drier areas the scrub opens out to give a lower, 1.5 to 2-m-high, open scrub or savanna known as the 'seridó' with a short grass layer with flowering culms to 20–30 cm high, and scattered spiny shrubs, 1.5–3 m high, together with low clumps of cacti.

Outliers of both the cerrado and coastal forest also occur within the caatinga, as well as outliers of the caatinga being present within the cerrado and coastal forest. A detailed description of the environment of the caatinga and its plant communities is given by Andrade-Lima (1981). Cole (1960, 1986) neatly assigns the distribution and type of caatinga and cerrado vegetation according to the geomorphology and soils, an interesting landform relationship that appears to have been overlooked by later researchers.

By Old World standards the semi-arid vegetation of the caatinga is extremely luxuriant. This can be attributed to the caatinga being surrounded on three sides by the more mesophytic cerrado woodland and on the remaining side by the coastal Atlantic forest, both of which must contribute to maintaining a relatively high humidity within the caatinga, and the absence of any dry desiccating winds. The caatinga thus appears to be an anomaly among the arid and semi-arid regions of the world, the only common factor being a semi-arid climate. Vegetation formations physiognomically similar to, and with strong taxonomic affinities with the caatinga, especially in the Leguminosae, also occur in more humid regions of South and Central America (Andrade-Lima 1981; Lewis, 1989, pers. comm.). Cabrera (1955) originally regarded the Caatinga Province of the Brazilian extra-Amazonian Domain as phytogeographically distinct from the Chaco Domain. However, he later recognised the close affinities of the two floras and placed the Caatinga and Chaco Provinces within the Chaco Domain (Cabrera and Willink 1973).

Schnell (1966), cited in Walter and Breckle (1986). lists the following aspects of the vegetation:

1. On deep soils with large water retention capacities the dominant vegetation is deciduous, thorny shrubs and trees that are often leafless for long periods.

2. On stony soils there are particularly large succulents, especially columnar cacti and large, prickly, terrestrial Bromeliaceae which can survive drought periods of more than a year without taking up water.
3. Bottle trees with succulent trunks, such as *Cavanillesia arborea* and *Ceiba pubiflora* (Bombacaceae), are frequent and grow to heights of 10–15 m.
4. Poikilohydric species, such as *Selaginella convoluta*, also occur.

The vegetation is largely of a closed, often impenetrable, mainly deciduous (semi-deciduous on some sandstone plateau tops in Piauï), broad-leaved and succulent shrubs, 3–5 m tall, often, but not always, with scattered low trees up to 6-8 m tall whose crowns overtop the scrub layer (Eiten and Goodland 1979). Both the trees and shrubs are particularly well represented by members of the Leguminosae subfamily Mimosoideae. Many of the leguminous trees and shrubs bear vicious thorns.

According to von Martius (1824) and cited by Schimper (1960) the characteristic species of the caatinga include *Acacia* spp., *Annona obtusifolia, Bombax* spp., *Caesalpinia* spp., *Capparis* spp., *Cavanillesia arborea* (syn. *Pourretia tuberculata*), *Chlorisia ventricosa, Galphimia brasiliensis* (syn. *Thryallis brasiliensis*), *Jatropha* spp., *Mimosa* spp. and *Spondias tuberosa**. Andrade-Lima (1981) gives the evergreen species as including *Albizia polyantha* (syn. *Pithecellobium multiflorum*), *Caesalpinia ferrea* (facultatively evergreen), *Capparis* spp., *Croton campestris, Licania rigida, Maytenus rigidus, Mimosa acutistipula* (facultatively evergreen), *Sideroxylon obtusifolium* (syn. *Bumelia obtusifolia*), *Tabeluia impetiginosa* (syn. *Tecoma caraibea*), *Ziziphus joazeiro, Z. cotinifolia* and members of the Palmae, such as *Copernicia prunifera** and *Syagrus coronata* (syn. *Cocos coronata*).

3.11 Western and South-Eastern Latin America

Under the influence of the northward-flowing cold Humboldt Current, the land along the west coast of South America, from Ecuador southwards to Patagonia, is either arid or semi-arid. Relatively small areas of semi-arid lands have also been mapped along the Caribbean coast of Colombia and Venezuela (UNESCO 1979). These semi-arid areas, 'guajira', named after the Guajira Peninsula in Colombia, are mainly found in the lee of the offshore islands, where the heat from the islands induces an upward movement of the air, leading to the formation of cloud towers above the islands and to dry winds descending on the coast (Walter and Breckle 1985).

Two major desert ecosystems are recognised: the coastal Peruvian-Chilean ecosystem, which includes the Sechura and Atacama Deserts and, east of the Andes, the Monte-Patagonian ecosystem, consisting of the tropical Monte region and cool temperate the Patagonian Desert. A useful, but incomplete, compilation of the economic plants of arid and semi-arid Peru, Chile and Argentina has been prepared by Gastó et al. (1985) and by Aronson (1991) for Chile.

3.1.1
Peruvian-Chilean Desert Ecosystems

The Peruvian-Chilean desert ecosystems extends for ca. 2400 km as a narrow strip along the Pacific Coast, from ca. 8°S in northern Peru to ca. 28°S in northern Chile, and lies in the rain shadow of the Andes. Starting in northern Peru, the 160-km-wide coastal plain, partly lying in the great bend of the Piura River and consisting mostly of an expanse of shifting sands, is known as the Sechura Desert (Meigs 1966). This narrows abruptly into a 16 to 150 km wide coastal plain, La Costa, which extends more or less uninterrupted to the south. To the east, in Peru the land rises sharply to the Andean Cordillera, and is dissected by transverse valleys which, fed by melting snows from the Andes, bring water to the desert from November to about June. In Chile the coastal plain is separated from the interior by a narrow range of mountains, the Cordillera de la Costa, rising in places to 1800-2100 m. To the east of this range lies the plain of the Atacama Desert, ascending from ca. 300 m to the Andean foothills at ca. 1200 m.

The deflection of the Humboldt Current brings cold water to this coast, resulting in a relatively cool climate. The Current is too cold to produce much water vapour, so that the oceanic winds reaching the mainland are exceptionally cold and dry for these latitudes. As these air masses pass over the hot land surfaces, their capacity for moisture retention increases and possibilities for precipitation decrease to rare or never. While a summer climatic instability provides relatively predictable and perennial precipitation above 900 m in central Peru, the lower limit in northern Chile is ca. 300 m. The result is a broad desert virtually free of significant precipitation and vegetation and extending inland from the coast for 80–100 km ((Rundel and Mahu 1976; Walter and Breckle 1986).

While the rainfall along the coastal strip is minimal, for several hours at night the relative air humidity can be as high as 80%. What little rainfall does fall also fluctuates from year to year, so that mean values are not very useful. In the north of the Peruvian arid region the rain falls in the summer, in the central region at any time of the year and in northern Chile it is during the winter months, if at all.

There are rare exceptions, such as when there is a stronger air circulation over the tropics, so that near Cabo Blanco the cold Humboldt current is replaced by a warm equatorial current from the Gulf of Guayaquil; extremely heavy rain then falls on the coast, from Ecuador to far into the desert. These rare El Niño (meaning Christ Child) years can produce disastrous floods; the rainwater being unable to infiltrate the hard surface soil crusts.

Although the Humboldt Current is too cold to produce much rain, an advective fog (garua or camanchaca), which forms above the current, is blown inland at night by the south-westerly winds. The fog increases both in density and size of water droplets as it rises over the coastal range, giving a maximum wetting effect at ca. 700 m; above 800–1000 m the rising wind will have lost all its humidity, which in Peru results in a rock desert devoid of vegetation. The condensation of fog water takes place when the fog passes over solid objects, the larger the surface of the object the greater the condensation, trees are particularly effective. Continuous condensation is clearly dependent on the air movement due to the light south-westerly winds. These fogs occur during the winter months, from May to October, especially in July and August, while in summer the land warms up too rapidly for there to be much condensation (Rauh 1985). The most dramatic fog moisture gradients are found where steep, high land masses closely approach the coast. Such conditions exist at Paposo, northern Chile, where the coastal mountains rise abruptly within a kilometre of the coast and reach 3000 m in height within 10–15 km from the shore (Rundel and Mahu 1976).

In Peru, the mean monthly temperatures for the warmest month (March) and the coolest month (September) for the coastal town of Callao are 21.6 and 17.9 °C, and for Lima, 30 km inland, 23 and 16.6 °C, respectively (Walter and Breckle 1986), while for Antofagasta, on the coast of Chile, the mean monthly temperatures for the warmest month (January) and coolest month (July) are 20.5 and 14.0 °C, respectively, latitude having little effect on the temperature.

Walter and Breckle (1986) recognise the following two vegetation subdivisions:

1. A coastal desert consisting of a flat coastal plain and the western slopes of the coastal mountains; this region is affected by the fog, producing the small and widely scattered patches of 'loma' vegetation (loma means hill in Spanish).
2. An inland, extreme desert lying between the coastal mountains and the lower western slopes of the Andes. The fog rarely reaches this area. There is little to no rainfall, the vegetation receiving subterranean water draining from the Andes, 1.5–40 m or more below the surface.

Geomorphologically the inland Chilean plain of the Atacama Desert and its northern extension, the Pampa de Tamarugal, represent a down-faulted marine depression filled with post-Tertiary deposits. The plain consists of a series of undrained basins, 40–80 km wide, the floors of which consist of alluvial fans from eroded material washed down from the Andes. The soils are varied, often calcareous or volcanic in origin. Generally, the soils are thin, saline or alkaline, often rich in exploitable minerals, especially nitrate, copper, iron and manganese, and to a lesser extent calcium borate, sodium chloride, sulphur and asbestos. The sites of former large lakes are recognisable as salt depressions or 'salares', often with extensive deposits of sodium nitrate.

3.11.1.1 Coastal Plain

In the coastal plain of the Sechura Desert the precipitation influenced by the El Niño is sufficient to support a sparse desert vegetation, grading into open woody scrub from Piura northwards, with *Prosopis chilensis* (*P. juliflora sensu* Rauh 1985) and *P. laevigata* (syn. *P. dulcis*), both of which have edible pods and are important sources of fuelwood (Meigs 1966). Other species present include the evergreen *Capparis angulata* and *Cryptocarpus pyriformis*; both species often trap sand to such a degree that only the leaves of the young branches emerge from the sand. *Cercidium praecox* is remarkable in that it flowers during the dry winter; the green bark of its stem and branches photosynthesise during the long dry season. There are a number of other species which are either evergreen or have green bark (Rauh 1985).

South of Olmos (6°S), in the rare years when rain falls, the desert is dominated by the rapidly maturing, pantropical, annual forage grasses *Aristida adscensionis*, *Cenchrus echinatus*, *Chloris barbata* (syn. *C. inflata*) and *Eragrostis ciliaris*, etc. According to Cabrera (1955) and Rauh (1985) other herbaceous forage plants are *Apodanthera biflora*, *Portulaca oleracea** and *Proboscidea peruviana* (*P. altheaefolia sensu* Rauh 1985). The halophytic vegetation, wherever brackish water is present, consists of *Batis maritima*, *Sarcocornia fruticosa* (syn. *Salicornia fruticosa*), *Sesuvium portulacastrum* and the grasses *Distichlis spicata** and *Sporobolus virginicus* (Rauh 1985).

South of the Sechura Desert, the sandy beach of the Peruvian coastal desert is barren except on the dunes, where apparently rootless rosettes of *Tillandsia* spp. occur. They are arranged in rows, at right angles to the direction of the wind, apparently so orientated as to obtain maximum benefit from the condensation of the fog. Transverse sections of the basal parts have shown that the roots are contained within the outer cortex and sheathed by the leaf bases, seldom reaching outside; they are consequently functionless. These species of *Tillandsia* are the only known true homeohydrous fog plants that can survive without uptake of water through roots. The plants obtain sufficient water from the moisture that condenses on the leaves and is absorbed through trichomes which function like valves and are typical of the Bromeliaceae (Walter and Breckle 1986).

Higher upslope but still among the *Tillandsia* spp. there are stands of a low, creeping cactus, *Haageocereus australis,* rooting on the lower surface, and with lichen-covered, spherical or short, vertical columns orientated towards the sea. Moisture is probably obtained from the water droplets formed by the fog condensing on the spines and falling to the ground. This cactus, together with globose forms of *Neoporteria* spp. (syn. *Islaya, Neochilenia*), occupies specialised habitats, where the dry, night winds blow from the mountains and locally disperse the fog. The narrow strip of the northern Chilean coastal desert is extremely dry, for there is no coastal fog and the rain falls on very rare occasions during the winter. On such rare occasions, the desert becomes carpeted with flowers; Kohler (1967) describes such an event.

Somewhat higher in the coastal range begins the true loma vegetation, where the condensing fog can produce an equivalent of ca. 120–1240 mm precipitation, depending on the ability of the vegetation to trap moisture. The natural vegetation on the lower loma slopes consists entirely of annuals, such as *Nolana* spp. and various members of the Solanaceae and Convolvulaceae. Grasses do not occur until higher up the slopes, between 200 and 1000 m, and appear during August to November, when the condensing fog can be expected to provide the necessary moisture, but even then the herbs are dominant, forming a thick carpet, 60 cm high (Cabrera 1955; Rauh 1985).

Woody plants are found where the fog is thickest, at 450–600 m; they are usually isolated specimens, growing among the rocks. Annuals and geophytes predominate in the herbaceous layer. The fog vegetation fades out at ca. 800–1000 m, with colourful crustose lichens covering the rocks. Higher up, often to almost 2000 m, the mountain slopes are almost completely barren (Walter and Breckle 1986). Further south, in northern Chile, the now broader coastal plain forces the loma vegetation higher up the Cordillera, at 1000–1200 m. The effects of the fog extend into the Pampa del Tamarugal, a northern extension of the Atacama Desert, where there are large stands of the endemic *Prosopis tamarugo**.

3.11.1.2
Atacama Desert

Separated from the coastal desert by the Cordillera de la Costa, the coastal fog rarely reaches the Atacama desert; consequently there is no loma vegetation, apart from some earth cacti, *Neoporteria* spp. Lichens and blue-green algae dominate the landscape, relying on the nightly dew for their source of water. Higher plants are absent; the few scattered species reported by Rauh (1985) in erosion gullies are characteristic of the pre-puna vegetation (see Sect. 3.11.1.3) and could well be expected to occur in the ectone between the Atacama and the pre-puna rather than in the desert proper (Wickens 1993).

Indeed, the Atacama, which represents an uplifted sea-bed, rich in minerals and highly saline, is extremely arid and hostile to life.

3.11.1.3
Andean Foothills

Villagrán et al. (1982), in an east-west transect survey, from the Sierra de Huaylillas (1540 m) to Cerro Guane-Guane (5200 m), on the western flanks of the Andes, recorded a plant cover of only 0.1% on the border with the Atacama Desert to 70% at 4000 m, in the transitional zone between the precordillera and the altiplano. The following three formations were recognised: desert (pre-puna), tola or shrub (puna) and tussock grass (high Andean).

3.11.1.3.1
Desert Formation (Pre-Puna Belt)

Desert formation (pre-puna belt) lies between 1540 and 3500 m and is characterised by a sparse cover of desert shrubs and subshrubs and is rich in columnar cacti. The shrubby species include *Ambrosia artemisioides* (syn. *Franseria meyeniana*) and *Atriplex imbricata* subsp. *imbricata* (syn. *A. microphyllum*), and among the cacti are the spiny, arborescent *Browningia candelaris* and *Opuntia* spp., many of which also have edible fruits. In the valley of San Pedro de Atacama there are areas of irrigation, where the major wayside trees are *Geoffroea decorticans**, *Prosopis chilensis* and *Schinus molle*, the *Prosopis* is also grown in plantations. Also present in the area is cardón, *Echinopsis atacamensis* (syn. *Helianthocereus atacamensis*), a columnar or sparingly branched cactus, and a valuable local source of timber.

Bordering the great salt lakes, such as the Salar de Atacama, are open stands of *Aloysia triphylla* (syn. *Lippia triphylla*). Also to be found are the shrubby *Atriplex atacamensis, Ephedra chilensis* (syn. *E. andina*) and *Tessaria absinthioides*, and closely grazed meadows of *Distichlis spicata** and *Nitrophila atacamensis,* with the rhizomatous rush *Juncus balticus* also present (Rauh 1985; Wickens 1988).

3.11.1.3.2
Tola or Shrub Formation (Puna Belt)

Tola or shrub formation (puna belt) lies between ca. 3000–4500 m and characterised by a sparse cover of small, often cushion-like, xerophytic shrubs and subshrubs up to 1 m high. The Argentinean puna, which occurs on the eastern slopes of the Andes, has a number of in common, or closely related species

(Cabrera 1955). The majority of the species are eaten by the camelids (Cajal 1989) and to a lesser extent, by other livestock.

3.11.1.3.3
Tussock Grass Formation (High Andean Belt)

Tussock grass formation (high Andean belt) occurs from 4000 to 5000 m, and is dominated by tussock grasses, with local aggregations of cushion plants; above 4900 m, in the subnival belt the grasses are replaced by cushion plants. The densely ramified and cushion-like habits of many of the puna and high Andean flora are regarded as morphological adaptations for restricting transpiration and possibly also adaptations to potent radiation and high winds, both being reduced by the dense branching (Oppenheimer 1960). Compared with more succulent perennial herbs the lignified cushion habit must also offer better survival advantages against the weight of winter snow. The lignification of these xeric shrubs and subshrubs, in what are regarded as mainly herbaceous families – Caryophyllaceae, Compositae, Solanaceae, Umbelliferae, Verbenaceae, etc. (Hutchinson 1959) – may perhaps be interpreted as an alternative to turgidity in fleshy leaves. Furthermore, the leaves of these cushion plants often exhibit considerable reduction in surface area, resulting in a linear shape and, in some species, decurrent leaves; pubescent, coriaceous or glutinous leaf surfaces are the norm, such morphological characters being accepted means of reducing water losses under xeric conditions.

3.11.2
Monte-Patagonian Ecosystems

3.11.2.1
Monte

The arid to semi-arid Monte, to the south of the Gran Chaco, constitutes the steppe and scrub of the much dissected pre-Andean foothills, occupying the intermontane bolsones (flat and short valleys) where the moist air currents rarely penetrate. Rainfall ranges from 100 to 250 mm, falling during the summer. The mean annual temperature for the region is 13 to 15 °C, the climate becoming progressively cooler southwards. Cabrera (1961) for Plaza Huincul gives a mean maximum temperature of 32 °C for January and -4.4 °C for July, and a recorded minimum of -15.7 °C; the annual precipitation is 150 mm, the driest month being December with only 5.4 mm.

Much of the vegetation cover of the Monte consists of extremely xerophilous shrubs, with many aphyllous and small-leaved species. Trees, where present, are confined to the river banks, and consist either of *Prosopis alba* or

Salix humboldtiana, usually in conjunction with thickets of *Baccharis salicifolia*. According to Morello (1958), the main plant communities are: subdesert scrub with species of *Bulnesia, Larrea, Monttea, Plectrocarpa*, etc.; succulent cactus scrub with species of *Cereus* and *Echinopsis* (syn. *Trichocereus*); xerophilous open woodland with *Acacia, Celtis, Geoffroea* and rock associations with the terrestrial bromeliads; and also psammophilous, gysophilous and halophilous associations. Due to overgrazing the perennial grasses are limited to moist sites. The dominant shrub species include the evergreen *Capparis atamisquea, Senna aphylla* (syn. *Cassia aphylla*), *Larrea cuneifolia, L. divaricata* subsp. *divaricarta**, *L. nitida, Lycium* spp., *Monttea aphylla**, *Prosopis alba, P. alpataco, P. flexuosa, P. strombulifera, Schinus* spp. and *Ximenia americana*, plus creeping cacti belonging to the genera *Cereus, Echinocactus sensu* Cabrera (according to Hunt 1992 the latter genus does not occur in Argentina) and *Opuntia* (Cabrera 1955, 1961). *Pappophorum subbulbosum, Piptochaetium napostaense and Stipa tenuis* are important fodder grasses of the area (Distel and Fernández 1986, 1987).

In the grass plains of the Province of Salta the moving dunes are dominated by *Sporobolus rigens**, with *Gomphrena martiana* and *Heliotropium mendocinum* dominating the interdunal ground (Cabrera 1955). The xerophytic mixed algarrobal woodland of the arid plain of Mendoza is dominated by *Prosopis flexuosa* includes the shrubs *Atriplex lampa**, *Junellia aspera* (syn. *Verbena aspera*) and *Lycium chilense* (Guevara et al. 1976; Colmer et al. 1991). In the land-locked, saline depressions the more typical species are the chenopods *Allenrolfea patagonica, Atriplex lampa**, *A. sagittifolia, Heterostachys olivascens, H. ritteriana, Sarcocornia perennis* (syn. *Salicornia ambigua*) and *Suaeda divaricata*, the composite halophytic shrub *Cyclolepis genistoides* and *Frankenia patagonica* (Cabrera 1955).

3.11.2.2
Patagonian Desert

Lying to the south of the Rio Colorado is the dissected plateau of Patagonia with its shrubby steppe vegetation. The cold temperate climate is characterised by strong westerly winds and frost throughout most of the year, with snow falls during the winter months. Annual precipitation varies between 100 and 200 mm. The mean annual temperature reaches 30 °C in the north and 5 °C in the south (McGinnies et al. 1977; Robinson 1977; Gastó et al. 1985). According to Meigs (1966), cited by McGinnies et al. (1968), the Patagonian Desert is the only high altitude, east coast desert and has no counterpart among the earth's deserts. It owes its aridity to the barrier formed by the Andes preventing rain-bearing winds from the Pacific, and to the cold Falkland Current off the east coast. UNESCO (1979) classes the western part of Patagonia, the area bordering the Andes, as semi-arid and the eastern part as arid.

The vegetation is basically a low shrubby steppe, with grasses interspersed with low shrubs; the grasses predominate in the low lying areas. The 100 to 150 km wide plain flanking the lower slopes of the Cordillera, from Mendoza south to the Santa Cruz River, is dominated by coarse grasses belonging to the genera *Bromus, Festuca, Poa, Stipa*, etc., together with shrubby herbs such as *Adesmia campestris, A. trijuga, Berberis heterophylla* (syn. *B. cuneata*), *Mulinum spinosum, Nassauvia axillaris, N. glomerulosa** and *Senecio filaginoides*. This vegetation type is considered to constitute the best sheep grazing in Patagonia.

Sala et al. (1989) investigated the hypothesis: (1) grasses take up most of the water from the upper layers of the soil and utilise frequent and short duration pulses of water availability; (2) shrubs take up most of the water from the lower layers of the soil and utilise infrequent and long-duration pulses of water availability. Grasses and shrubs were selectively removed and plant performance and availability of soil resources monitored. Plant variables were evaluated for *Poa ligularis* and *Mulinum spinosum*, but grass primary production was assessed for the entire life form.

The results supported the overall hypothesis that grasses and shrubs in the Patagonian steppe use mainly different resources. Although the removal of shrubs did not alter grass production, the removal of grasses resulted in a small increase in shrub production which was associated with an increase in deep soil water and in shrub leaf water potential. However, the increase in production by the shrubs did not compensate for the loss in total production resulting from the removal of the grasses. The water resources freed by the removal of the grasses were inefficiently utilised by the shrubs, which utilised water exclusively from the lower soil horizons. Grasses, with 54% of their root biomass located in the upper 10 cm of the soil, took up most of their water from the upper layers but were also capable of absorbing water from deep layers. Rainfall events, which in the presence of grasses did not penetrate the deeper soil horizons, did so when the grasses were removed. There were no significant differences in the nitrogen content in the leaves of shrubs and grasses following removal and the control. There were, however, significantly high differences between the nitrogen content of the shrubs and the grasses. This pattern of water partitioning together with the lack of response in leaf nitrogen to the removal treatment suggested that shrubs may be at a disadvantage compared to grasses regarding nutrient capture and led to questions about the role of nitrogen circulation, leaching and nitrogen fixation in the steppe.

Soriano (1961) had noted that many species of grasses and shrubs dominant in the area lacked the regulating mechanisms commonly ascribed to desert plants. In his study of the perennial grass *Stipa neaei* he noted that the seeds germinated over a wide range of temperatures. Since the rainfall lacks any real periodicity, this characteristic can seriously hinder regeneration since plants do not produce good seeds every year. Rapid and total germination could deplete the soil seed bank at a time when soil moisture is insufficient for seedling establishment. The seed-water relations neatly counterbalance this behaviour. Apparently, moisture tension for germination needs to be low, the required minimum inhibition is rapidly attained and germination then proceeds rapidly, so that establishment is likely to be ensured.

In the drier, central part of the Patagonian Desert, the total annual precipitation from rain and snow is only 100–150 mm; strong, persistent winds are frequent. The mean annual temperature ranges between 8 and 13 °C. The soils are low in humus. The vegetation cover is poor and does not exceed 35% cover. Mesetas (plateaux) and low sierras are distinguished. These ranges are most productive in years when there is much snow; the moisture from the snow then has enough time to enter the soil. The occasional spring and summer rains are mostly ineffective due to the winds and high temperatures resulting in excessive evaporation.

The phenology of six species within the shrubby steppe were studied by Bertiller et al. (1991). The shallow-rooted arido-passive deciduous shrub *Lycium chilense* and the grasses *Poa ligularis, Stipa speciosa* and *S. tenuis* exhibited biological activity early in the growing season, during winter late spring. This was the period when such shallow-rooted species were best able to exploit the high moisture and available nitrogen in the upper 80 cm of soil. Phenological development was arrested in late spring and summer due to the depletion of available moisture in the topsoil. Any further reproductive development either failed to start or failed shortly after initiation. However, the two evergreen and deep-rooted, arido-active species, *Chuquiraga avellanedae* and *Larrea divaricata**, continued their development during the dry summer and late summer, when their deep roots were able to draw on the available moisture and nitrogen in the subsoil after the top soil had been depleted. The roots of *Larrea divaricata** penetrated to a depth of 1.7 m and the plant maintained photosynthetic activity and vegetative growth throughout the entire growing season irrespective of air temperature. The shallower roots of *Chuquiraga avellanedae* descended to 1.2 m. Active growth occurred during the spring and mid-summer periods of high temperature, with a longer-lasting reproductive growth than in *Larrea*. Green tissues were maintained during the period of vegetative inactivity from late summer to early winter.

CHAPTER 4

Ecophysiology, Nitrogen Fixation and Plant Stress

4.1 Ecophysiology

Botanists have long been aware of the interrelation between ecology and physiology, e.g. Clements (1907), Schimper (1960), but it was apparently Billings (1957) who first formalised the union with the terms 'physiological ecology' and 'ecological physiology'. The term 'eco-physiology', however, was first introduced in 1962 at UNESCO's Montpellier Symposium where it was defined by Eckardt (1965) in the English summary of his paper as follows:

"*Eco-physiology, in the acceptance of the term adopted in the proceedings of the Montpellier Symposium, deals with all relationships existing between living beings and their physical and biotic environment. It thus embraces not only the study of the adaptive structural and functional features which link the individual plant or animal to its specific environment, but also that of all forms of transfer and transformation of energy and mass connected with the dynamics of the ecosystem.*"

For the purposes of this present study ecophysiology is interpreted as the morphological, anatomical, physiological and biochemical adaptations of a plant that ensure its survival in a particular environment and the description of that environment. It seems appropriate here to quote the conclusions of Walter (1963b): "The adaptation of perennial desert plants seems to be chiefly a morphological one rather than a physiological one. The chief adaptation is the reduction of the transpiring surface, the better development of the main root system, and the ability to reduce water loss during prolonged drought periods to nearly nil".

The ecophysiological behaviour shown by the species growing in the different climatic zones is so great that it is not possible to identify any one type as typical for the zone. This is in agreement with the general observation that the water supply to individual plants in humid and arid areas does not differ greatly because there is a reduction of the total transpiring surface per square metre or hectare of ground surface which is in proportion to the decrease in rainfall. Furthermore, it would be helpful if, in ecological investigations, transpiration and production rates could be expressed according to not only leaf surface area but also to ground area. This would make it possible to evaluate such data in relation to rainfall and climate. A careful examination of the root system and soil water content and the depth of the root tips should also be made, so that the available dry season water supply could be better assessed (Walter and Breckle 1986).

4.2 Nitrogen Fixation

Although the symbiotic fixation of nitrogen by plants is not confined to the arid and semi-arid ecosystems, in such ecosystems it does give a competitive advantage to those plants that are able to fix atmospheric nitrogen when, as is often the case, soil nitrogen levels are low. However, nitrogen fixation in arid and semi-arid ecosystems may apparently fail to occur in species that are elsewhere either known or expected to fix nitrogen. For example, species of the mimosoid legumes *Acacia* and *Prosopis* which are normally expected to nodulate have been reported not to do so in truly arid environments. This lack of nodulation is compensated for by deep root systems to obtain nutrients and water and is considered to make better use of plant resources than those required to fix nitrogen. Nevertheless phreatophytes may find conditions near the water table suitable for their roots to nodulate and fix nitrogen to support growth. The nitrogen is eventually returned to the soil as litter and mineralised to form nitrates in the soil. However, in arid ecosystems the annual rainfall is insufficient to wash these nitrates down to the nodulating zone, where they would inhibit nitrogen fixation.

In semi-arid ecosystems the situation is somewhat different. In a Mediterranean-type climate where the growing season coincides with winter rainfall, the nitrogen cycle is likely to broadly resemble that of temperate wet seasons, although low temperatures rather than water deficit may limit the biological processes. In summer rainfall regions, such as the savanna woodlands of Australia's Northern Territory, dry season fire may lead to a loss of 90% or more of the available nitrogen from litter, thereby biasing the nitrogen cycle in favour of nitrogen-fixing species. Grass regrowth in the following year may be reduced by as much as 50% until the soil nitrogen has been built up again by nitrogen-fixing legumes. In parts of semi-arid Western Australia cycads, such as *Macrozamia riedlei* becomes established following fire, regenerating from boles and fixing nitrogen by means of the cyanobacteria-infected coralloid roots, thereby providing as much as 18.8 kg N ha^{-1}; *M. riedlei* would thus appear to play an important role in the nitrogen economy of the community. Eastern Australia provides yet another example of the value of nitrogen fixation where, following the clearance of pure stands of *Acacia harpophylla* for pasture, there were sufficient nitrogen reserves remaining in the soil for about a decade, after which nitrogen fertilisers had to be applied in order to overcome the rapidly falling pasture productivity (Sprent 1985; Sprent and Sprent 1990). An interesting observation by Professor P. Sprent was the regeneration of *Eucalyptus*-dominant forests following wildfire. The early stages of recovery were often dominated by many nitrogen-fixing species of *Acacia* and *Casuarina**. Although it has often been assumed that these nitrogen-fixing species were replenishing the lost nitrogen, the available data does not always support this view (Sprent and Sprent 1990).

Desiccation, heat stress, drought, reduced oxygen diffusion resulting from low photosynthate availability, the inhibiting effect of high soil nitrogen, low soil phosphate and non-availability of an appropriate strain of inoculate can inhibit nitrogen fixation (See Sprent and Sprent 1990 for more details). In the arid and semi-arid regions nitrogen fixation occurs in members of the Leguminosae but also in a number of other families, see Table 1. The ability of C_4 members of the Gramineae as *Brachiaria decumbens, Digitaria decumbens* and *Zea mays* to fix nitrogen with the aid of the bacterium *Azospirillum lipoferum* is of obvious potential interest to the grazer. However, whether the nitrogen is immediately available after fixation to the host grass or whether it only becomes available after the death and breakdown of the bacterial cells is not yet understood. Neither are consistent improvements obtained from the pasture after appropriate inoculation (Sprent 1985). The presence of rhizosheaths may also have a nitrogen-fixing function although the actual nitrogen uptake by the plant remains unproven (see Sect. 10.1.3).

Vesicular-arbuscular (VA) mycorrhizas, as distinct from nitrogen-fixing bacteria, are where a fungus lives between the cortical cells and forms temporary hyphal projections that penetrate the cortical cells. Such projections may take the form of swollen vesicles or finely branched masses known as arbuscules. VA mycorrhizas may occur in both legume and non-legume, annual and perennial plants; indeed, they may be found in most plants. Plants infected by VA mycorrhiza growing in impoverished soils can exhibit improved phosphorus uptake, especially from an insoluble source, compared with non-infected plants. Water uptake too may be increased due to the VA mycorrhiza providing a low resistance pathway for radial water flow across the cortex in addition to possible changes in the permeability of the membrane to water due to the improved phosphorus metabolism. In experimental studies it has been demonstrated that plants infected with both *Rhizobium* and VA-mycorrhizal fungi may interact to produce an effect greater than the sum of their individual effects on a plant (Sprent and Sprent 1990; Danin 1996; Jacobson 1997).

Table 1. Familes and genera capable of nitrogen fixation in arid environments. (Sprent 1985; Sprent and Sprent 1990)

Family	Genus	Nitrogen-fixing agent
Angiosperms: Dicotyledons		
Betulaceae	*Alnus*	*Frankia*
Casuarinaceae	*Allocasuarina**, *Casuarina*	*Frankia*
Coriariaceae	*Coriaria*	*Frankia*
Datiscaceae	*Datisca*	*Frankia*
Elaeagnaceae	*Elaeagnus**, *Hippophae**, *Shepherdia*	*Frankia*
Leguminosae		
Subfamily Caesalpinioideae	Sporodic	*Rhizobium*
subfamily Mimosoideae	Fairly common	*Bradyrhizobium,*
subfamily Papilionoideae	Nearly 100% of genera	*Azorhizobium,* etc.
Myricaceae	*Comptonia*	*Frankia*
Rhamnaceae	*Ceanothus, Colletia, Discaria, Kentrothamnus. Retanilla, Talguenea, Trevoa*	*Frankia*
Rosaceae	*Cercocarpus, Chamaebatia, Cowania, Dryas, Purshia*	*Frankia*
Angiosperms: Monocotyledons		
Gramineae	*Brachiaria (B. decumens), Digitaria (D. decumbens), Zea mays*	*Azospirillum*
Cycadophyta		
Cycadaceae	*Cycas*	Cyanobacteria
Zamiaceae	*Encephalatos, Macrozamia*	Cyanobacteria

4.3 Restraints and Adaptations of Plants to Arid Environments

Traditionally the term *xerophyte* has been used to denote a plant restricted to a dry locality while *xeromorphic* is used to denote plants which, from the appearance of their external morphology and internal histology, give the impression that they will only be found in dry localities. However, xeromorphic plants may not, in fact, be always restricted to dry localities; more correctly the so-called xerophytic histological characters should be considered as xeromorphic (Metcalfe 1983).

The major constraint on plant life, especially in an arid or semi-arid environment, is undoubtedly water. Plants need water not only to stay alive but also to grow and develop; they need to be able to survive drought. Walter and Breckle (1985) distinguish two groups of plants with fundamentally different types of water economy:

1. *Poikilohydrous plants*: where the water economy is completely dependent on that of atmosphere and where the plants must therefore be able to survive complete drying out. Such plants are mainly found among the terrestrial algae, bacteria, fungi and lichens and rarely among the higher plants (see Chapter 5).

2. *Homeohydrous plants*: where the water status of the cell protoplasts is maintained and is largely independent of the water status of the environment. Halophytes, with their special adaptation to a saline environment, are included in this group.

Other important constraints are high temperature, low temperature and salinity. The physiological adaptations of plants to these constraints are defined by Street and Öpik (1984) as:

1. *Stress resistance*: the ability to endure an externally applied stress.

2. *Stress avoidance*: the ability to prevent an externally applied stress from producing an equivalent internal stress in a plant.

3. *Stress tolerance*: the ability to endure an internal stress induced by externally applied stress.

In addition to the plant's physiological adaptations to water, temperature and salt stress there are also anatomical and morphological adaptations, which are dealt with in Chapter 10. While it is convenient to consider these stress adaptations separately, it must be borne in mind that all three are interconnected and functioning to ensure survival in a stressful environment.

CHAPTER 5

Water Stress

The quantification of external water stress is expressed in terms of the water status of the environment, i.e. external water potential, relative humidity or as vapour pressure. Tissue dehydration may also be measured by changes in tissue water potential or changes in tissue water content (percentage of tissue weight). However, individual tissues differ widely in water contents at full turgidity. Thus, a fully turgid leaf may contain 85% water, whereas a fully imbibed seed storage cotyledon may contain only 60%; dehydration to 55% water would mean a relatively much greater water loss by the leaf than by the cotyledon. Consequently, the relative water content (RWC) is sometimes used in comparisons of water stress resistance in order to indicate the degree of water deficit:

$$\text{RWC} = \frac{\text{tissue water content}}{\text{tissue water content at full turgidity} \times 100}$$

$$= \frac{(\text{fresh weight}) - (\text{dry weight})}{(\text{fully turgid fresh weight}) - (\text{dry weight}) \times 100.}$$

The RWC thus indicates that proportion of the saturation water content still retained by the cell. An alternative measurement is the water saturation deficit (WSD), or the proportion of the saturation water lost from the tissue: WSD = 100 - RWC.

Under severe dehydration, as in air-dry material and poikilohydrous species, when the water content is below ca. 10%, the water potential values also become extremely low and difficult to measure accurately. In such circumstances the water status of the plant tissue and its dehydration resistance may be expressed as the equivalent relative humidity with which it is in equilibrium (Street and Öpik 1984).

According to Levitt et al. (1960), whose classification can be variously interpreted, drought resistant and drought tolerant plants, i.e. xerophytes, can be classified (with modifications) according to their mechanism for drought tolerance as follows:

1. *Drought resistant species*; plants able to survive a reduction in water content:
   a) *Drought avoiding species*; drought escaping species or therophytes which complete their life cycle before the onset of extreme drought by means of desiccation tolerant seeds.
   b) *Drought evading species*; water spenders that have the ability to obtain large amounts of water during drought.
   c) *Drought enduring species*; plants that are able to reduce water loss to a minimum.

2. *Drought tolerant species*; plants able to withstand desiccation without permanent damage to the protoplasm; poikilohydrous or resurrection plants.

5.1 Drought-Resistant Species

Levitt et al. (1960) have defined drought resistance as the ability of a plant to survive a reduction in water content, while Parker (1968) considers drought resistance as being a plant characteristic that is the sum of the many parts of the plant, from whole organs to the fine structure of the cytoplasm, from the seed to the whole plant.

Evenari (1985) recognises three main plant survival strategies for hot deserts, based on plants' main morpho-anatomical and eco-physiological adaptations. Such plants are termed *poikilohydrous, arido-passive* and *arido-active homeohydrous*. Poikilohydrous plants have the ability to enter a condition of suspended animation (*anabiosis*) so that do not die when dehydrated and become active again when wetted. The arido-passive plants have their drought-surviving parts metabolically inactive (quiescent, dormant) during the dry season, whereas arido-active plants remain, at least in part, metabolically active throughout the year. It should be noted that these terms are not synonymous with drought evading or drought tolerant plants sensu Levitt et al. (1960). Within the arido-passive annuals and perennials Gutterman (1993) recognises:

1. Rainy season annuals and ephemerals.
2. *Synanthus geophytes* which survive the dry season as subterranean storage organs, i.e. bulbs, corms or rhizomes.
3. *Hemicryptophytes*, whose renewal buds are close to the soil surface and who also possess one or more main storage roots, or cormlets.

The arido-active plants can be divided into:

1. ***Bi-seasonal annuals***, which include species such as *Salsola inermis* whose seeds germinate during the Mediterranean winter and which remains as a

rosette plant until the summer, when the life cycle is completed by the development of flowering stems and the maturing of seeds.

2. ***Hysteranthous geophytes*** which, unlike the synanthus geophytes, complete their life cycle either during the dry season or soon after.

3. ***Heterophyllous shrubs*** that survive the dry season by a reduction in the number, size and surface area of the leaves (see Sect. 10.1.1).

4. ***Phreatophytes***, which are trees and shrubs that have access to water throughout the year by means of their roots tapping the water-table (see Sect. 5.1.2).

5.1.1 Drought-Avoiding Species

Included in the category of drought avoiding species are the *therophytes*, the ephemeral and annual plants which are not resistant to drought and can only survive the onset of drought, i.e. the dry season, as desiccation tolerant seeds. *Ephemerals* are plants with a short life cycle that take advantage of the occasional favourable light rainfall. They are capable, under more mesophytic conditions, of producing more than one generation in a year. Should a favourable rainfall not occur their seeds are capable of remaining dormant for a number of years. The life cycle of some ephemerals can be very short indeed. For example, in the Sahara species of *Boerhavia* and *Tribulus* have been reported as being capable of passing through the stages of germination to fruiting in 8–15 days (Ozenda 1977).

Some ephemerals and annuals produce large quantities of seed, a survival strategy to ensure that at least some will survive animal and insect predation, which Gutterman (1993) regards as an important source of loss from the soil seed bank. In favourable sites in Qatar, Batanouny (1981) has recorded densities of ephemeral seedlings of up to 2000 m^{-2}, while Archibold (1995) cites observations in the Sonoran Desert by Reichman (1984) of average seed densities between 4000 and 104 000 seeds m^{-2}, and Kemp (1989) with seed densities ranging from 400 to 7700 seeds m^{-2}. However, with such densities mortality is extremely high due to root competition for moisture in the top 20–30 cm of soil. The combination of differences in temperature and moisture appears to regulate the germination of the various species of ephemerals, thus to some extent reducing competition between species.

Less opportunist than ephemerals, *annuals* are capable of only one generation a year, completing their life cycle during the rainy season and surviving the dry season as seed. The ability of an ephemeral or annual to respond rapidly to rainfall is clearly an advantage, often enabling the plant to utilise soil moisture and nutrients before its competitors, as is the case with the annual grass *Bromus tectorum**. Depending on the season's water availability some perennials behave as annuals, completing their life cycle within the first rainy season. Such facultative behaviour may be found in the Namibian grasses *Centropodia glauca* and *Stipagrostis ciliata* (Jacobson 1997).

As a protection against failure to survive because of insufficient rainfall many therophytes produce heterogeneous seeds (*heterochory*), which respond to different moisture regimes and are governed by chemical inhibitors or the permeability of the seed coat. Furthermore, factors such as daylength during the season in which the seeds develop and mature may also result in the production of seeds with differing abilities to germinate (*heteroblasty*). Such heterogeneous seeds affect the number germinating in the first season and allow other seeds to remain dormant until the following year (Danin 1983; Gutterman 1993). Some seeds of *Salsola volkensii*, for example, may germinate after 5 years, while the annual *Pteranthus dichotomus* produces bunches of prickly fruit, each comprising of one to seven one-seeded fruits, each fruit with different germination capacities. Thus, a reservoir of seed remains in the soil and the survival of the species is thereby ensured (Walter and Breckle 1986; Gutterman 1993).

Seeds may be retained in the capsule until the capsule is sufficiently wetted to open, as in *Blepharis ciliaris** where, as an additional safeguard, only a proportion of the seeds are explosively ejected at a wetting and are then anchored to the soil by means of a mucilaginous layer. Other examples of hydroscopic capsules are to be found in *Aizoon canariense**, *Mesembryanthemum crystallinum** and *M. forsskålii** (Gutterman 1993). Mucilaginous seeds occur in a number of families, including the Acanthaceae (*Lepidagathis bandraensis*), Aizoaceae*, Cistaceae (*Helianthemum* spp.), Compositae (*Seriphidium sieberi*), Cruciferae (*Anastatica hierochuntica**, *Carrichtera annua*), Labiatae (*Salvia* spp.) and Plantaginaceae (*Plantago coronopus**) (Bhandari 1990; Gutterman 1993).

The mature seeds of most angiosperms can dry out until their water content is in equilibrium with atmospheric humidity (relative humidity 20–50%), bringing their water content down to 5–20%, corresponding to water potential values of -100 to -10 MPa. During germination this desiccation tolerance is lost, although seeds in the imbibition stage of germination can be dried back to the air-dry condition again without damage (Street and Öpik 1984), which is a useful survival mechanism should insufficient rains fall to complete germination.

5.1.2 Drought Evading Species

Some drought-evading woody plants in dry regions have massive underground organs but only annual or short-lived shoots above ground, e.g. the arido-passive legume *Andira laurifolia* var. *laurifolia* (syn. *A. humilis*) from the cerrado of Brazil, which produces short leafy shoots up to 30 cm above ground level, thereby keeping transpiration to a minimum, and with storage roots in the upper 2 m of soil and absorbing roots descending for 18 m to the water-table (Ferri 1961, Parker 1968). The shrubby, multistemmed, mallee *Eucalyptus* spp.* from the Australian shrublands also possess large, woody, underground parts, which have developed from two swellings produced between the angles of the cotyledons growing down over the hypocotyl and giving rise to a swollen rootstock or *lignotuber* composed of accessory strands of meristematic tissue capable of producing numerous stems – *lignotillers*. The lignotubers are protected from lethal temperatures by being underground. It appears that lignotuber development is accentuated in soils of low fertility, especially where phosphorous is limiting, suggesting a wider adaptive function than as a regenerative base following fire. Because of the regular, but not necessarily annual, shedding of mallee bark during the main summer growing season (December to April), the remaining periderm is quite thin and offers little protection against fire. It is the shed bark, some of which may be caught up in the branches, that constitutes a fuel hazard. White (1976) gives other examples of these geoxylic suffrutex species from Africa. The stems of others have a corky layer within the xylem, e.g. *Artemisia* spp., and it is believed that such layers occurring between the annual rings help to retain stem moisture.

Some perennial grasses, such as *Astrebla* spp.*, have the ability to take advantage of the slightest rainfall and, because of their extremely rapid growth, are able to complete vegetative growth within 2 weeks and produce ripe seed after 6 weeks (Walter and Breckle 1986). Species known as *phreatophytes* (see Sect. 10.1.3) have deep roots tapping the water-table which, in evergreens, permits ample transpiration throughout the year, e.g. *Larrea divaricata**. However, not all are evergreen. Thus, although deciduous during the dry season, deep-rooted *Acacia* spp. in the Negev, Israel were found to carry out uninterrupted cambial activity throughout the year. In the Suez area, roots of *Acacia tortilis** were found down to a depth of 30 m (Oppenheimer 1960). In Arabia the roots of *Prosopis cineraria** may penetrate as deep as 36 m, while some American species of *Prosopis* descend to a depth of 80 m (Leakey and Last 1980). *Tamarix* too may have roots descending to 30 m, with superficial laterals spreading over a radius to 50 m (Baum 1978).

In addition, other phreatophytes, such as *Acacia nilotica* subsp. *nilotica** and *Faidherbia albida**, can also tolerate waterlogging for several months of

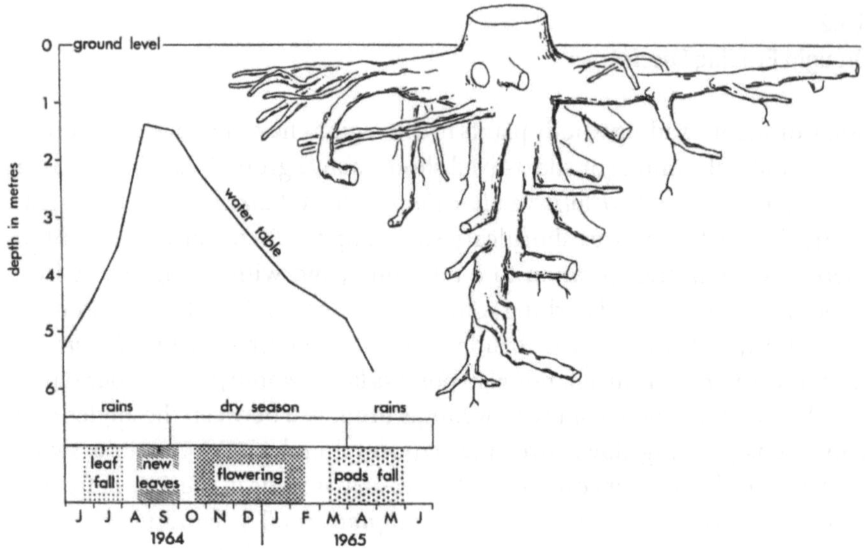

Fig. 15. Root system of *Faidherbia albida* growing on middle terrace soils of Wadi Aribo, Zalingei, Sudan. Seasonal behaviour is shown in relation to changes in level of water-table and waterlogging. Water-table levels were recorded from a nearby borehole. (Based on observations by the author and recompiled from Hunting Technical Services 1968)

the year. The respiration pathway of the root system under such anaerobic conditions has not been investigated. The latter species also has the unusual habit throughout its distribution range of bearing leaves during the dry season and shedding them during the rainy season (Fig. 15). *Acacia caven** behaves in a similar manner, but only for part of its distribution; the reason for such behaviour is not understood.

Roots have definite limits to the amount of water that they can retain above the lethal point in cool periods. Small roots, especially root tips, shrink, change colour and may even die during a drought. When, for example, the growing roots of *Pinus halepensis* var. *halepensis** become dormant during a drought, the outer six to eight layers of cortical cells collapse and die, or turn brown, the colour change being related to the total suberisation of the endodermal cell walls and not to the production of suberin. When growth is resumed, new white root tips burst out of the suberised layers.

In his survey of drought resistance mechanisms, Parker (1968) lists a number of leaf characters that aid water retention in desert plants. These include greater thickness of leaf veins per leaf area, greater number of stomatal openings per unit surface area, smaller size of stomata, epidermal and mesophyll cells, denser and shorter indumentum and thick outer walls to the epidermis and cuticle. In the past there has been an argument as to whether stomatal densities decreased in desert plants to give reduced transpiration, or increased to provide evaporative cooling, as suggested by Parker (1968); in fact, both situations occur, accompanied by an adjustment in stomata size (see Chap. 8).

There appear to be at least four related oxygenated sesquiterpenes responsible for stomatal closure. Apart from abscisic acid (*ABA*), the mechanism by which they function is not entirely clear. When under drought stress the ABA is released from the mesophyll chloroplasts and moves to the guard cells, where it overrides the normal diurnal pattern of stomatal opening and closure and activates the stomata to close, thereby reducing transpiration losses. The evidence that drought resistant plants contain large amounts of this important hormone is largely circumstantial. It has been found, for example in wheat plants, that ABA can increase as much as 40-fold within 4 h of wilting; and the level continues to rise in such osmotically stressed plants for at least 48 h after wilting induction. When the water supply is replenished the ABA drops to its former low level, being stored in an inactive form within the leaf (Harborne 1988).

Many non-succulents have high transpiration rates, some may even increase their rate with increasing water stress. Unlike succulents, non-succulents are unable to tolerate very high daytime temperatures and must therefore avoid any heat build-up in their tissues by way of evaporative cooling through high transpiration rates. Because of the generally flat leaf area in non-succulents, the diffusion resistance, due to the boundary layer of air surrounding the leaf, becomes more important than stomatal resistance by the time the stomata are half open. Many small stomata are consequently required for maximum gas exchange, rather than a few, relatively large ones. Although an individual small stomata would be less efficient than an individual large one in allowing gas exchange per unit pore area, considerably more small stomata can be packed into a leaf unit area, thereby providing greater total transpiration and consequently greater evaporative cooling potential. The efficiency of stomata operations is demonstrated in the evergreen *Zuccagnia punctata** where Morello (1955) calculated that with the stomata open the shrub would spend all its water in less than 1 h while with the stomata closed it would take 7 days and 9 h.

Plants may also adapt to drought stress by an accumulation of the 5-carbon imino acid *proline*, and its increase may account for as much as 30% of the free amino acid pool. In legumes, proline is replaced by the aliphatic polyol *pinitol*. It is possible that the increase in proline or pinitol could be a symptom of some more fundamental adaptation to water stress. It is also possible that the special osmotic properties of proline enable it to contribute directly to the retention of water within the plant (Harborne 1988).

The accumulation under moisture stress of large quantities of proline in the leaves of many higher plants is due either to the inhibition of proline oxidation or to the breakdown of protein and, in some plants, to the more rapid biosynthesis of proline from its precursors. The physiological significance of proline accumulation is controversial, some workers maintaining that it is pathological, while others suggest that, at high concentrations, proline acts as a solute for intracellular osmotic balance, and provides a store of nitrogen and carbon for subsequent utilisation after release from stress. It has also been suggested that free proline synthesised from glutamate serves as an energy donor during environmental stress (Mohammed and Sen 1987).

The ability of some desert plants, mainly in coastal areas, to trap moisture from dew or fog is well documented (Gindel 1970, Sundberg 1985). The moisture is concentrated on the plant surface and falls to the ground as leaf drip and stem flow, which can significantly augment the water available to the plant. Adaptations by the plant to maximise the trapping of the dew or fog include divaricate branching patterns and rosette growth form with upright leaves and spreading spines (Rundel 1982). It has been suggested that some plants can absorb some of this moisture through the leaves; however, significant and unambiguous direct foliar uptake of moisture from the atmosphere by arid zone flowering plants has yet to be demonstrated (Rundel 1982). However, Walter and Breckle (1986) cite the bromeliads *Tillandsia latifolia, T. purpurea, T. recurvata* and *T. werdermanniii* from the Peruvian-Chilian Desert as being the only known true homeohydrous fog plants, being able to survive without water uptake through the roots by absorbing the water condensing on the leaves through valve-like sucking scales.

The reaction of semi-arid plants to changes in water and light can be surprisingly rapid. Ferri (1961) recorded that when the water supply to *Spondias tuberosa** was cut off, initial transpiration was reduced by more than 50% in only 2 min and complete stomata closure occurred in ca. 5 min. Parker (1968) also considers the role of leaf fall and size reduction in moisture retention. Leaf abscission during drought is believed to be a means of reducing transpiration. For example, many trees and shrubs shed their leaves at the start of the dry season and bear leaves and flowers during the rainy season, e.g. the legumes *Acacia tortilis** and *Anagyris foetida*. Although generally associated with trees and shrubs, deciduous leaves may also occur in some perennial herbs, such as *Anthyllis cytisoides* (Alados et al. 1996).

The rocky heaths of the Mediterranean and Middle East abound in species with superficial root systems, without access to deep soil water. These plants possess large leaves during the rainy season and replace them with smaller leaves during the dry summer months, e.g. *Thymus serpyllum* and *Salvia fruticosa**, while in *Zygophyllum dumosum* (Fig. 16) and the legume *Caragana* spp.* the leaflets are shed, but the petioles persist, thereby reducing the tran-

Fig. 16. *Zygophyllum dumosum.* Flowering and fruiting branch showing pairs of succulent leaflets borne on succulent petioles

spiring surface by ca. 85%. The petioles of the former act as water reservoirs, becoming shrivelled and shrunken when dehydrated but swelling two to three times their size 2 days after the first effective winter rain (Danin 1983).

Some shrubs may even replace their leaves with *brachyblasts* (short lateral shoots), or have stipules which persist after leaf fall and grow turgid under low soil moisture conditions. In many evergreen species of the largely Australian *Acacia* subgenus *Heterophyllum* adapted to dry climates, e.g. *A. aneura** and *A. saligna*, the pinnate lamina is reduced and the functions of photosynthesis taken over by a *phyllode* (see Sect. 10.3.1) formed by the dilation of the petiole and leaf rachis (Brenan 1959, Harder et al. 1965). The ability of leaves to roll or fold under water stress is another means of reducing transpiration, especially in the Gramineae, e.g. *Andropogon gerardii**, *Panicum turgidum**, *Plectrachne schinzii**, *Stipa tenacissima** and *Triodia pungens**.

Hairs are also believed to reduce transpiration, but their removal has also been demonstrated to reduce transpiration! The trichomes on *Crassula* sp., for example, can readily absorb surface water droplets. Therefore it would not be surprising if, in some cases, the removal of hairs reduced the total transpirating surface. However, the situation is not straightforward. For example, *Seriphidium tridentatum* has both live hairs filled with protoplasm and dead, air-filled cells. Furthermore, hairs may perform other functions, such as scattering incoming radiation or in breaking up soil-reflected radiation, preventing insect attack and helping to keep the leaf cool by increasing the total radiating surface.

A number of succulent plants have thick fleshy stems adapted to water storage and they contract or expand, depending on their water content, e.g. *Adansonia digitata** and *Carnegiea gigantea** (Crosswhite 1980, Wickens 1982b). It is assumed that very little water can be transpired through the bark and consequently much of the water in the trunk should be available for transpiration in times of drought. Yet when the actual amount of water used per day by an average forest tree is estimated to be ca. 90 l, it is difficult to envisage how the trunk water can act as an adequate reservoir (Parker 1968).

5.1.3
Drought-Enduring Species

Drought endurance is the ability of plants to continue to function, albeit at a slower rate, during the dry season and is usually found in plants with small, evergreen leaves. In general, the photosynthetic potential of their leaves tends to be substantially lower than that of deciduous plants, but this is compensated for by an increased leaf area duration and lower annual investment in replacement of leaf tissue (Fitter and Hay 1987). True *evergreen trees* have their shoots and leaves unreduced and physiologically active throughout the year. For example, the summer transpiration rates of *Arbutus andrachne**, *Laurus noblis** and *Pistacia lentiscus** are low, not exceeding 500 mg g^{-1} h^{-1} fresh weight. Summer osmotic pressures are also low, with values of 19.6, 27.2 and 26.6 atm. respectively. The leaves of *Laurus noblis** are also capable of withstanding a 60 to 65%-water deficit as well as a sublethal water content of 44–49% dry weight.

5.2
Drought-Tolerant Species

Drought tolerance or desiccation resistance is defined by Parker (1968) as the lowest relative humidity at which a plant can come to equilibrium without injury. This not only varies from species to species but also between provenances and individuals of the same species. Since the protein production by a living cell is dependent on nucleic acids bound to protein, any injury to this mechanism by desiccation, or any other means, is obviously an important factor in cell injury, since the means of producing replacement enzymes is destroyed. As water is withdrawn there is probably some interference with enzymatic reactions and phosphorylation (the combination, for example, of sugar with phosphoric acid) is practically inhibited as viscosity increases. There can also be changes in enzymatic reactions to favour one enzyme more than another, or shifts in pH values which could cause, for example, a tendency for starch to be changed to sugar. There is little doubt that sugar increases occur in many plants subjected to drought.

Many desert ferns, mosses, liverworts, lichens and algae dehydrate during the dry season but do not die. They remain in an anabiotic state of desiccation and resume activity following wetting. Among poikilohydrous plants or resurrection plants Danin (1983) records three ferns from Sinai, while Lazarides 1992) lists some 15 ferns from Australia. In the poikilohydrous angiosperm *Myrothamnus flabellifolia* from the Namib, the mitochondria (and possibly the plastids) are separated from the rest of the cell contents during desiccation by barriers which are perforated on rehydration (Mabberley 1987); it revives in about 30 min after rain, changing from a brown, broom-like plant to a spreading, bright green bush (Walter and Breckle 1986). Conversely, rapid dehydration can be fatal, as is the case with the Australian monocot *Borya nitida*; the species is also *poikilochlorophyllous*, i.e. it loses its chlorophyll on dehydration (Bewlag and Krochko 1982). It is also believed to endure desiccation for 2.5 years; the Australian grass *Eragrostiella bifaria* is also poikilochlorophyllous (Lazarides 1992), while Zohary (1955) cited by Danin and Ganor (1997) reported that a dried specimen of *Poa eigii* produced new leaves after 7 years in the herbarium. The leaves of the African perennial grass *Eragrostis nindensis* also have the ability to tolerate dehydration and loss of chlorophyll, with the replacement of the latter 48 h after rewetting (Gaff and Ellis 1974).

Other examples of poikilohydrous angiosperms occur in the Cyperaceae (*Fimbristylis dichotoma*), Gesneriaceae (*Boea hygroscopica*), Gramineae (*Eragrostiella bifaria, Micraira* spp., *Poa bulbosa, Sporobolus* spp., *Tripogon loliiformis*), Scrophulariaceae (*Lindernia intrepida*) and Velloziaceae (*Vellozia elegans*, syn. *Talbotia elegans*). According to Bewlag and Krochko (1982) poikilohydrous plants occur in 27 genera of angiosperms, with 39 species from South Africa and none from North America, while Lazarides (1992) lists 22 angiosperms indigenous to Australia, of which only *Fimbristylis* spp. are regarded as valuable pasture plants.

The annual herb known as rose of Jericho or resurrection plant, *Anastatica hierochuntica**, has erroneously been classified in the past as poikilohydrous. It is the dead plant which acts as a tumble weed, opening out when wet and closing on drying, thereby dispersing the seeds, i.e. it is hygrochastic. During periods of drought, the leaves of resurrection plants naturally reach air-dryness [relative humidity (RH) 50-30%], corresponding to water potentials of ca. -95 to -160 MPa, the most tolerant survive drying in air of RH 0%; by comparison, the desiccation tolerance of many mesophytic crop plants falls in the range of RH 98-85%, or water potentials of ca. -3 to -22 MPa. A small amount of water, nevertheless, still remains bound up in the plant tissues even when the external humidity is at zero (Street and Öpik 1984).

The ability of resurrection plants to revive is due to the presence of the non-reducing disaccharide trehalose, which is also present in fungi, generally replacing sucrose in non-chlorophyllous plants as well as being the principal carbohydrate of insect haemolymph (Sharp 1983). There is considerable interest in the possible use of trehalose to preserve blood plasma and inoculations as an alternative to freezing, a method which would be of considerable value in developing tropical countries where refrigeration could be a problem in rural areas. There is also the possibility of using genetic engineering to introduce trehalose into cereals in order to maintain dormancy under conditions where the plants would otherwise die, as for example during severe drought, and thus permit recovery when moisture conditions improve. Current research at the Quadrant Research Foundation, Cambridge is considering the possibility of its use in resurrecting dehydrated foods without denaturing. Thus, a rehydrated dried grape would be revived as a grape and not as a raisin!

CHAPTER 6

Temperature Stress

The genetic and physiological understanding of resistance in plants to heat stress is still only imperfectly understood, especially in the absence of efficient and accurate laboratory tests to select the physiological traits for heat tolerance. This can be illustrated by the breeding of certain tomato cultivars, where one high temperature tolerant cultivar, 'Nacarlang', was also found to set fruits better at temperatures too low for normal fruit set, while another cultivar, 'Cold Set', developed for the cool temperate, short-season regions, was found to be heat tolerant. Subsequently it has been found that some of the best high-temperature tolerant lines had been developed in and for cool, short-season regions. The explanation of this apparent paradox is that the basic physiological mechanisms are involved in stress tolerance rather than temperature tolerance. An addition mechanism is the greater thermostability of certain enzymes that may serve a genotype at both high and low temperatures (Arnon 1992). An added complication could be the photosynthetic pathway followed, although it would be premature to suggest any definite relationship. As Nobel (1996) has pointed out, Crassulacean acid mechanism (CAM) plants such as the agaves and cacti are extremely tolerant of high temperatures, although CAM plants in general are not particularly tolerant of low temperatures.

6.1 High Temperature Stress

The basic metabolism of living cells is generally heat sensitive, since many of the molecules involved in the intermediary pathways, e.g. ATP (adenosine triphosphate), are unstable at temperatures above 40 °C. This is especially true of the macromolecules DNA, RNA and the enzymes, which often become denatured and inactive with rising temperature. The response of plants to *heat shock* (HS) is the synthesis of two groups of HS proteins. These HS proteins persist for some time and would appear to be responsible for the plant's ability to withstand increases in temperature. Their synthesis is based largely on laboratory observations and it is still unclear as to how far this process is important in the field (Harborne 1988).

The temperature control by the plant of its tissues is discussed by Cloudsley-Thompson and Chadwick (1964) and Street and Öpik (1984). The former's classification of the mechanisms for heat control are as follows:

1. *Heat tolerance*; by plants with protoplasm capable of withstanding relatively high temperatures without injury.
2. *Heat avoidance*; by keeping the temperature either below that which would seriously endanger the tissues or at least that which would have been experienced in the absence of the avoidance factor.
   a. Insulation against heat absoption.
   b. Loss of heat through evaporative cooling.
   c. Reduction in the radiation absorbed by increasing the amount reflected or transmitted.
3. *Plant evasion*; by avoiding the hot season by producing heat tolerant seed or fruit.

From a study of resistance to high temperatures by local Mediterranean maquis trees and shrubs it was found that the highest temperatures that leaves can withstand under natural conditions are considerably lower than those found to be lethal under experimental conditions. Not unexpectedly, it was also found that lethal temperatures are lower for younger than mature leaves and that deciduous plants are less heat tolerant than evergreen. From these results Zohary (1962) concluded that heat in itself is not a damaging factor in maquis plants under natural conditions.

While heat injury is less of a problem than drought or frost damage, high temperatures can cause problems indirectly through water stress. Indeed, in hot climates high temperatures are a limiting factor for survival, but since the hottest regions are also arid, it may be difficult to dissociate the effects of heat stress from those of water stress in such habitats (Street and Öpik 1984). The lethal temperature, the *thermal death point* of a plant tissue, varies with the period of exposure; higher the temperature the shorter the period of endurance. Dry seeds can withstand higher temperatures than other plant organs. The thermal death point for organs of temperate plants generally follow exposure for a few hours at temperatures between 45 and 55 °C, while desert plants, such as agaves and cacti, survive air and tissue temperatures exceeding 55 °C (Nobel 1996) and have thermal death points at over 60–65 °C.

As well as affecting soil water evaporation, soil temperatures are of interest for their direct effect on seed germination and seedling establishment. In Jerusalem, Ashbel (1942), cited by Oppenheimer (1960), recorded a June soil surface temperature of 70 °C. In El Obeid, Sudan the author recorded a late

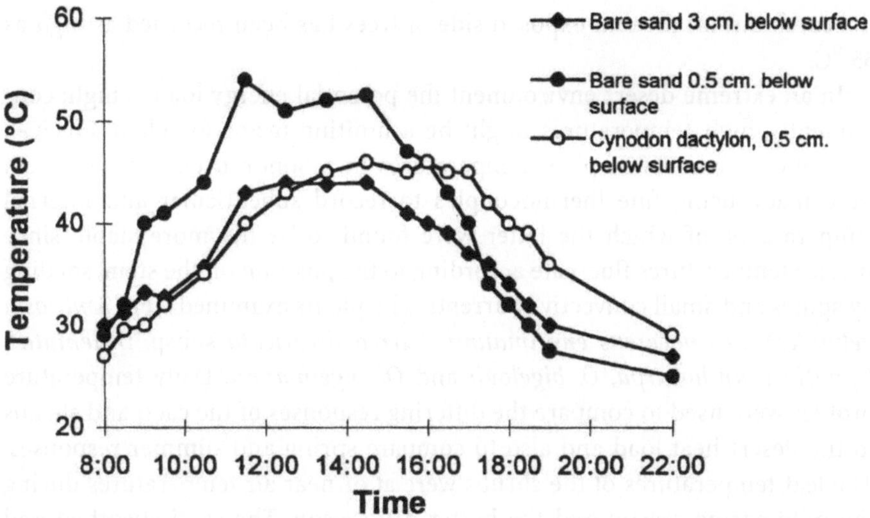

Fig. 17. Soil temperatures at El Obeid, Sudan on 13 April 1963. Maximum and minimum air temperatures at El Obeid Meteorological Station were 35 and 18 °C respectively (Redrawn from Huntin Technical Services 1964)

morning sub-surface maximum temperature on bare sand of 54 °C and at 3 cm depth a maximum of 45 °C, yet the maximum air temperature was only 35 °C (Fig. 17, Hunting Technical Services 1964). Since an air temperature of 46.1 °C has been recorded for El Obeid (Ireland 1948), even higher soil surface temperatures can be expected. Indeed, in Wadi Halfa, Sudan a daytime surface temperature of 84 °C has been recorded; such an extreme surface soil temperature must surely seriously affect the chances for a natural recovery of the vegetation following desertification. For much of the arid and semi-arid regions the period during which a plant is in daily contact with potentially lethal temperatures is clearly insufficient for the thermal death point to be reached. Diurnal and seasonal fluctuations in soil temperatures decrease with depth and at about 20–30 cm they almost disappear and according to Shachak (1975) cited by Gutterman (1993) the temperature is about the daily average of the surface maximum and minimum.

In temperate climates, where the air temperatures are unlikely to exceed 40 °C, it might be expected that heat injury would be unlikely. However, the temperatures of plant tissues may well be above ambient, especially where the organs are bulky and unable to dissipate absorbed heat. In the sun, even thin leaves may attain temperature 6–10 °C above air temperature despite transpirational cooling. Temperature of 40–50 °C have been recorded internally for fleshy leaves with an air temperature at 20–30 °C, while the temperature of the cambium on the sun exposed side of trees has been recorded as high as 55 °C.

In an extreme desert environment the potential energy load is high; consequently, high temperatures might be a limiting factor for plant survival. Field measurements of plant temperatures in a Sonoran Desert ecosystem were made using fine thermocouples to record subcuticular and internal temperatures, of which the latter were found to be the more stable since surface temperatures fluctuate according to the position on the stem, shading by spines and small convective currents. The plants examined were *Ambrosia deltoides**, *Echinocereus engelmannii, Larrea divaricata* subsp. *tridentata**, *Opuntia acanthocarpa, O. bigelovii* and *O. engelmannii*. Daily temperature profiles were used to compare the differing responses of the cacti and shrubs to the desert heat load and also to compare spring and summer responses. The leaf temperatures of the shrubs were at or near air temperatures during the mild spring season and the hotter dry season. The cacti absorbed and stored heat so that their temperatures were often above air temperature. The energy absorbed is largely determined by the orientation of the plant and the surface area exposed. The actual energy absorbed by the plants was extracted from energy diagrams.

The flat stems of *Opuntia engelmannii* were oriented so as to receive maximum sunlight without long periods of continuous heating. The spines of *O. bigelovii* reflect heat and absorb much of the environmental energy load, thereby protecting the thick succulent stems from overheating. The smaller, cylindrical stems of intricately branched, hemispherical, up to 2-m-tall *O. acanthocarpa* dissipate heat most effectively by their large surface area exposed to connective air currents. In *Echinocereus* the heat flow during the day began with absorption of light by the upper parts of the cylinder, followed by a temperature rise. The upper part appears to absorb more radiant energy than the basal portion. Later during the day the basal portion slowly heats up, receiving heat from the upper irradiated surface and from the hot soil by conduction. The time lag of heating by conductance demonstrates the poor thermal conductivity of succulent stems. The slow rate of cooling and retention of heat into the night should create a relatively high internal metabolic rate at night while the epidermal layers are cooled for efficient uptake of CO_2. The leaves of the two desert shrubs, *Ambrosia deltoides* and *Larrea divaricata* remain closer to air temperatures than do the succulent stems of the cacti because their large surface area to volume ratio allows them to dissipate heat by convection. *Larrea* lost a little heat by reradiation while *Ambrosia* dissipated more heat by reradiation (Gibbs and Patten 1970).

Plants growing in hot climates show some adaptations for heat avoidance, i.e. for keeping their temperatures near or below air temperature. Desert plants commonly have highly reflective external surfaces in order to reduce heat absorbtion. *Transpirational cooling* is a means of reducing shoot temperature and, in favourable circumstances, leaves may be cooled to 10 °C below air temperature. However, transpirational cooling is greatly affected by environmental conditions and also varies from species to species. Significant transpirational cooling is dependent on an ample water supply, whereas many desert plants are water conservers. Once leaf temperature has risen above that of air temperature, heat convection results in heat loss, and for small leaves this effect may be important (Street and Öpik 1984).

Some plants take advantage of the insulating effect of the soil for protection against high atmospheric temperatures by having the minimum of aerial parts exposed to the atmosphere. For example, in the coastal desert of Chile there are the unusual 'earth cacti' belonging to the *Thelocephala* group of *Neoporteria* (Hoffmann 1989), which are sunken into the soil. The aerial parts are small and brown, barely visible except when in flower or in times of extreme drought when the underground portion shrinks and a gap appears between the plant and the soil. The swollen subterranean storage tubers have thin root necks from which arise fine lateral roots, which radiate out at 1.5 cm below the surface, absorbing any condensed fog (Rauh 1985).

6.2 Low Temperature Stress

Low temperature stress is experienced mainly by plants during the long winters in the cold deserts of the world, such as the arid regions of Central Asia. The lowest temperature recorded for plant ecosystems is -70 °C, although experimentally certain plants and plant structures can survive even lower temperatures. As an extreme example, the bark cells of a *Morus* twig can survive at -196 °C for 160 days. Even in warm deserts, such as the Sahara, there may be dramatic diurnal changes in temperature, such as that experienced at Bir Mighla, southern Tripolitania, from -0.5 °C to 37.5 °C (Cloudsley-Thompson 1977; Street and Öpik 1984). In the Sonoran and Chihuahuan Deserts of North America the northern limit for various cacti species is the -1 °C annual isothem. It is the seasonal low temperature that mainly limits the cultivation of *Opuntia ficus-indica** in the 20 or more countries where it is grown. The influence of freezing temperatures on cell viability and plant survival has not yet been fully investigated for many of the cultivated CAM species (Nobel 1996).

Many species originating in warm climates are unable to endure temperatures below 5 °C, sometimes not even below 15 °C. This is believed to be due to a gelling of the cellular membranes, and is known as *cold shock* or *chilling injury* (Street and Öpik 1984).

The resistance of plants to subfreezing temperatures has been classified by Street and Öpik (1984) as follows:

1. *Low temperature resistance*; survival at subzero external temperature.
   a) *Low temperature avoidance*; maintenance of high internal temperature.
   b) *Low temperature tolerance*; survival at subzero external temperature
2. *Freezing resistance*; the lowest temperature at which a plant can survive.
   a) *Freezing avoidance*; avoidance of ice formation in tissues;; depression of freezing point, supercooling.
   b) *Freezing tolerance*; survival if ice formation in tissues; prevention of intracellular freezing; dehydration tolerance.

Most plants do not have any specific mechanism for temperature regulation; even inside bulky organs, the temperature is close to air temperature, except in direct sunlight. On cold nights, leaf and bud temperature often falls below ambient. However, some adaptations for keeping the temperatures of plant organs above air temperature have been recorded for high altitude plants in Africa and South America, where the night temperatures throughout the year may fall to ca. -10 °C. The plants produce giant rosettes, which form diurnal '*night buds*', where the freeze tolerant adult leaves close tightly over the central meristem. A large volume of mucilaginous fluid in the bud also acts as a thermal buffer, and the meristem temperature is thereby maintained appreciably above the air temperature. These are, however, exceptional cases.

While plants have many adaptations for avoiding water stress, low temperature tolerance is the main resistance mechanism. The positioning of perennating organs such as buds, rhizomes and bulbs below ground level, where the temperature does not fall as low as at the surface, has been regarded as a kind of freezing avoidance. These organs are, however, at near temperature equilibrium with their immediate environment. The ability to survive internal subzero temperatures is known as *freezing resistance* (Street and Öpik 1984).

The effects of cold depend not only on the lowest temperature reached but also on the rates of cooling and warming, the time for which the minimum temperature persists and the subsequent treatment after warming. Not surprisingly, species in regions with cold winter climates are more resistant to cold than species from warm climates. The cold resistance of perennials of temperate and Arctic zones undergoes striking seasonal changes, being very much higher in the winter months than in the summer. The increase in hardiness occurs in the autumn with the falling temperatures and shortening daylength, without which the plants would be unable to survive the rigours of winter. The relative rapid rise in temperature in the following spring results in a loss of hardiness.

Artificially, hardening can be induced at any season by chilling, and resistance destroyed in mid-winter by warming. In the field, cold injury is most frequent in spring when late frosts act on non-hardy tissues. Greater resistance is developed in severe winters than in mild ones. Resistance also varies greatly between plant organs. Underground structures are much less resistant than aerial parts; they also have a more uniform resistance throughout the year because soil temperatures show less annual temperature variation and never fall as low as air temperatures. Freezing damage depends on ice formation within the tissues; injury could, therefore, be regarded as a result of mechanical disruption caused by ice crystals. However, as the ice crystals form, dehydration of the tissue occurs as water passes from the cells into the growing crystals; freezing damage could, therefore, also be a form of dehydration damage caused by the denaturation of the cell proteins, especially those of the cell membranes.

Freezing resistance of a tissue is clearly dependent upon the amount and state of the water in its cells. The greatest cold resistance is encountered in dry seeds, a factor that is utilised by seed banks for the long-term storage of seeds. During the period of frost hardening of perennials, there is a tendency for a decrease in total water content of the tissues, and an increase in the proportion of bound water, due to an increased hydration capacity of the cell colloids. Nevertheless, many frost-hardy tissues contain very appreciable amounts of freezable water.

Avoidance of freezing may be achieved by a lowering of the freezing point by the presence of solutes; frost hardening is associated with an increase in soluble carbohydrates, most commonly sucrose; other sugars or polyhydric sugar alcohols may also be present. The freezing point of the cell contents is, thereby, usually lowered by just a few degrees (Street and Öpik 1984). The role of the soluble carbohydrates is at least two-fold. Firstly, their osmotic effect reduces the amount of ice formed in the vacuole. Secondly, they achieve an additional protective role by their metabolic effect, by being converted within the protoplasm into other constituents. Whether these changes involve conversion of the sugars into polyhydric alcohols or whether a more complex biochemistry is involved is uncertain (Harborne 1988). In halophytes, freezing points as low as -14 °C have been recorded owing to the cryoscopic effect of the high salt levels. Much lower temperatures may be endured without ice formation in tissues capable of deep supercooling, such as in the xylem ray parenchyma cells and flower primordia in the buds of some temperate zone woody perennials; other tissues in these species are able to tolerate ice formation. Water can be supercooled to -38 °C, at which temperature ice forms spontaneously; for cell solutions the temperature is slightly lower. The northern growth limit for species relying on deep supercooling is set by the -40 °C winter minimum isotherm, for when the supercooled cells do freeze, they freeze very rapidly and intracellularly. The capacity for deep supercooling is only achieved during the winter months; it is not known how this is achieved.

Tolerance of freezing enables tissues to endure even lower temperatures than supercooling, down to -196 °C, though such tolerance is not shown by all tissues that can withstand freezing. Freezing tolerance depends on ice formation being confined to outside the cell membranes and on the ability of the cells to survive desiccation and compression by the extracellular masses of ice. Freeze hardening may involve an increase in the permeability of cells to water, making it easier for water to leave the cells and thereby lessening the risk of intracellular freezing. Protoplasts of hardy cells have been reported to withdraw more easily from the cell walls when water is lost, lessening the risk of mechanical strains as the protoplasts shrink. The formation of soluble sugars may also help to stabilise colloids against dehydration damage as well as lowering the freezing point. The frost hardening of perennial tissues is associated with increases in soluble protein content, as in the bark of *Robina pseudoacacia** (Street and Öpik 1984).

CHAPTER 7

Salt Stress

Salinity in arid and semi-arid lands is becoming an ever increasing problem with 222 750 ha of irrigated lands going out of production annually. Salt accumulation, as well as producing toxic concentrations that stress plant growth also reduces water availability to the plant by increasing the osmotic potential of the rhizosphere. In addition, high sodium ions in the soil can negatively affect the physical soil characteristics, for example, sodic soils, which have a high concentration of exchangeable sodium but a low concentration of total salts, can seriously affect soil structure and become impermeable, thereby further reducing water availability (Briggs 1996).

A saline habitat is defined as one where the NaCl concentration equals or exceeds 0.5%, and plants capable of growing normally in such concentrations are classed as halophytes (sea water contains 3–5% NaCl, ca. 0.5 mol l^{-1}). Salinity is usually measured in terms of specific electrical conductivity (EC_e or EC_i) of saturation extracts at 25 °C in decisiemens per metre (dS m^{-1}), where the criteria for salinity is EC_e or EC_i = 7.8 dS m^{-1}, if not considerably lower; for sea water EC_i = 50–80 dS m^{-1}. The electrical conductivity of underground brackish water typically varies from as low as 1.3 dS m^{-1} to as high as 10 or even 20 dS m^{-1} (Aronson 1989). Salt sensitive plants will show reduced growth with levels as low as 2 dS m^{-1} while salt tolerant plants will show no ill effects at levels as high as 10 dS m^{-1}. Most plants, however, will show some signs of slight to moderate sensitivity at levels of 5–10 dS m^{-1} (Briggs 1996). Data on the maximum recorded salinity tolerance by plants is given in Aronson (1989).

Many inhabitants of saline habitats are facultative halophytes, but there are also obligate halophytes, requiring at least 0.5% NaCl, such as *Salicornia europaea**, *Suaeda maritima* and the mangroves *Avicennia germinans* and *Rhizophora mangel*. Halophytes have to not only satisfy their internal water requirements, but also regulate the salt concentration. In order to absorb water from the saline soil solution with a low osmotic potential, the osmotic potential of the cell sap in the vacuoles must be even lower (Walter and Breckle 1985). This is achieved by the plant first absorbing salt and water until the osmotic forces of the soil water and cell sap are in equilibrium. Excessive salt must not be taken up since it will then have a toxic effect upon the cells. Older roots are almost impermeable to salts so that almost pure water is absorbed from the saline soil water, and this under very high cohesive tension in the conducting vessels.

Stress due to high salinity may be caused by either as a water stress (the water potential of sea water is ca. -2.5 MPa) or to the toxic effects of the Na^+ and/or Cl^- ions, the latter being considered the major stress factor. Some species may achieve a limited resistance to salinity by avoiding the uptake of excessive amounts of salt, the exclusion mechanism depending on the properties of the cellular membranes. Sometimes, although salt is absorbed into the root system, very little is transported into the shoot. The halophytes capable of growing in salinity levels of sea water, or even above, accumulate high levels of Na^+ and Cl^-, which enables them to maintain a low internal water potential and to avoid water stress deficits. For example, the perennial grass *Leptochloa fusca** has the ability to sequest high levels of Na^+ and Cl^- in the sheath, away from the blade. *Salicornia europaea**, when grown at a soil salinity with a soil water potential of -1.6 MPa, can maintain a shoot water potential of -6.5 MPa.

Although the high NaCl uptake by halophytes solves the water stress problem, it does necessitate other physiological adjustments. To avoid the susceptibility of the cell enzymes to salt inhibition the ions are segregated away from the cytoplasm, in vacuoles and cell walls. For example, it has been found by X-ray analysis that in the halophytes *Frankenia grandifolia, Suaeda maritima* and *Tamarix aphylla** most of the Cl^- are in the vacuoles and cell walls. The presence of high salt concentrations in the vacuoles and cell walls, while enabling the plant as a whole to absorb water would, nevertheless, lead to a dehydration of the cytoplasm if this did not contain osmotically active solutes to keep it at the same water potential as the vacuole and wall. Osmotic balance is provided by the cytoplasmic synthesis of organic solutes, including amino acids, especially the protein imino acid proline and the quaternary nitrogen compound glycinebetaine, and sometimes sugars. Although a few halophytes do not exhibit enhanced proline levels, the concentration of proline in others may reach ten-fold the level present in the free amino acid pool of non-halophytes. The precise mechanism by which proline functions in salt resistance is not known, but its presence is presumed to enable halophytes to withstand the high osmotic forces to which their cells would otherwise be subjected. A function of proline in drought resistance has already been briefly discussed in Section 5.1.2, and there may well be a common mechanism in adaptation to salt and drought stress. In some ways glycinebetaine may be more important than proline in the adjustment of halophytes to salt stress, and may even function in a similar manner. The protective role of both compounds is further emphasised in that neither has any inhibitory effect on enzyme activities, even at high concentrations.

Fitter and Hay (1987) cite the work of Stewart and Lee (1974) on the use of proline in halophytes for osmotic balance, who found proline accumulation was directly related to the salinity of the growing medium and comprised on average 54.6% of the amino acid pool in halophytes, 2.9% in calcicoles, 2.4% in calcifuges and 4% in ruderals. They found proline to have no effect on the activity of nine enzymes extracted from *Triglochin maritima* at concentrations where NaCl was extremely inhibitory and that the concentrations used of up to 700 mM were comparable with measured internal NaCl levels. Since the ratio of proline to Na^+ and Cl^- in the vacuole had previously been demonstrated to be maintained constant in three halophytes, the role of proline in maintaining osmotic balance is an attractive one.

Other common halophyte characters are succulence, which is believed to be a means of diluting the salt with large volumes of water, and the possession of salt glands. Salt glands enable many halophytes, especially non-succulent halophytes growing in saline habitats, such as in the Compositae (*Nassauvia glomerulosa**), Convolvulaceae (*Cressa cretica**), Frankeniaceae (*Frankenia palmeri**), Gramineae (*Aeluropus litoralis*), Plumbaginaceae (*Limonium* spp., *Statice* spp.) *and* Tamaricaceae (*Reaumuria**, *Tamarix**), to actively eliminate excess salt and thereby prevent the accumulation of high and possibly toxic concentrations of salt in the shoot tissues (Pollak and Waisel 1970; Street and Öpik 1984; Harborne 1988).

Greenwood (1986) made the following observations on the water use of trees and shrubs and their effect on saline groundwater:

1. The uptake of water increases with rooting depth.
2. Phreatophytes transpire more water than non-phreatophytes. Because of their deep root system, phreatophytes have access to a more readily available water supply provided it is not too saline (it is not unusual for aquifers to be less saline lower down than at the surface, where the salt is concentrated by evaporation).
3. The uptake from the saturated zone decreases as the depth and salinity of the water-table increases. It has been demonstrated in tank experiments, using the phreatophyte *Tamarix ramosissima*, that the uptake of low salinity water decreased by 510 mm $year^{-1}$ for each additional metre to the water-table. In contrast, using high salinity, the decline in water uptake was 800 mm m^{-1} $year^{-1}$. Further investigations are necessary to discover whether these rates continue at even greater depths since they imply zero uptakes at depths of 7 and 4 m respectively.
4. Halophytes transpire more water than non-halophytes in saline soils; under non-saline conditions the reverse is true. Transpiration from ungrazed

Table 2. Proline µg g-1 fresh weight) content in non-stressed and stressed plants. (After Mohammed and Sen 1987)

Species	Non-stressed	Stressed	Increase over non-stressed %	Category non-stressed
Category 0 (proline absent)				
Aerva javanica				
Calotropis procera				
Corchorus depressus				
Crotalaria burhia				
Portulaca oleracea				
Category I (increase <100%)				
Cenchrus biflorus	1.94	2.90	49	II
Farsetia stylosa (syn. *F. hamiltonii*)	29.00	37.34	28	III
Category II (increase <200%)				
Convolvulus prostratus (syn. *C. microphyllus*)[a]	0,42	1.16	177	I
Indigofera cordifolia	1.32	3.57	171	II
Trianthema triquetra	3.89	8.16	109	III
Category III (increase >200%)				
Citrulus colocynthis	0.50	3.47	594	I
Convolvulus prostratus[b]	0.50	4.31	755	I
Dacyloctenium aegyptium	1.30	9.90	661	II
Ocimum americanum	4.95	15.45	212	III
Tephrosia purpurea[c]	2.34	9.10	288	III
Tephrosia purpurea[d]	5.07	53.96	964	III
Zaleya redimata	2.10	31.61	1404	III
Category IV (proline estimated only under stressed conditions)				
Calligonum polygonoides		8.61		
Capparis decidua		1		
Leptadenia pyrotechnica		1.63		
Prosopis cineraria		4.86		
Salsola imbricata		2.74		
Salvadora oleoides		0.5		
Salvadora persica		1.3		

[a] White-flowered
[b] Pink-flowered
[c] Long-leaved
[d] Short-leaved

*Atriplex vesicaria** plantations growing on a salt flat with a water-table at 1.2 m and an NaCl concentration of 24 x 03 mg l^{-1} lowered the water-table by ca. 10 cm on a site where no recognised agricultural crop plant could survive.

5. Transpiration increases with leaf area where the water supply is plentiful. Grazing, fallow and the felling of forests reduce transpiration.

Mohammed and Sen (1987) examined 65 species of the Indian Desert for proline content; the results for 19 out of the species discussed are shown in Table 2. Since those species classified as belonging to category IV have deep taproots that can absorb water from depth, it is not possible to classify them as being stressed or non-stressed. Table 2 demonstrates the wide range of increase in proline under stress, and also that a number of well-adapted desert plants such as *Aerva javanica**, *Calotropis procera* and *Crotalaria burhia* show no proline under any conditions, while *Capparis decidua*, *Leptadenia pyrotechnica** and *Salvadora oleoides* of category IV contained insignificant amounts of proline. Mohammed and Sen (1987) concluded that such insignificant amounts in well-adapted desert plants indicate that proline is not necessarily an indicator of stress. A more likely conclusion is that osmoregulation is achieved by other means; indeed another amino acid, glycine-betaine, has been found to be of greater significance than proline in a number of species (Fitter and Hay 1987).

CHAPTER 8

Photosynthesis

The three photosynthetic pathways, C_3, C_4 and CAM, are briefly summarised below. The biochemical processes involved are described in greater detail by Harborne (1988), among others. There are considerable variations to be found in the C_3 and C_4 pathways; to date these have been most widely investigated in the Gramineae – the CAM pathway is not known in the Gramineae (Hattersley and Watson 1992).

8.1 C_3 and C_4 Photosynthetic Pathways

The terms C_3 and C_4 recognise major differences in the processing of the internal products of plant photosynthesis. The first product of CO_2 fixation by photosynthesis in most higher terrestrial plants is 3-phosphoglyceric acid, a three-carbon acid, hence the designation C_3 plant. Most plants found in the temperate regions are C_3 plants. They are relatively inefficient photosynthetically compared with C_4 plants, generally having lower CO_2 fixation rates and higher compensation points than C_4 plants, the compensation point being the minimum level where the photosynthetic uptake of CO_2 is exactly balanced by its respiratory release, i.e. the rate of synthesis of organic matter is equal to breakdown by respiration (Tootill 1984).

Other species follow a modified carbon cycle where, in addition to the C_3 cycle, there is a second CO_2-fixing pathway (Hatch-Slack cycle) and the first step in photosynthesis is a four-carbon decarboxylic acid, oxaloacetic acid. Such plants are known as C_4 plants, most of which occur in the tropics (Noggle and Fritz 1983). They possess the following characteristics:

1. The primary initial products of photosynthesis are the C_4 decarboxylic acids oxaloacetate, malate and aspartate. Carbon is donated from the C_4 acids into the RPP (ribulose 1,5 biphosphate) pathways of photosynthesis. These events, unlike in CAM plants, occur simultaneously in the light, albeit the first fixation is separated in space from the other two processes.

2. Generally there are two photosynthetic cell types; mesophyll and large, thick-walled Kranz cells (from the German word *Kranz* meaning wreath). Anatomical examination shows the two photosynthetic cell types arranged in concentric layers, forming a wreath around the vascular system; anatomical examination for the presence of Kranz cells is the quickest way of identifying C_4 plants; however, their absence does not necessarily mean that it is not a C_4 plant. Furthermore, the grass *Alloteropsis semialata** contains morphologically indistinguishable races with either non-Kranz (C_3) or PS Kranz variant (C_4) leaf anatomy. In C_4 photosynthesis there is a division of labour between the mesophyll and sheath. In grasses the PS variant achieves this by an elongation of the chlorenchyma cells around the secondary bundles. In the MS variant minor bundles are interpolated in the mesophyll located under bulliform cells. The selection of PS or MS types according to environmental stress is problematical (Clayton and Renvoize 1986; Hattersley and Watson 1992).
3. Apart from some noticeable exceptions, C_4 plants have an optimum photosynthetic efficiency at about 10 °C above that of C_3 plants; their photosynthetic efficiency is also higher (Le Houérou 1984).
4. C_4 plants produce more glucose per unit leaf area than C_3 plants and consequently grow more quickly (Tootill 1984).

Experimental claims that C_4 plants are more efficient users of water than C_3 plants have proved to be mainly erroneous. Since C_4 plants in their natural environment function at much higher temperatures, their water demand is also considerably higher. Thus, at 15 °C higher temperatures, water demands are three times higher; consequently, C_4 plants are generally not better water users than C_3 plants; however, there are exceptions, including some cultivars of maize, sorghum and pearl millet (Le Houérou 1984). This increase in water demand is in accordance with Van't Hoff's rule where, in the absence of other limiting factors, for every 10 °C increase in temperature between -5 and 35 °C the transpiration rate approximately doubles (Curtis and Clark 1950; Le Houérou 1984)

The C_4 synthetic pathway is found in the following families: Acanthaceae, Aizoaceae, Amaranthaceae, Boraginaceae, Capparaceae, Caryophyllaceae, Chenopodiaceae, Compositae, Cyperaceae (tribes Cypereae and Fimbristylideae; Raynal 1973), Euphorbiaceae, Gramineae (subfamilies Arundinoideae, Chloridoideae, Eragrostoideae and Panicoideae; Clayton and Renvoize 1986), Nyctaginaceae, Polygonaceae, Portulacaceae, Saxifragaceae and Zygophyllaceae. Many of the world's worst weeds are C_4 plants, as are also the tropical economic grasses sorghum, maize, millets and sugar cane (Edwards and Walker 1983). Winter (1981) observed that the C_4 pathway had rarely been observed in arborescent species, having only been previously reported from members of the Euphorbiaceae from the Hawaiian rainforest. It is now known from the Chenopodiaceae, where *Haloxylon ammodendron**, *H. aphyllum** and *H. persicum** certainly attain the status of trees.

8.2 Crassulacean Acid Metabolism (CAM)

Plants with the CAM pathway are very conservative in their water usage due to their metabolic adaptation during photosynthesis whereby the CO_2 is fixed at night into malate. During the day the malate is decarboxylated and the released CO_2 assimilated. This allows the stomata to be closed during the day, when the potential water loss is high due to high day temperatures. The CAM pathway differs basically from the C_4 pathway in that all the three processes of initial fixation, decarboxylation and refixation occur in the same cells, except that the initial fixation occurs in the dark and the other two processes in the light.

At a given stomatal resistance a CAM plant can conserve more water by fixing CO_2 at night than during the day. The potential for water loss from a leaf depends on the differences between the water vapour concentration in the leaf versus that in the atmosphere. Thus, in a hot, arid environment with a high leaf temperature and low atmospheric humidity, the potential for water loss is very high. With high irradiance and low temperatures, the leaf temperature can easily exceed the air temperature (the heat absorbed when water is converted from a liquid to vapour during transpiration removes heat from the leaf, a process known as evaporative cooling).

Besides being adapted to water conservation, CAM plants conserve carbon and minimise respiration losses of CO_2. When photosynthesising in the CAM mode they may go through acomplete 24-h cycle without net efflux of CO_2 at any time. At night there is a net uptake of CO_2 and any respired CO_2 may be refixed through PEP (phosphoenol pyruvate) carboxylase. The stomata are closed during the day and the CO_2 released by C_4 acid decarboxylation or respiration is assimilated through the RPP (ribulose 1,5 bisphospate) pathway (Edwards and Walker 1983).

CAM plants, like C_4 plants, are widely distributed in nature and both forms of specialisation may be present in different species of the same family. For example, in *Euphorbia* the CAM pathway occurs in species from tropical Africa, while the C_4 pathway is present in the species from subtropical North America (Harborne 1988). Species with the CAM pathway are five times as numerous as C_4 species (Nobel 1991), and the CAM pathway is found in the following angiosperm families: Agavaceae, Aizoaceae, Asclepiadaceae, Bromeliaceae, Cactaceae, Compositae, Crassulaceae (where it was first discovered), Cucurbitaceae, Didiereaceae, Euphorbiaceae, Liliaceae, Orchidaceae, Portulacaceae and Vitaceae, as well in the gymnosperm and pteridophyte families Welwitschiaceae and Polypodiaceae. It is also suspected to occur in the Geraniaceae, Labiatae, Oxalidaceae and Piperaceae, with inconclusive evidence for its occurrence in some species of Butaceae, Caryophyllaceae, Chenopodiaceae, Convolvulaceae and Plantaginaceae (Edwards and Walker 1983). It has not been found in the Gramineae (Clayton and Renvoize 1987).

Although usually associated with succulents, there are many succulents, e.g. halophytes, which do not exhibit CAM; there are also CAM plants which are best described as semi-succulent. Walter and Breckle (1986) report that succulents with the CAM pathway do not occur in the Negev Desert except for *Caralluma negevensis*, although *Mesembryanthemum forsskalei**, which also occurs in the Negev, uses CAM when under stress from drought or salinity, but converts to the C_3 pathway when the stress is removed (Danin 1983).

CAM plants often have large mesophyll cells in the leaf, several times larger than those found in C_3 and C_4 plants; the very large vacuole of the cell, which is the site of malate storage, is surrounded by a layer of cytoplasm, although the leaf anatomy of CAM plants can vary considerably. However, anatomical studies alone cannot distinguish between CAM and C_3 plants. CAM plants are particularly well adapted to semi-arid conditions, their metabolic pattern being such, that under drought conditions they conserve respiratory CO_2 and assimilate external CO_2 in the dark, with little water loss. Some succulents may even carry out C_3 photosynthesis as an additional means of growth when water is plentiful. The transpiration to photosynthesis ratios (gram water transpired to gram CO_2 fixed) are about 50:100 in CAM plants (in CAM mode), 250:300 in C_4 plants and 400:500 in C_3 plants.

Ecologically CAM can be described as an adaptation to water stress and water conservation. CAM plants, according to Edwards and Walker (l983), have a higher water-use efficiency than C_3 or C_4 plants but a lower productivity. However, although many CAM plants are slow growing, such as the short stemmed cacti and many of the epiphytes from other families, the energetics of CO_2 fixation demonstrate that this is not necessarily true for all plants following the CAM photosynthetic pathway.

While photorespiration in C_3 plants can release some of the CO_2 fixed and also has an energy cost, photorespiration is much less in C_4 and CAM plants. CAM plants can actually perform most CO_2 fixation 15% more efficiently than C_3 plants and with only 10% less efficiency than C_4 plants (Nobel 1991). However, when soil water is readily available the CAM species *Agave deserti** can switch to the C_3 photosynthetic pathway, fixing CO_2 during the day. Although net photosynthesis is about the same for either pathway, the latter allows for greater productivity when days are long and nights are short.

CHAPTER 9

Plant Metabolites

Many of the phytochemicals present within a plant have important metabolic functions that respond to environmental stress and thus ensure the plant's survival through one generation to the next (Table 3). *Gums* consist of complex, highly branched polysaccharides which swell in water to form a gel or sticky substance. Some gums are of economic importance, such as gum arabic from *Acacia senegal** and tragacanth gum from *Astragalus gummifer*, *A. microcephalus**, etc.; these are glassy, hard gums which are formed from acidic polysaccharides, often as the result of injury (Tootill 1984).

Plant growth substances are known as *auxins*; the most widely occurring being indole acetic acid, while others are based on the indole ring. When present in low concentrations they have the ability to promote the elongation of shoots, as in *Jatropha curcas*, and of roots (Tootill 1984). Synthetic auxins,

Table 3. Biochemical responses of plants to various forms of environmental stress. (Harborne 1988)

Stress	Biochemical response in plant
High temperature	Long term: C_4 photosynthesis Short term: heat shock proteins[a]
Drought	CAM photosynthesis Increase in abscisic acid Increase in proline or pinitols[a]
Low temperature	Accumulation of sugars and/or polyols[a] Synthesis of antifreeze proteins[a]
Salt	Accumulation of cytoplasmic osmotica (polyols, quaternary ammonium compounds, etc.)
Selenium	Accumulation of selenium as non-protein amino acid
Heavy metals	Modification of enzymes in root surface Binding to sites in cell walls of root Accumulation/detoxification as peptide complex[a] Accumulation/detoxification as organic acid complex

[a] Similar responses have been observed in micro-organisms and/or animals

such as 2,4-D, MCPA and 2,4,5-T, are now widely used as weedicides. Some metabolites, such as proline, oxaloacetic acid, malate and aspartate, have already been mentioned (See Sect. 5.1.2, Chap. 7, Sect. 8.1) others will be discussed elsewhere in the text (see Chap. 11). However, what of the *secondary compounds*, those that were formerly believed to be the waste products of the primary metabolism? In a pioneer study on the coevolution of plants and butterflies, Ehrlich and Raven (1964) concluded that these secondary plant substances play the leading role in determining patterns of evolution, a role that now appears to be true for all phytophagous groups.

During the course of evolution, plants have produced a number of compounds that are not directly involved with the basic metabolism but, by chance, serve to enhance, reduce or destroy the plant's palatability to the phytophagous members of the animal kingdom – chiefly insects. All angiosperms tend to accumulate concentrations of at least one type of secondary compound, whether it be alkaloid or flavonoid or terpenoid (Table 4), but rarely concentrations of different classes of secondary compounds.

Such defensive phytochemicals offer an explanation of what Feeny (1975) cited by Harborne (1988) considered a conspicuous non-event as to how the plant kingdom had managed to survive and dominate the earth against the predations of the numerically greatly superior phytophagous insects. For example, members of the Cruciferae commonly contain a glucosinolate-myriosinase system that principally acts as a defence against herbivore insects (Blau et al. 1978; Chew 1988).

Plant toxins are of obvious interest to graziers. The intensity of toxicity may vary within a species; plants may even be innocuous when eaten alone and require another species to trigger the toxicity, as with *Eremophila maculata** which requires an enzyme present in *Acacia georginae** to liberate prussic acid. Local immunity to toxicity can sometimes be acquired, for example, in South Africa locally bred cattle are unaffected by *Moraea* spp. which is lethal to introduced cattle.

Other secondary compounds reduce competition by affecting the growth of other plants within the immediate vicinity, implicating either those belonging to the same species (*auto-allelopathy* or *auto-toxicity*), as in *Parthenium argentatum**, or different species, as in *Eucalyptus camaldulensis**, *Gutierrezia sarothrae** and *Juglans regia**. These deleterious or advantageous biochemical reactions between higher plants are known as *allelopathy*, although the term is sometimes extended to include reactions between higher plants and microorganisms. Such allelopathic substances or toxins are either volatile terpenes or else phenolic compounds. The former compounds are more usual among the aromatic plants of the arid and semi-arid habitats, such as *Artemisia californica* and *Salivia leucophylla* of the Californian chaparral, while the latter compounds tend to be found among plants of more humid habitats

Table 4. Major classes of secondary plant compounds involved in plant-animal interactions (Harborne 1988)

Class	Approximate number of structures	Distribution	Physiological activity
Nitrogen compounds			
Alkaloids	6500	Widespread in angiosperms, especially in root, leaf and fruit	Many toxic and bitter tasting
Amines	100	Widespread in angiosperms, often in flowers	Many repellent smelling; some hallucinogenic
Amino acids (non-protein)	400	Relatively widespread, especially in legume seeds	Many toxic
Cyanogenic glycosides	30	Sporadic, especially in fruit and leaf	Poisonous (as HCN)
Glucosinolates	75	Cruciferae and ten other families	Acrid and bitter (as isothiocyanates)
Terpenoids			
Monoterpenes	1000	Widespread, as essential oils	Pleasant smells
Sesquiterpene lactones	1500	Mainly in compositae, but increasingly in other angiosperms	Some bitter and toxic, also allergenic
Diterpenoids	2000	Widespread, especially in latex and plant resins	Some toxic
Saponins	600	In over 70 plant families	Haemolyse blood cells
Limonoids	100	Mainly in Meliaceae, Rutaceae and Simaroubaceae	Bitter tasting
Cucurbitacins	50	Mainly in Curcurbitaceae	Bitter tasting and toxic
Cardenolides	150	Especially common in Apocynaceae, Asclepiadaceae and Scrophulariaceae	Toxic and bitter
Carotenoids	500	Universal in leaf, often in flower and fruit	Coloured
Phenolics			
Simple phenols	200	Universal in leaf, often in other tissues as well	Antimicrobial
Flavonoids	4000	Universal in angiopserms, gymnosperms and ferns	Often coloured
Quinones	800	Widespread, especially in Rhamnaceae	Coloured
Other			
Polyacetylenes	650	Mainly in Compositae and Umbelliferae	Some toxic

Table 5. A classification of disease resistance factors in plants (Harborne 1988)

Class	Description
Pre-infectional compounds	
1. Prohibitins	Metabolites which reduce or completely halt the in vitro development of micro-organisms
2. Inhibitions	Metabolites which undergo post-infectional increase in order to express full toxicity
Post-infectional compounds	
1. Post-inhibitins	Metabolites formed by the hydrolysis or oxidation of pre-existing non-toxic substrates
2. Phytoalexins	Metabolites formed de novo after invasion by gene depression or invasion of a latent enzyme system

(Tootill 1984; Harborne 1988). The essential oil of the latter species yields appreciable quantities of five terpenoids, camphor, 1,8-cineole, isothujone, ù-thujone and artemene ketone. The camphor is the most toxic, followed by 1,8-cineole, while artemene ketone is the least toxic. The amounts and composition of the essential oil change with the season, increasing in late summer and autumn and with less in the winter. Isothujone largely replaces ù-thujone and camphor becomes more abundant in late summer and autumn (Halligan 1975).

Yet another group of secondary compounds are responsible for disease resistance at the pre-infectional or post-infectional stage. The distinction is somewhat arbitrary since pre-infectional compounds can undergo significant post-infectional changes. Their classification is shown in Table 5. While the phytoalexins are most active against fungal infection, they are also able to react to bacterial and viral infections. In addition, they can also be formed abiotically under stress conditions, including temperature shock and wounding (Harborne 1988). Plants of the arid and semi-arid regions, like the rest of the plant kingdom, are subject to herbivory and infestation, and are equipped with the appropriate secondary biochemicals to overcome such stresses with varying degrees of efficacy. The terpenoid resin produced by *Grindelia camporum** is believed to function as an antifeedant and is produced in sufficient quantities to be considered a potential source of biocrude (Timmermann and Hoffmann 1985). Some of these anti-predatory and allelopathic phytochemicals will be mentioned, as appropriate, elsewhere in the text (Chapt. 11).

CHAPTER 10

Anatomical and Morphological Adaptations

Metcalfe (1983) lists the following anatomical and morphological characters as being xeromorphic:

1. Leaves may be small, sometimes with the margins inrolled, as in *Erica cinerea* and *E. tetralix*.
2. Characteristically, in many xerophytes the leaves become detached, and water loss is thereby reduced.
3. In some taxa, such as *Salicornia* spp., leaves are absent and photosynthesis is carried out by the stem. In some cases the stem chlorenchyma may then consist of palisade cells.
4. Water loss may be controlled by the stomata, as in *Flourensia thurifera*. The stomata are often numerous and frequently protected by being depressed, surrounded by flanges of cuticle, etc.
5. The leaf cells, especially the epidermal cells, are often conspicuously smaller than corresponding cells in mesophytes and hydrophytes. However, there are exceptions, such as, for example, the enlarged water-storage cells in the epidermis or hypodermis of *Peperomia incana* and *P. pereskiaefolia*.
6. The cuticle is often above average thickness, as are also the walls of the epidermal cells, especially the outer periclinal walls. This is well developed in members of the Cactaceae.
7. Many xerophytes have hairy leaves and are generally to be found in members of the Caryophyllaceae, Chenopodiaceae, Compositae, Ericaceae and Solanaceae. In some xeromorphs these are air-filled or of a distinctive type. It was once believed that the indumentum gave a physiological advantage to xerophytes by shielding the stomata and thereby reducing transpiration loss. In many cases it has been experimentally demonstrated that such an hypothesis is false.
8. The proportion of assimilatory tissue in either leaf or stem is often higher than usual in xerophytes and is commonly achieved by an increase in the number of layers of palisade cells. Additional palisade cells also occur when leaves become isobilateral, as in *Nerium oleander**, instead of dorsiventral in structure.

9. The surface area is sometimes reduced in proportion to the leaf volume. Thus, the adaxial surface may be reduced to a groove or even eliminated altogether, thereby producing centric leaves, which are strictly circular in cross-section, albeit the surface may be ridged or furrowed. In such terete leaves the chlorenchyma is usually restricted to the peripheral tissues and the component cells are often, but not always, of the palisade type and radiately arranged around the cross-section, as in *Hakea sulcata*, *Isopogon* sp. and *Salsola* sp.
10. In some xerophytes the normal mesophyll structure is reversed, so that the palisade tissue is restricted to the abaxial side of the leaf. The network of veins is then often greatly compacted and bundle sheaths, which may consist of either water storage or mechanical tissue, are often developed.
11. In leaves with the highly distinctive Kranz-type structure (see Sect. 8.1) the bundle sheaths are of a special type.
12. Cells with sclereidal idioblasts in the mesophyll and sclereids at the vein endings are believed to prevent tissue damage from desiccation.

10.1 General Anatomical and Morphological Adaptations

10.1.1 Leaf Adaptations

The shedding of leaves is a common means of reducing evaporation. Since the majority of broad-leaved species are deciduous any xeromorphic adaptation would appear unnecessary (Eiten and Goodland 1979). For many trees and shrubs the leaves are *deciduous*, falling either at the end of the rainy season or *drought deciduous*, falling at the onset of drought, as in *Retama raetam** and *Zygophyllum dumosum*. *Fouquieria splendens** may even produce three or four sets of leaves each year, depending on the distribution of the rainfall. It has already been noted that in many xeromorphic plants a reduction in transpiration under dry conditions is obtained by a reduction in actual leaf area (see Sect. 5.1.2). The greatly reduced leaves of *Parkinsonia florida* and *P. microphylla** are compensated for by as much as 50% of the annual photosynthetic output being carried out by the green bark; this occurs similarly in the Chenopodiaceae (*Haloxylon* spp.*) and Euphorbiaceae (*Euphorbia* spp.*)

Other plants are *heterophyllous*, bearing *dimorphic leaves*, with relatively large leaves during the rainy season and smaller leaves and often with increased pubescence during the dry season. This characteristic is to be found in a number of families, including the Chenopodiaceae (*Atriplex hymenelytra**, *Nitraria retusa**, *Noaea mucronata*, *Rhagodia baccata**, *Salsola kali*), Compositae (*Artemisia monosperma**, *Encelia farinosa**, *Seriphidium herba-alba**),

Labiatae (*Coridothymus capitatus**, *Hyptis emoryi**, *Salvia fruticosa**, *Satureja thymbra**) and Leguminosae (*Alhagi graecorum**).

The leaf *indumentum* is believed to shade the chloroplasts, reduce the heat load and lower leaf temperature, as for example in *Chrysothamnus nauseous**. Leaf pubescence has also been shown to have an important influence on the spectral properties of the leaves of the three Mediterranean shrubs *Cistus incanus*, *C. salviifolius* and *Salvia fruticosa**. Since transpiration cannot always overcome the plant's heat load, the three shrubs have apparently developed leaf characteristics that help to cool them. These include a reduction in surface area by producing smaller leaves and folding the leaf margins, changing leaf properties by differences in indumentum densities, and the ability to expose the lower leaf surface under conditions of water stress. Despite their similar reflectance, the more pubescent leaves of *Cistus incanus* absorbed more energy in the 0.4- to 0.5-μm waveband than the less pubescent leaves of *C. salviifolius* (Gausman and Quisenberry 1990). Similarly, a narrow-leaved, more hairy form of *Aerva javanica** occurs in harsh environments and a broader, less hirsute form under less harsh conditions (Lekhak et al. 1983). The presence of a waxy leaf surface, as in *Copernicia prunifera**, effectively reduces water losses during the dry season and, in the above example, appears to be a response to low humidity since it is not formed if the humidity is too high. The leaves may inroll in order to protect the stomata, as for example in *Cutandia dichotoma**, *Helianthemum lippii** and *Hippophae rhamnoides**, or fold, as in *Andropogon gerardii** and *Distichlis spicata**.

The steep angling of the leaves of *Atriplex hymenelytra** during the summer months increases reflectance and thus reduces midday solar inception (Mooney et al. 1977), and similarly for the phyllodes of *Acacia* spp. such as *A. craspedocarpa*. The north-south orientation of the terminal pad of *Opuntia basilaris** likewise ensures maximum absorption of solar radiation during the winter months (Turner et al. 1995). In the Cactaceae the stem surface to volume ratio of *Lophocereus schottii** is correlated with rainfall, with the thicker stem and lowest surface to volume ratio occurring under lower rainfall (Felger and Lowe 1967). In the halophytic members of the Chenopodiaceae such as *Atriplex* spp.*, the bursting of salt-excreting, vesicular hairs deposit salt on the leaf surface, where their reflectance is believed to reduce absorption of solar radiation (Danin 1983; Metcalfe 1983; Archibold 1995).

Sundberg (1985) reported low stomatal frequencies associated with succulents (32.64 mm^{-2}) and high frequencies with non-succulents (165.03 mm^{-2}). Since stomatal values in most plants range from ca. 100–300 mm^{-2}, it is clear that succulents are characterised by a reduced number of stomata per unit surface area, which effectively reduces water losses from transpiration. As transpiration in succulents is highest at night and most, if not all, succulents are CAM plants, the relatively few stomata are mainly open at night anyway.

Also, according to Sundberg (1985), the stomata of succulents are relatively large in order to counter their low density. Furthermore, because of the generally rounded shape of a succulent's transpiring surface, the boundary layer resistance is greater than the resistance to gas diffusion provided by the stomata; consequently the large stomata are effective regulators of gas exchange over their entire range of surface apertures. The stomata densities of non-succulents fall within the typical range of densities for more mesophytic plants and include some of the highest densities recorded: *Buddleja saligna* 580 mm^{-2}, *Guaiacum coulteri** 523 mm^{-2}, *Pithecellobium flexicaule* 457 mm^{-2}, and *Ambrosia deltoidea** 440 mm^{-2}. As a consequence, the stomata would be expected to be materially smaller, but in practice they were less than expected. Sunken stomata are a fairly common method of reducing transpiration losses, and are presumably even more effective when combined with a thick, overhanging cuticle, as in *Boscia senegalensis** and *Chilopsis linearis**.

10.1.2 Stem Adaptations

Species with small, vestigial or caducous leaves (*apophyllous*), the photosynthetic function being undertaken by the stem, are widely present in the arid and semi-arid regions and are to be found in a number of families including Asclepiadaceae (*Leptadenia pyrotechnica**, *Sarcostemma viminale*), Asparagaceae (*Asparagus* spp.), Burseraceae (*Commiphora* spp.), Cactaceae, Casuarinaceae (*Allocasuarina decaisneana**, *Casuarina cristata*) and Chenopodiaceae (*Arthrocnemum macrostachyum**, *Halocnemum strobilaceum**, *Haloxylon* spp., *Maireana aphylla**, *Salicornia europaea**, *Salsola imbricata*, *Sarcocornia pilansiae*). In other examples, such as the deciduous *Geoffroea decorticans**, the young stem and shoots are initially green and photosynthesising and later develop a bark which 15–25 years later peels away to reveal a green and photosynthesising underbark.

Metcalfe (1983) cites anatomical studies in South America by Böcher and Lyshede (1968, 1972) where four groups of apophyllous species were recognised:

1. Transverse sections of the stem show a continuous chlorenchymous cylinder formed by many layers of parenchyma and uninterrupted by sclerenchyma. The stomata are contained in pits in the thick cuticle. This occurs in the Scrophulariaceae (*Monttea aphylla**, Fig. 18).
2. The chlorenchyma cylinder is interrupted by the sclerenchyma. This is found in members of the Compositae, Leguminosae subfamily Mimosoideae and Verbenaceae, e.g. *Neosparton aphyllum* (Fig. 19).

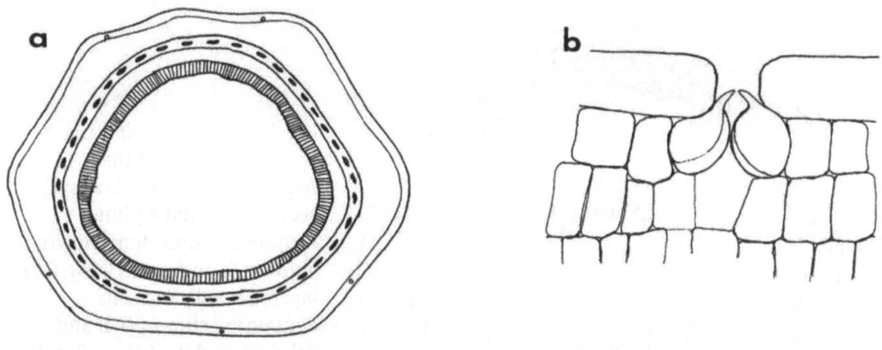

Fig. 18. *Monttea aphylla*, an aphyllous shrub of the Argentinian Monte. **a** transverse sectionthrough a photosynthetic young stem exhibiting typical structure for Scrophulariaceae showing epidermis with stomata, assimilative cortical parenchyma, pericycle with bundles of fibres, phloem and xylem cylinders and large central pith. **b** Detail of a stomata on a young branch protected by thick cuticle, large front cavity to stomata and palisade cells of outer cortex. (Redrawn and modified from Cabrera 1961, plate xxv, A–B)

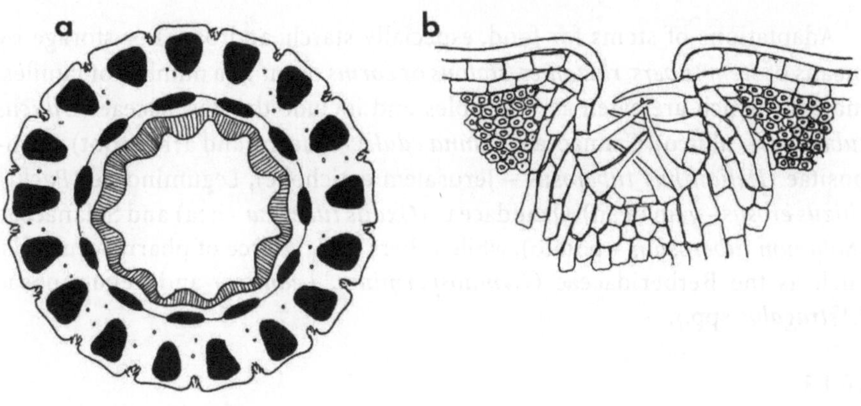

Fig. 19. *Neosparton aphyllum*, an aphyllous shrub of the Argentinian Monte. **a** Transverse section through a photosynthetic branch showing numerous shallow, longitudinal furrows alternating with bundles of sclerenchymous fibres; other bundles occur near phloem cylinder. Thick cuticle is present between furrows and absent in furrows. **b** Details of furrow with stomatal pit and unicellular trichomes. (Redrawn and modified from Cabrera 1961, Plate xxiv, A–B)

3. The stem is furrowed and the sclerenchyma confined to the furrows. This occurs in members of the Casuarinaceae, Celastraceae, Compositae and Leguminosae, e.g. *Senna aphylla* (syn. *Cassia aphylla*) (Fig. 20)
4. The stems are flattened and either form cladodes with a leaf-like structure or they are winged. This occurs in the Compositae (*Baccharis* spp.) and Leguminosae (*Genista*).

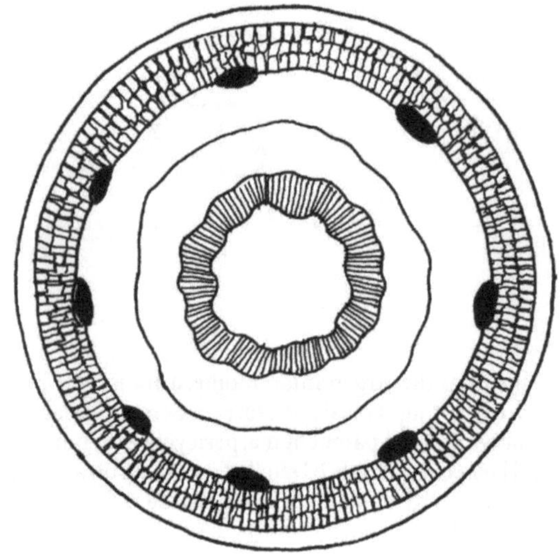

Fig. 20. *Senna aphylla*, an aphyllous shrub of the Argentinian Monte. Transverse section through a photosynthetic young stem showing epidermis and underlying three layers of pallisade cells containing chlorophyll and calcium oxalate druses. Bundles of sclerenchymous fibres are present between inner and outer cortex. Phloem and xylem are continious and enclose a pith with red pigment. (Redrawn and modified from Cabrera 1961, plate xxvi, H)

Adaptations of stems for food, especially starch, and/or water storage by means of *stem tubers, rhizomes, stolons* or *corms* occur in a number of familes, many of which are eaten as vegetables and include the Basellaceae (*Ullucus tuberosus* – ulluco), Cannaceae (*Canna edulis* - Queensland arrowroot), Compositae (*Helianthus tuberosus* – Jerusalem artichoke), Leguminosae (*Pachyrhizus erosus* - yam bean), Oxalidaceae (*Oxalis tuberosa* - oca) and Solanaceae (*Solanum tuberosum* – potato), while others are a source of pharmaceuticals, such as the Berberidaceae (*Gymnospermium, Leontice*) and Leguminosae (*Astragalus* spp.).

10.1.3 Root Adaptations

In arid regions *phreatophytes* rely on their roots reaching the water table for their survival. The access to soil water allows fairly rapid rehydration following diurnal transpiration losses, thus permitting photosynthetic activity to be maintained during dry periods. Nevertheless, many plants, such as *Acacia* spp. and *Prosopis* spp., are deciduous during the dry season, while a few, such as *Conocarpus lancifolius**, are evergreen. Except for phreatophytes, deep tap roots are generally the exception. For example, the swollen, carrot-like taproot of *Adansonia digitata** is short-lived, the tree relying on extensive lateral roots for stability, nutrients and water (Wickens 1982b). However, a distinction should be drawn between obligate phreatophytes which continue to tap the perennial water-table throughout the year and facultative phreatophytes whose deep roots are in contact with deep but seasonally fluctuating ground water.

The former favours a more mesophytic canopy and the latter a deciduous or more xerophytic foliage. Only the rigorous confirmation of water uptake by roots from the capillary fringe of a permanent water-table should qualify a plant as being a true phreatophyte (Smith et al. 1997). The ability of non-phreatophyte shrub species with deep roots and grasses with shallow roots to utilise the soil moisture from infrequent, long duration pulses and frequent, short duration pulses of water availability, respectively, and the effect on their phenology is discussed in Section 3.11.2.2.

In the Chenopodiaceae roots exposed by deflation are believed to be safeguarded against desiccation by the scattering of the xylem and phloem bundles, so that even when the outer bundles cease to function the inner ones will continue to function, as in *Anabasis articulata* and *Cornulaca monacantha** (Danin 1996), while a thick corky bark protects the exposed taproot of *Convolvulus lanatus** and *Cornulaca monacantha** from desiccation (Danin 1983).

The bulbous *Hymenocallis litoralis* (?syn. *H. amancaës*), which may be cultivated as an ornamental, exhibits a pronounced *root dimorphism*, considered by Rauh (1985) as typical of many plants of the arid zone. Besides normal roots, which function solely as contractile roots, negatively geotropic roots develop inside the old dead bulb scales. They eventually penetrate the scales and grow below and parallel to the soil surface. It is only these roots, spreading through the upper 2 cm of soil that are able to take up water from soil moistened by nightly fog water. Root dimorphism is also exhibited by members of the soil cacti, as for example in *Neoporteria napina* (syn. *Neochilenia napina*) with its large napiform storage root supporting a narrow, subsurface collar and the extremely shallow lateral roots lying a mere 10-20 mm below the surface (Hoffmann 1989; von Willert et al, 1992).

The mucilaginous secretions from persistent root hairs and culms cause sand particles to agglutinate and form a *rhizosheath*. They are reported for a number of species growing on sandy soils, especially among the Gramineae [perennial species: *Aristida purpurea**, *Centropodia forsskaolii*, *Dichanthium foveolatum*, *Elymus farctus*, *E. lanceolatus* (syn. *Agropyron dasystachyum*), *Hilaria rigida*, *Lasiurus scindicus**, *Oryzopsis hymenoides**, *Panicum turgidum**, *P. urvilleanum*, *Pennisetum divisum*, *Phragmites australis*, *Stipa comata**, *S. pellita*, *Stipagrostis acutiflora*, *S. ciliata*, *S. hirtigluma*, *S. lanata*, *S. obtusa*, *S. plumosa*, *S. pungens**, *S. scoparia**; annual species: *Avena barbata* subsp. *wiestii*, *Bromus tectorum**, *Corynephorus divaricatus*, *Cutandia memphitica*, *Rostraria smyrnacea*, *Stipa capensis*, *Trisetaria linearis*, *Vulpia pectinella* (syn. *Ctenopsis pectinella*)]. Other members in the Monocotyledons are Asphodelaceae (*Asphodelus ramosus*), Cyperaceae (*Carex pachystylis*, *Cyperus arenarius*, *C. capitatus*, *C. conglomeratus*, *C. macrorrhizus*, *Scirpus holoschoenus*) and Juncaceae (*Juncus acutus*, *J. bufonius*). The Dicotyledons are less well represented, possibly because they have not been sought for, and include Boraginaceae (*Tiquilia plicata*), Chenopodiaceae (*Atriplex polycarpa**, *Haloxylon aphyllum**, *H. scoparium**, *Sarcobatus vermiculatus**), Compositae [*Launaea residifolia, Senecio glaucus* subsp. *glaucus* (syn. *S. joppensis*)], Convolvulaceae (*Calystegia soldanella*), Polygonaceae (*Polygonum palaestinum*) and Tamaricaceae (*Tamarix ramosissima*) (Danin 1996). In the Gramineae Danin (1996) found rhizosheaths on all perennial grasses inhabiting desert dunes; they appear to be restricted to plants growing on sandy soils.

Their function is still subject to speculation and requires further research. Why rhizosheaths are confined to plants growing on sandy soils and not those on heavier textures also requires investigation. Using acetylene reduction, Wullstein et al. (1979) found a relatively high rate of nitrogen fixation within the rhizosheaths of *Aristida purpurea**, *Elymus lanceolatus, Oryzopsis hymenoides** and *Stipa comata**, although the actual uptake of nitrogen was not demonstrated. Rhizosheaths are also believed to protect roots from abrasion by sand grains when exposed by deflation (Danin 1996). In the Chenopodiaceae it is suggested that they enable the root hairs to continue to function during drought (Oppenheimer 1960), while Volkens (1887) and Price (1911) believed that the mucilage concentrated moisture around the sheath, thereby promoting water uptake in addition to providing protection against desiccation; certainly Wullstein et al. (1979) and Wullstein (1991) demonstrated the water content of rhizosheaths to be four times that of control sand samples, the rhizosheaths providing a nano-climate wherein nitrogen-fixing bacteria are able to survive.

A somewhat similar situation occurs in numerous South African grasses and in the herbs *Commelina* spp. and *Gerbera* spp., where the inner cortex of the root eventually become spongy due to the perforation of the disintegrating cell walls and forms dead sleeves; on contact with water the roots of *Bromus rubens* were immediately able to absorb 12% of their weight by capillary action, while in the rhizomatous perennial grass *Lygeum spartum** the dead roots are strongly hydroscopic and capable of absorbing 100% of their original weight from moist air (Oppenheimer 1960).

The ability to produce *adventitious roots* on stems buried by shifting sands occurs in a number of families. including Anacardiaceae (*Pistacia lentiscus**, *P. terebinthus**), Asclepiadaceae (*Leptadenia pyrotechnica**), Chenopodiaceae (*Sarcocornia pilansiae*), Compositae (*Artemisia monosperma**), Labiatae (*Coridothymus capitatus**, *Satureja thymbra**), Rhamnaceae (*Rhamnus lycioides*), Rosaceae (*Sarcopoterium spinosum*) and Tamaricaceae (*Tamarix ramosissima*).

Some plants have the ability to trap sand either within the plant canopy or on their leeward side and to grow up through the accumulated sand, known as *nebkas*, either by producing adventitious roots on the buried branches or maintaining their original rhizosphere. Such nebka-building plants are widely distributed both globally and taxonomically. They have been reported in the Chenopodiaceae (*Atriplex halimus**, *Halocnemum strobilaceum**, *Salsola nollothensis, Sarcobatus vermiculatus**), Compositae (*Artemisia monosperma**), Cyperaceae (*Cyperus macrorrhizus*), Ephedraceae (*Ephedra strobilacea*), Euphorbiaceae (*Croton californicus*), Gramineae (*Aeluropus lagopoides**, *Ammophila arenaria**, *A. breviligulata, Cutandia memphitica, Elymus farctus, Hilaria rigida, Panicum urvilleanum, Stipagrostis acutiflora, S. ciliata, S. obtusa, S. pungens**, *S. scoparia*), Leguminosae (*Retama raetum**), Loasaceae (*Petalonyx thurberi*), Plumbaginaceae (*Limoniastrum guyonianum, Limonium axillare*), Polygonaceae (*Calligonum polygonoides* subsp. *comosum**), Rhamnaceae (*Ziziphus lotus**, *Z. spina-christi**), Tamaricaceae (*Tamarix* spp.) and Zygophyllaceae (*Nitraria retusa**, *N. schoberi, Zygophyllum album*) (Le Houérou 1986; Walter and Breckle 1986; Danin 1996).

The shallow feeder roots or rain roots of *Citrullus colocynthis, Maireana pyramidata** and *M. sedifolia** cease to function during the dry season and are shed. Adaptations of roots for food and/or water storage by means of swollen taproots and *root tubers* are to be found in the Anacardiaceae (*Spondias tuberosa**), Bombacaceae (*Ceiba acuminata**), Caricaceae (*Carica corumbaensis*), Chenopodiaceae (*Beta* spp. - beet), Compositae (*Tragopogon porrifolius* - oyster plant), Cruciferae (*Brassica* spp. - turnips, rutabagas, etc.; *Raphanus* spp. - radishes), Cucurbitaceae (*Cucurbita foetidissima**, *Marah gilensis**), Euphorbiaceae (*Manihot glaziovii*), Leguminosae (*Dolichos fangitsa, Luetzelburgia auriculata, Tylosema esculentum**), Pedaliaceae (*Harpagophyton procumbens**, *Proboscidea peruviana*) and Umbelliferae (*Daucus* spp. - carrots).

10.2 Cushion Plants

Cushion plants are mainly dwarf, somewhat woody plants with a much branched, compact, hemispherical habit where, in extreme forms, such as *Azorella compacta** (Umbelliferae) the tightly packed branching presents a firm, apparently unbroken, hemispherical surface reminiscent of a cauliflower (Fig. 21). However, they are not a very well defined group owing to the degree of compactness and shape, with intermediate forms with closely packed, long stems, as in *Artemisia,* and more laxly branched, subglobose forms, as in Andean members of the Compositae (*Parastrephia lucida*) and Verbenaceae (*Junellia seriphioides*, syn. *Verbena seriphioides*), merging with other life forms, and they may even be confused with nebka-forming species.

The cushion habit occurs largely in cold temperate regions, including high montane regions; for example, Went (1971) recognises over 40 genera with

Fig. 21. A young plant of *Azorella compacta* (Ilareta) forming a cushion 30 cm in diameter, growing at 5200 m near Ollague, El Loa Province, Chile. Mature plants may grow to form a cushion 4.5 m in diameter and live for 3000 years

cushion plants in the Patagonian lowlands, the Andes and New Zealand Alps. The cushion habit occurs less frequently in tropical deserts, such as the Saharan *Fredolia aretioides**. It is to be found in a number of families, including the Caryophyllaceae, Chenopodiaceae (*Anabasis articultata*, *Fredolia aretioides**, *Halocnemum strobilaceum**), Compositae, Cruciferae, Leguminosae (*Astragulus microcephalus**), Malvaceae, Primulaceae, Rosaceae, Scrophulariaceae, Umbelliferae (*Azorella* spp.) and Verbenaceae (Metcalfe 1983). It has been suggested that the cushion habit offers protection against the wind and high solar radiation. In cold deserts the compact structure and cushion habit can be expected to provide mechanical strength against compaction by heavy snow falls. Ralph (1978) considered that the cushion habit of the high Andean *Azorella compacta** was an adaptation to an alpine desert environment which, by reducing the plant's surface to volume ratio, reduces water vapour loss and conserves heat - on sunny days the plant's interior was decidedly cooler to the touch. The moisture in the damp interior could act as a reservoir as well as enhancing the insulating properties of the plant's biomass. It is possible that the resin content of the plant's tissue could act as an antifreeze.

The densely ramified and cushion habit are regarded as morphological adaptations for restricting transpiration and possibly also as an adaptation to potent radiation and high winds, both being reduced by the dense branching (Oppenheimer 1960). Compared with more succulent perennial herbs, the lignified cushion habit must also offer better survival advantages against the weight of winter snow. The lignification of these xeric shrubs and subshrubs, in what are regarded as mainly herbaceous families, Caryophyllaceae, Compositae, Solanaceae, Umbelliferae, Verbenaceae, etc. (Hutchinson 1959), may perhaps be interpreted as an alternative to turgidity in fleshy leaves. Furthermore, the leaves of these cushion plants often exhibit considerable reduction in surface area, resulting in a linear shape and, in some species, decurrent leaves; pubescent, coriaceous or glutinous leaf surfaces are the norm, such morphological characters being accepted means for reducing water losses under xeric conditions.

10.3 Non-succulents

10.3.1 Leaf Adaptations

For some plants the leaf blade is either reduced or absent and photosynthesis undertaken by *phyllodes*, i.e. modified petioles. Phyllodes are particularly prevalent among Australian species of Compositae, Labiatae, Leguminosae, Myrtaceae, Pittosporaceae, Proteaceae, etc., those of *Acacia* in Australia being particularly well documented (Figs. 22 and 23). Stipular spines are usually absent and the phyllodes are usually accompanied by a complete loss of the rachis and leaflets, although in some species the leaflets are present in subgenus *Heterophyllum* on juvenile plants or vestigial as in *A. victoriae** (Pedley 1981; Maslin 1988). As a general rule, thin, cylindrical phyllodes predominate in arid areas, while in more mesic areas the phyllodes tend to be relatively broad and flat. However, both forms may occur in the same habitat or even on the same bush in relation to their position to sun or shade (Walter and Breckle 1986). *Acacia willardiana** from the Sonoran Desert is of particular interest as being the only phyllodenous *Acacia* in the New World, although it is somewhat anomalous in occasionally having bipinnate leaves, with the rachis as well as the petiole phyllodenous, perhaps an unfulfilled evolutionary trend. Alternatively, the stems may be flattened and leaf-like in appearance and are termed *cladodes*, as in *Euphorbia enterophotra* and *Opuntia ficus-indica**, or they may be winged, as in species of *Baccharis* and *Genista*.

Selective aphylly or *branch dimorphy* occurs in the Saharan composite *Launaea arborescens*, where the outer 'sun branches' are aphyllous and inner

Fig. 22. *Acacia holosericea.* Flowering branch and details of a mature phyllode and fruits. (Adapted from drawing in Simmons 1981, p. 279)

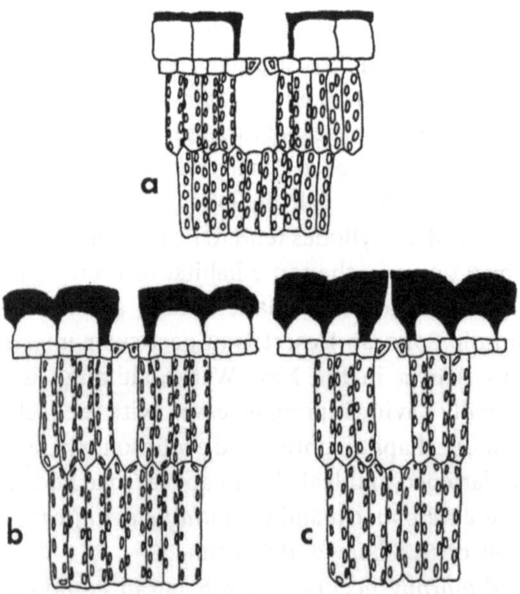

Fig. 23. Transverse sections through phyllodes of *Acacia craspedocarpa*. **a** Young and unlignified. **b** Mature, green. **c** Mature, yellowish-green. Note increase in cuticle thickness and development of rugose adaxial surface with increase in age, and development of epistomatal cavity in c. Guard cells are attached to hypodermal cells, and chloroplasts in pallisade cells are large and relatively few. (Hellmuth 1969)

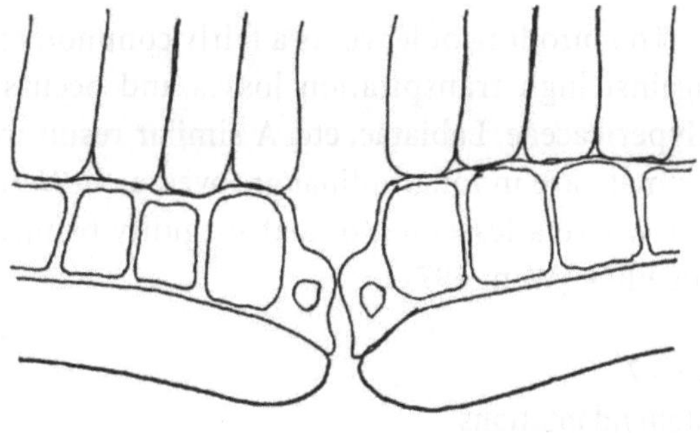

Fig. 24. *Larrea divaricata* subsp. *divaricata*. transverse section of lower leaf epidermis showing thick, glutionous coating to single layer of epidermal cells, unicellular trichomes and external stomata with a small, outer atrium. Homogenous mesophyll has two to three layers of pallisade cells (Redrawn from Cabrera 1961, Plate xxvi, M)

branches bear leaves. Similarly, in the spinose and intricately branched, Andean shrub *Adesmia spinosissima*, the outer branches are likewise bare and the leaves and flowers confined to the inner branches. The leaf rachi or spinescent stipules may persist after the lamina has fallen and serve to protect young growth in the following growing season, as in the legumes *Astragulus* spp., *Caragana* spp. and *Halimodendron halodendron**. This appears to be a relatively common characteristic of cold desert species.

The tough, leathery leaves of *sclerophytes* have long been believed to have arisen in response to drought; consequently they could be considered as xerophytes, but with the additional implication of being particularly characteristic of areas with hot, dry summers and cold winters, such as the Mediterranean region. The leaves of sclerophytes show a number of what might be considered xeromorphic features, such as thick outer walls to the epidermal cells and overlying cuticle, although the stomata are not always depressed. For example, epidermal cells with thick outer walls, thick cuticles and high stomata density and abundant sclerenchyma have been found in Mediterranean maquis species of *Olea*, *Pistacia* and *Styrax*. On the other hand, the plants do not always show the expected high proportion of typical palisade tissue and congested veins. The stomata are not depressed in *Larrea* spp., but it has been suggested that the oily coating to the leaf would flow into the stomata if they were depressed (Fig. 24); spongy tissue is also absent and otherwise obvious xeromorphic characters appear to be lacking (Smith et al. 1997).

Although historically the Mediterranean sclerophytes were among the first to receive attention, sclerophytes are now known from other parts of the world. For example, they occur in some 20 families in Australia, the more important being Epacridaceae, Leguminosae, Myrtaceae and Proteaceae, and less numerously among the Dilleniaceae, Euphorbiaceae and Rhamnaceae. It is now evident that sclerophytes also occur in more mesophytic habitats and should not be regarded as xerophytes, although they can, because of their structure, be regarded as xeromorphic (Metcalfe 1983).

The inrolling of leaves is a fairly common means of protecting the stomata against high transpiration losses and occurs in the Cistaceae, Gramineae, Hypericaceae, Labiatae, etc. A similar result may also be achieved where the stomata are in longitudinal grooves, as in *Nerium oleander** and *Nolina parryi*, where a lessening of leaf turgidity brings the lips of the grooves closer together (Went 1971).

10.3.2
Stem Adaptations

In extreme cases of drought intercalary cork may be formed between the stem segments, as in members of the Chenopodiaceae (*Atriplex torreyi**, *Haloxylon persicum**, *H. salicornicum**, *Noaea mucronata*), Compositae (*Rhanterium epapposum*, *Seriphidium herba-alba**) and Rosaceae (*Coleogyne ramosissimus**), thereby creating multiple stems so that actual limbs may be shed, thus enabling the plant's water resources to be concentrated in a section of the canopy (Gintzburg 1963; Danin 1983).

In others the leaves may persist as rudimentary scales, as in *Ephedra*, and the plant aquires a broom-like habit. In such *switch-plants* the function of photosynthesis may be transferred to other organs, such as the stem; the usually cylindrical stems of these *aphyllous* species may be longitudinally ridged or furrowed, with stomata confined to the protection of the ridges or furrows. In the Didiereaceae* the photosynthetic function of the soon deciduous leaves of the long shoots is replaced by the short shoots as the photosynthetic organ (Rabesandratana 1984). The stems of aphyllous plants are often mechanically supported by axially aligned radiating plates or groups of fibres; alternatively the fibres may be confined to the crests of the longitudinal ridges (Metcalfe 1983).

Selaginella convoluta has the ability to retract its side shoots during the dry period and extend them again when the rain comes (Cabrera 1955). When dehydrated, the water content of this species was found to be only 9% of that when fully saturated. According to Andrade-Lima (1981) and Smith et al. (1997) a number of *Tillandsia* spp. are capable of taking up atmospheric moisture through their scale-like hairs.

10.4
Succulents

Succulent is a term that is widely applied to plants which have a high proportion of parenchyma tissue in which they are able to store water and have a lowered surface to volume ratio, thereby reducing water loss. Such plants are generally considered xeromorphic and are associated with arid or saline environments, but not necessarily so. For example, in the past numerous Cactaceae have been successfully cultivated at the Jodrell Laboratory, Kew, with their lower portions immersed in bottles of tap water and have remained perfectly healthy for a number of years (Metcalfe 1983).

For non-halophytes succulence is a means by which plants can survive in habitats that endure periodic droughts. Such plants have been defined by von Willert et al. (1992) as follows:

"*A succulent (or succophyte) is a plant possessing at least one succulent tissue. A succulent tissue is a living tissue that, besides possible other tasks, serves and guarantees an at least temporary storage of utilisable water, which makes the plant temporarily independent from external water supply when soil water conditions have deteriorated such that the root is no longer able to provide the necessary water from the soil.*"

While water-storage tissue is always present in succulents, the cellular organisation within the plant, whether it be a leaf or stem, is not uniformly sited but varies according to taxonomic affinities. Thus, in many members of the Aizoaceae, the epidermal cells are greatly enlarged, but it is more usual to find the water-storage tissue in the central core of the organ, as in *Didierea madagascariensis*. The translucent tissue is often surrounded by a transitional zone between itself and the chlorenchyma (Metcalfe 1983).

Columnar stem succulence is found in the families Cactaceae, Didiereaceae, Euphorbiaceae and Fouquieriaceae. The scattered presence of columnar succulents in an otherwise treeless, 'science fiction' landscape requires some explanation. Their scattered distribution can be assumed to indicate the plant's soil water requirements as being the limiting factor. Height would appear to be an a dvantage as a means of reducing evapotranspiration due to air temperatures decreasing with height above ground level. The aerodynamics of the bristly appearance presented by numerous short shoots could be a further means of reducing water losses. The argument could be further theorised to suggest that in arborescent growth the canopy is due to competition for light.

10.4.1 Leaf Succulence

The function of a succulent is to store utilisable water, with the functioning of the water-storage tissue being determined by the growth form of the plant. Thus, short-lived annuals or deciduous plants generally have leaves in which, apart from the vascular tissue, stomata and parts of the epidermis, all the parenchyma cells store water and photosynthesise. There is no separation of cell function. The leaves usually do not show cyclic refilling of the tissue. Their utilisable water is mostly used irreversibly and its withdrawal leads to the death of the plant. Such plants, e.g. *Mesembryanthemum sladenianum* (syn. *Prenia sladeniana*), have what are termed *all-cell succulent leaves*. Like many other members of the Aizoaceae, the ephemeral *Mesembryanthemum crystallinum** has epidermal idioblasts or bladder cells which are used for water storage (von Willert et al. 1992). Anatomical strengthening may be present to prevent collapse on loss of turgidity, thus the presence of thick sclereids in the internal palisade sclerenchyma of *Arthrocnemum macro-stachyum** (Walter and Breckle 1986).

Non-deciduous and evergreen plants may have leaves that contain only one or rarely two succulent tissues, such as the epidermis, central or peripherous parenchyma. There is a separation of water storage and photosynthetic functions. Their utilisable water is rechargeable. Such plants, e.g. *Gasteria pillansii* have what are termed *partially succulent leaves* (von Willert et al. 1992).

Leaf succulence is a means of reducing the transpiration area; thus *Agathophora alopecuroides** is able to dramatically reduce its transpiration surface by changing from cylindrical leaves to globose (Danin 1983). In the halophytic members of the Chenopodiaceae the leaves may be caducous, vestigial or succulent; succulent leaves occur in *Kalidium foliatum**.

In the living stones, *Lithops* spp.*, the entire plants are embedded in the soil, thereby providing some insulation against surface temperatures. The leaf apices of the succulent leaf pair are often provided with transparent 'windows', the clarity of which causes significant variations in leaf temperature and is regarded by Turner and Picker (1993) as probably being the sole thermal adaptation to a hot environment. In the Zygophyllaceae the cylindrical petioles may persist for up to 3 years; during the dry summer the petioles are small and wrinkled but swell two- to three-fold following an effective rain, as in *Zygophyllum dumosum*.

10.4.2 Stem Succulence

Stem succulence is relatively common, occurring in a number of families including the Anacardiaceae (*Pachycornus discolor*), Apocynaceae (*Adenium obesum**, *Pachypodium* spp.), Bombacaceae (*Adansonia* spp., *Cavanillesia arborea, Ceiba parvifolia, C. pubiflora, Chorisia glaziovii, C. insignis*), Burseraceae (*Bursera hindsiana**, *B. microphylla**, *Commiphora monstruosa*), Euphorbiaceae (*Cnidiscolus urens*, syn. *Jatropha urens, Schinziophyton rautanenii**), Fouquieriaceae (*Fouquieria* spp.*), Hernandiaceae (*Gyrocarpos americanus*), Icacinaceae (*Pyrenacantha malvifolia*), Passifloraceae (*Adenia globosa**) and Xanthorrhoeaceae (*Xanthorrhoea thorntonii**), while *Haloxylon aphyllum** (Chenopodiaceae) is reported to have an extremely succulent bark which acts as a water reservoir (Schimper 1960).

CHAPTER 11

Selected Arid Land Plants, their Ecophysiology and Uses

There are probably at least 20 000 economic species in the arid and semi-arid regions of the world. The plants mentioned in this chapter are those whose adaptations or ecology have been discussed in the previous chapters. Many well-known plants in the economies of the developed countries have been omitted while many of what may appear to be of little economic importance have been included. However, it should be borne in mind that items that people of the developed world take for granted, such as string, soap, toothbrushes, medicines and pesticides, are often only available as raw plant products by peoples of the less developed arid and semi-arid lands. Some economic plants have been omitted either because of lack of information or because their form of ecophysical behaviour and adaptations appear sufficiently similar to others already described. Also, not all the economic plants mentioned earlier in the text are discussed here for reasons of space limitations.

As far as possible an attempt has been made to include plants from all the arid and semi-arid regions, even when the regions and their plants are poorly known. The abbreviations ZB and ZE are used for zonobiome and zonoecotone, respectively, sensu Walter and Breckle (1985). They are only given for plant distributions lying within the arid and semi-arid ecosystems; montane plant distributions outside the arid and semi-arid regions are not shown.

ANGIOSPERMS

Dicotyledons

ACANTHACEAE

Blepharis ciliaris, syn. *B. persica, B. edulis*: an annual or perennial herb, the leafy stems of which become coarse and spiny at maturity. Widely distributed across Sahelian Africa to Arabia (ZB and ZE II-III, III, IIIsr), it is eaten by livestock but less readily when dry. The persistent and densely bracteose inflorescences retain the seeds in the unopened capsule, enabling the seeds to remain protected from weather and predators for many years without loss of viability. Provided with a double fail-safe system, continuous wetting by rain or flood waters (dew is ineffective) is necessary for the bracts to part and expose the capsule, while the wetting of the capsule allows the tension previously built up in the septum of the capsule during ripening and drying to be released, causing the capsule to explode and eject the dry seed. When wetted, the adpressed, multicellular hairs coating the seed become erect in such a manner as to orientate the seed at a 30–45° angle to the soil surface, with the radicle end touching the soil. On contact with the soil the primary walls of the apical cells of the multicellular hairs rapidly produce a mucilaginous layer, which anchors the seed to the soil and aids germination. The dry seeds falling on the wet ground normally germinate within 50 min of being exposed to moisture, the radicles attaining ca. 49 mm within 24 h. However, with excessive wetting the cell walls of the multicellular hairs swell to such an extent as to fill all the spaces between the hairs, thereby acting as an oxygen barrier and thus preventing germination; on drying the seeds will germinate immediately. Thus, riverbed ecotypes develop a mucilaginous layer several millimetres thick within a few minutes of wetting, which both prevents germination and protects the embryo from damage by the flood waters.

Only a portion of the capsules from the mother plant eject their seed at a wetting, thereby ensuring a succession of seed to compensate for any earlier failure in establishment. Ecotypes with genotypic differences in germinability also exist under a wide range of environmental conditions (Gutterman 1967, 1972, 1982, 1983, 1990, 1993; Gutterman et al. 1969; Witztum et al. 1969; Danin 1983).

AIZOACEAE

Aizoon canariense: a prostrate, thickly pilose and tough annual or perennial herb. It is widely distributed from the Atlantic islands through Africa to Baluchistan (ZB and ZE II–III, III, IIIsr, IIIwr, III–II, III–IV, IV, IV–III). The Bedouins grind and cook the seeds as a gruel. The capsules are hydrostatic, opening after wetting, a strategy of seed dispersal in the following rainy season. At least a few seeds are dispersed by raindrops at each opening, on drying the capsule closes again (Osborn 1968; Gutterman 1993).

Lithops spp. (living stones): glabrous succulents embedded in the hot surface soils of south-western Africa (ZB and ZE III, III-II). They are well camouflaged in their stony surroundings, the colour variations being due to the orange-red chromoplasts in the epidermal and subepidermal cells as well as betalain pigments in the latter (Mabberley 1987). The plants are widely grown as ornamentals.

The species of *Lithops* produce annual growths of one or more obconical bodies composed of a pair of fleshy leaves on top of a dual column; leaf apices are flattened and often with transparent 'windows' where aqueous tissues of the body in the leaf meet the surface (Fig. 25). In areas of very high light intensity the plant apparently has a protective layer of calcium oxalate. The

Fig. 25. *Lithps lesliei* grown at Kirstenbosh, South Africa and at the royal Botanic Gardens, Kew. **a** Colony in flower and young fruit. **b** Plant before flowering **c** Diagrammatic, longitudinal section of a plant showing 'windows'. (From drawings in Kew archives of plants and Brown 1921, Figs. 110, 112)

solitary flower produced between the leaves terminates the season's growth, after which a new shoot develops in the axil of one or both leaves. The new shoots withdraw moisture from the old leaves, which then persist as withered sheaths.

Since the leaves are embedded in the soil, they are unable to utilise such common mechanisms for controlling leaf temperatures as orientating the leaf away from the sun or engage in self-shading. Instead, the thermal coupling of the leaf and surrounding soil combines the plant's thermal capacity with that of the soil and thus reduces daily variations in leaf temperature. The papery sheath surrounding *Lithops* has been suggested as providing insulation from the hot soil and thus keeping the leaf cool; however, it is now considered doubtful if such thermal decoupling could function in this way. The steep decrease in soil temperatures immediately below the surface keep the deeper, chlorenchyma-containing areas of the plant relatively cool. The variation in leaf colour found within the genus has little effect on leaf temperature. However, variation in window clarity causes significant variation in leaf temperature and is probably the sole thermal adaptation to hot environmental conditions (Turner and Picker 1993).

Mesembryanthemum crystallinum (ice plant): a sprawling, succulent, annual halophyte covered with translucent, glistening papillae; the epidermal idioblasts or bladder cells are used for water storage. It is a native of the Cape (ZE IV-III), now naturalised in the Mediterranean region, the islands of the Atlantic and California. The plant was formerly cultivated as a source of soda but is now grown as a bedding plant. The leaves may be eaten as a spinach (Bramwell and Bramwell 1974; Mabberley 1987).

A comparatively long vegetative phase is spent in accumulating water before beginning flowering at the start of the dry season; flowering and fruit production then continue until all the utilisable water has been consumed. The hydrochastic capsules open after wetting, and with each heavy rainfall a few seeds are dispersed by raindrops. Initially C_3, there is a transition in time to CAM; the regulation and sequence of steps of this reaction are still imperfectly understood (von Willert et al. 1992; Gutterman 1993).

Mesembryanthemum forsskålii, syn. *Opophytum forsskålii*: an erect, annual herb bearing fleshy, terete to subterete, linear leaves, distributed from Egypt to the Middle East and Arabia (ZB and ZE III, III–IV). In summer the Bedouin harvest the seeds by placing the dry plants in a container of water; the valves of the fruit then open and release the seeds, which fall to the bottom of the container. The ground seeds are made into a bread. The hydrochastic capsules open on wetting and release at least some of the seeds, which are then dispersed by raindrops. It is a C_3 plant which converts to CAM when under stress (Batanouny 1981; Danin 1983; Gutterman 1993).

AMARANTHACEAE

Aerva javanica, syn. *A. persica, A. pseudo-tomentosa*: a perennial, dioecious undershrub up to 1.5 m tall, branched from the base and bearing narrowly linear to suborbicular leaves up to 2 cm long. Lekhak et al. (1983) recognise two distinct ecological races of *A. javanica* in the Indian Desert: a narrow-leaved form that is widely distributed in sandy areas with harsh climatic conditions, and a broad-leaved form under the less harsh, semi-arid environment. Compared with the broad-leaved form, the narrow-leaved form showed a marked increase in anthocyanins, a thicker tomentum at maturity, both upper and lower surfaces exhibiting a two-fold increase in epidermal cells with a three-quarters' reduction in cell size, and a slight increase in the number of stomata accompanied by a slight reduction in size, all characters that are considered as aiding a reduction in transpiration losses. It is widely distributed throughout the drier parts of the Old World and adventive in Australia and elsewhere [ZB and ZE I-III, II-III, III, III(tI), III-II, III-IV, IV, IV-III]. The white, woolly seeds which are used for stuffing pillows and mattresses; the woolly, old inflorescences may be used as tinder (Danin 1983; Bhandari 1990).

ANACARDIACEAE

Pistacia lentiscus (mastic tree): a deep-rooted, evergreen, dioecious shrub or small tree up to 5 m tall. Stems buried by drifting sand have the ability to produce adventitious roots (Danin 1996). It is a Mediterranean species (ZB and ZE III–IV, IV, IV–III). As its name suggests, it is the source of mastic, an agreeable resin that has been used since Grecian times as a masticatory to quell halitosis, to preserve teeth and as a filler for dental caries; it is also used medicinally as a diuretic, stimulant and aphrodisiac. Other uses are for varnishes, especially for oil paintings, and to flavour date spirit (Oppenheimer 1960, Nuttonson 1961; Zohary 1962, Townsend and Guest 1980; Mabberley 1987).

It is a true evergreen tree, with shoots and leaves unreduced and physiologically active throughout the year. The summer transpiration rates and osmotic pressures are low (Zohary 1962). The sclerophyllous leaves contain xeromorphic features such as a thick cuticle, thick outer walls to the epidermal cells, high stomatal density and abundant sclerenchyma, characters that are also present in *Olea, Quercus* and *Styrax* (Metcalfe 1983).

Pistacia terebinthus, syn. *P. palaestina, P. terebinthus* subsp. *palaestina*: a winter deciduous, up to 6-m-tall shrub or small tree. Stems buried by drifting sand have the ability to produce adventitious roots (Danin 1996). It is widely distributed throughout the Mediterranean region (ZB and ZE III-IV, IV, IV-III), source of an oleo-resin known as Chian or Cyprus turpentine and formerly used as a stimulant and diuretic. The kernels are edible (Hedrick 1972; Townsend and Guest 1980; Wickens 1995b).

It has a relatively low osmotic pressure, although the summer transpiration rates are high. Under conditions of severe atmospheric and soil drought complete suspension of transpiration is possible. Mild winters considerably reduce bud dormancy, so much so that bud-break may sometimes start before the last leaves have been shed. The drought resistance of both species of *Pistacia* is due to the ability of their roots to penetrate porous or softened rock in their search for moisture (Oppenheimer 1960; Zohary 1962).

Spondias tuberosa (umbuzeiro or imbuzeiro): a low, spreading tree, up to 8 m high, with a shallow root system extending for a radius of 15–18 m from the trunk and bearing spongy, water-filled, spherical tubers of up to 20 cm in diameter, an interesting adaptation for the long dry season. It is endemic to north-eastern Brazil [ZB I(rIII)]. The root tubers, known locally as bucus, are a source of water for the thirsty traveller; the water can be under sufficient pressure to spurt out from the tuber when tapped. The tubers may also be eaten in times of famine. The fruits of umbuzeiro are edible and used in ice-cream, sherbets, jellies and jams, and also medicinally as a stomachic. The fruits are also a good source of vitamin C, especially when green. Although deciduous, the tree is reported to be a source of dry season forage (Mors and Rizzini 1966).

Ferri and Labouriau (1952) and Ferri (1954) remark on the complete absence of any anatomical feature of the leaves of *S. tuberosa* for protection against water loss, instead the plant relies entirely on physiological control. During the wet season the stomata open very early in the day, at about 9 a.m., but during the dry season they may sometimes not open for several days and show very low cuticular transpiration. The stomata are very sensitive to changes in both light and humidity. If a leaf is removed from the plant, thereby cutting off its water supply, the stomata will reduce transpiration by more than 50% within 2 minutes and close completely in 5 minutes; in darkness they close in 30 min. This photoactive movement is so efficient that under natural conditions leaves from different sides of the tree, and therefore exposed to different concentrations of sunlight, will show different degrees of stomatal opening.

APOCYNACEAE

Adenium obesum (desert rose): a succulent and very variable, deciduous shrub or small tree up to 5 m or more tall, the stem with a grotesquely swollen base up to 2 m in diameter with one to several ascending or decumbent, thin and semi-succulent branches, the leaves usually in terminal clusters; the latex clear or white (Fig. 26). It is distributed from eastern Africa to Arabia and probably introduced in West Africa [ZB and ZE II-III, III, III(tI)]. All parts are toxic and the latex is used as an arrow or homicide poison; wood chips are used as a piscicide. The liquid in which the pounded bark has been soaked is applied to swellings, painful joints, etc. The plant may be grown as an ornamental (Watt and Breyer-Brandwijk 1962; Miller and Morris 1988).

Nerium oleander (oleander): a robust, evergreen, leggy shrub up to 4 m tall. A native of the Mediterranean region (ZB and ZE III–IV, IV, IV–III), it is widely cultivated as an ornamental in many parts of the tropics and subtropics. The leaf and bark have been used as a rodenticide and as an insecticide and piscicide. All parts are highly toxic, one leaf being a potentially lethal dose, with death within 24 h. These toxic properties were also known to the ancient Greeks when, during the Persian campaign, Alexander's army lost horses which had fed on the shrub as well as some soldiers who had grilled their meat on skewers made from the wood. People have even been killed from drinking water in which a leaf or flower had fallen; the plant also imparts its poisonous properties to honey. It is used medicinally as an abortive, cardiotonic, febrifuge, parasiticide and sternutatory; it is also used as a rat poison.

The plant contains 0.08% of a digitalis-like glucoside oleandrin, which produces a cardiotonic action, causing a slower, stronger and more regular heart beat. Olearandrin has been successfully substituted for digitalis and ouabain (from *Acokanthera* spp.) in the treatment of cardiac deficiency and may even be tolerated when the latter two cannot. A number of other cardiac glucosides have been identified, including folinerin, cortenerin and cornerin; the latter has been successfully used in cardiac therapeutics. A homeopathic tincture of the plant produces a little less diuresis than that from *Solidago*. The leaf structure is isobilateral, thereby increasing the assimilatory surface (Standley 1924; Watt and Breyer-Brandwijk 1962; Boulos 1983; Metcalfe 1983; Chiej 1984; Mabberley 1987; Dowling and McKenzie 1993).

Fig. 26. Stem succulents growing in Socotra, Indian Ocean. Left *Adenium obesum*; right *Dendrosicyos socotrana*. (Adapted from a photograph by A. Radcliffe-Smith taken in 1967 in Kew archives

Pachypodium geayi (vontaka): a single-stemmed, bottle-shaped tree up to 5 m tall, having a trunk with spines in groups of three and branches bearing terminal clusters of caducous, long, coriaceous leaves. It is endemic to south-western Malagasy (ZE II-III), growing on calcareous soils. The functions of the water-storing trunk, coriaceous and caducous leaves are obvious adaptations for survival (White 1983; Rabesandratana 1984).

ASCLEPIADACEAE

Leptadenia pyrotechnica: a green stemmed and more or less leafless shrub or tree up to 5.5 m tall, with a short, stout trunk which is often buried in a mound of trapped sand. Branching close to the ground, the numerous, long, virgate, spinescent branches bear minute and soon deciduous, subulate leaves, although those leaves shaded within the interior may be retained for a longer period. From personal observation, branches buried by drifting sand may produce adventitious roots. The plant has a deep and extensive root system. It is widely distributed through the Sahara and the Sahel and through Arabia and the Middle East to India, growing in well-drained, sandy soils in areas with less than 100 to 450 mm annual rainfall (ZB and ZE II–III, III, IIIsr, IIIwr).

The stems are a source of khip fibre, a rot-proof fibre which has an economic potential for textiles, carpets, cordage and as raw material in the cellulose acetate industry. Water availability limits its commercial application unless an alternative method to retting can be found for fibre extraction. The slimy fruits and seeds, young shoots and flowers may be eaten and the tuberous root cooked as a vegetable. The plant is browsed by camels, although it is not greatly appreciated by other livestock. The stems may be used for thatching, matting, fuel and as an inextinguishable slow match. Propagated from seed, it may be grown for fencing, shelter belts and dune stabilisation. The branches, seeds and latex are also used in local medicine (Boulos 1983, 1989; IBPGR/RBG 1984; Ali 1986; Booth and Wickens 1988; Bhandari 1990).

In a study of *L. pyrotechnica* made by Batanouny and Abdel Wahab (1973) the roots of a bush 1.6 m tall and with a canopy spread of 1.35 m in diameter reached a depth of 11.5 m and extended laterally for 10 m. It was estimated that the root system occupied ca. 850 m^3 of soil containing 23000 kg of available water. The annual water output was 5700 kg and the authors calculated that the available water within reach of the root system could sustain the plant for 4 years without replenishment.

A drought resistant species, its xeromorphic, leafless habit implies a reduction in transpiration surfaces, with the young branches acting as the assimilating organs. The thin cuticle, sunken stomata and surface hairs are ineffective in reducing transpiration, while the deep root system and low transpiration rate ensures an effective water balance. The transpiration rate was found to be low during the winter months and rising at the onset of summer. This rise, however, was not comparable with the evaporation rate, maximum transpiration being generally attained earlier than maximum evaporation, especially in summer. The transpiration curve shows a rapid decrease after the maximum despite the increase in potential evaporation, thus demonstrating an effective stomata regulation. The osmotic pressure is relatively low and shows narrow daily and seasonal variations, e.g. the midday and midnight readings for June were 1.5 and 1.41 MPa, and for November 1.14 and 1.12 MPa.

Stomata frequency varies from plant to plant and with habitat, the shade leaves having a greater number on the lower surface while those in the sun having more stomata on the upper surface. Exposure to more xerophytic conditions is manifested by a decrease in the size of epidermal cells with a concomitant increase in the number of stomata per unit area, the decrease in cell size being due to the increased transpiration and the relative deficiency of water supply to the upper leaf compared with the lower. The young branches are the main assimilating organs and remain unsuberised for a considerable period, the epidermal cells having a thick cuticle and sunken stomata. Although the young shoots are hairy, these do not appear to reduce transpiration (El-Hadidi and Abdel Wahab 1964; Migahid et al. 1972).

BALANITACEAE

Balanites aegyptiaca, syn. *B. roxburghii sensu auct.* (desert date): a spinous, semi-deciduous tree up to 10 m tall, which, under heavy browsing, may not develop beyond a dense cushion shrub. Widely distributed through the Sahelian, Sudanian, Somalia-Masai and Zambezian regions of tropical Africa and in Arabia, in areas receiving 100–1000 mm summer rainfall [ZB and ZE I-II, I-III, II, II-III, III, III(tI)], it is a tree of wide ecological tolerance, including seasonal flooding.

The foliage is browsed by both domestic and wild animals. The young leaves, shoots and flowers may be cooked and eaten, and the pulp of ripe, ellipsoid, 2.5 to 4-cm long and 1.5 to 2.5-cm diameter fruit may be eaten raw, cooked or fermented to an alcoholic beverage; the fruit are laxative if eaten to excess. The fruit may also be eaten by donkeys. The seed kernels may be ground for making bread or used for soap. The mesocarp is a potential industrial source of steroidal saponins which, on hydrolysis, yield up to 4% diosgenin/yamogenin. An edible oil, zachon oil, may be extracted from the kernel; the expressed cake may be used for animal food as well as an additional source of steroids and saponins. The chemical composition of the fruit is shown in Table 6.

The saponins, which may also be obtained from the bark and root, are lethal to the freshwater snails *Biomphalaria glabratus, B. pfeifferi, Bulinus truncatus* and *Lymnaea natalensis*, and the miracidia and cercariae of schistosomes; they are also toxic to *Cyclops*, the alternative host of the Guinea worm, *Dracunculus medinensis*, the active principles being the three saponins balanitin 1, 2 and 3. Thus, the planting of along river and canal banks has been proposed for the control of schistosomiasis. Unfortunately, the fruits (and other plant parts) can also be used as a fish poison, making the planting of *Balanites* for schistosomias control unacceptable to local fishermen. Root and bark extracts have been shown to have antimicrobial activity against *Bacillus subtilus, Penicillium crustosum* and *Saccharomyces cerevisae.* The

Table 6. Chemical composition (%) of the fruit of *Balanites aegyptiaca*. (Al-Futuh 1983, 1989)

	Mesocarp	Kernel
Moisture	22.32	2.3–4.5
Crude protein	5.56	26.4–29.9
Fat	0.1	46.0–50.6
Carbohydrates	65.51	ND
Crude fibre	1.72	2.4–3.3
Ash	4.85	2.8–3.3
Vitamin C (mg 100 g^{-1})	0.02	

ND, no data

plant is also used in local medicine as an abortifacient, antimalarial, antiseptic, antisyphilitic, antiviral (against herpes zoster) and vermifuge.

The pale-yellow to yellowish-brown wood is fine-grained, hard and durable, resistant to insects and is used for cabinet work, local furniture, bowls, mortars, tool handles, etc. As a fuelwood it has a high calorific value and is reported as being smoke-free in West Africa, although in the Sudan it is recorded as burning with an oily smoke. The tree is readily propagated from seed and may be grown for shade, shelter, windbreaks and for live, stockproof hedges (Suliman and Jackson 1958; Sahni 1968; Al-Futuh 1983; Baumer 1983; IBPGR/RBG 1984; Henderson et al. 1987; Kloos and McCullough 1987; Booth and Wickens 1988; Boulos 1989; Hall and Walker 1991; Pettit et al. 1991).

BIGNONIACEAE

Chilopsis linearis (desert or flaming willow): a cold deciduous, rapidly growing, graceful shrub or small tree up to 9 m tall with a short trunk up to 30 cm in diameter and with linear leaves up to 14 cm long, the leaves more or less erect on young stems, spreading with age. A phreatophyte with roots descending to 16 m, it is distributed from southern California and Baja California to western Texas and from Sonora to Tamaulipas, Zacatecas and Durango (ZB III, VIIa).

The tough, slender branches are used for making baskets. The dark brown wood is soft, weak and close grained, with a specific gravity of ca. 0.50; it is used for fence posts. It may also be cultivated for its sweet-scented flowers and for erosion control, especially on unstable dunes where its sprouting ability enhances its ability to withstand burial. It is only browsed when more palatable species are not available. The flowers are sold in the local markets. The Seri people prepare a tisane from the flowers and pods for coughs and as a heart stimulant.

The leaves have a thick cuticle projecting over the stomata, thereby reducing water losses. However, over much of its range summer evapotranspiration is so great that the tree is drought deciduous and the leaves are renewed with the onset of the summer rains. Leaf fall normally occurs when winter temperatures drop to 5 °C. Although capable of maintaining positive net photosynthesis at all but the lowest levels of leaf water potential, photosynthesis decreases rapidly with relatively small decreases in tissue water potential (Standley 1926; Kearney and Peebles 1951; Felger and Moser 1985; Mabberley 1987; Turner et al. 1995).

BOMBACACEAE

Adansonia digitata (baobab): with its swollen trunk of up to 10 m or more in diameter, it is undoubtedly Africa's most conspicuous and easiest recognisable tree. It is widely distributed through the savannas and grasslands of tropical Africa and into southern Arabia and possibly India, in areas receiving 100 to 1500 mm, usually summer rainfall (ZB and ZE I–III, II–III, IIIsr). It prefers sandy topsoils overlying loam, however. it can tolerate poorly drained, heavily textured soils but not deep sands, which provide neither groundwater nor a strong enough rooting medium to support the tree. It is often associated with villages and an examination of its distribution, ecology and local history suggests that the villagers took advantage of its shade and economic benefits to site their villages rather than the baobab being a 'weed' of habitation.

The baobab's economic benefits, which are recorded in greater detail in Wickens (1982b), are numerous, with practically all parts of the plant being utilised. The roots, leaves, fruit pith and seeds may be eaten; the latter may also be eaten after germination. The naturally or artificially hollowed trunks may be used to store up to 9000 l of water, or used for tombs, prisons, stores, etc. The bark provides a strong and durable fibre used for cordage, bark cloth, etc. Unlike most trees the bark may be stripped from the entire circumference of the trunk without damaging the tree. The wood is light and spongy but susceptible to fungal attack and therefore unsuitable for carpentry and fuelwood. Although suitable for paper manufacture, the cost of extracting the moisture content fortunately makes the exploitation of this venerable tree an uneconomic proposition. The seedlings and saplings bear simple leaves before they are progressively replaced by combinations of simple, two-foliate and three-foliate leaves before attaining the five to seven leaflets of the adult trees. However, even when adult the annual flush commences with simple and early caducous leaves and progresses rapidly through the juvenile sequence to the adult leaves.

The grossly swollen trunk and branches are composed almost entirely of water-storing parenchyma upon which the tree depends for moisture during times of drought. So much so that during a severe drought the trunk will visibly shrink so as to leave a gap in the soil surrounding the trunk. According to Walter and Breckle (1986) the water content of the living wood is 31.5–41.6%, sufficient for the tree after it has shed its leaves to endure a prolonged drought without water uptake from the soil. The mechanical strength of the branches is also to some extent dependent on the water content of the parenchyma providing turgid support to the branches. In a very severe drought the branches may become so dehydrated that they may snap. Fallen trees may be found with surprising frequency and, provided some roots remain in the ground, are capable of producing erect growth from the prostrate trunk. Even more bizarre is the ability of the felled trees to produce a new shoot from the centre of the stump in addition to peripheral shoots (Wickens 1982b).

Ceiba acuminata (pochote): a fast growing, openly branched, drought deciduous tree to 20 m tall with large, blunt spines on the trunk. It is a native of western Mexico from Sonora, Puebla and Oaxaca (ZB and ZE II-III, III). The 20 cm-long kapok-filled fruit have been occasionally commercially harvested for their floss (*C. pentandra* is the primary source of kapok). The seeds are edible and have a nut-like flavour. At the end of the summer rains the large tuberous roots (*comotes*) are eaten roasted while they are still turgid and succulent. The young leaves are also eaten as a pot-herb. The species could have a potential as a perennial root crop for the hot, semi-arid regions. The tuberous roots and short duration of the foliage suggest high photosynthetic rates with substantial carbon allocation to the roots and that flower development, which occurs when the tree is leafless, draws on root and stem reserves (Nabhan and Felger 1985; Turner et al. 1995).

BURSERACEAE

Bursera hindsiana (copal or red elephant tree): a large, aromatic, drought deciduous tree or small tree up to 8 m tall with a pachycaul, red-barked trunk and large, simple, three- or five-foliate leaves on the same plant. It is a native of Baja California and coastal Sonora, growing in valleys, washes,on gentle slopes and rocky hillsides in areas of warm, maritime climates with bimodal or summer rainfall (ZB IIIsr, IIIwr, IIIswr). The relatively soft but firm wood does not split on drying and is excellent for carving and musical instruments. The dry wood is used for kindling and to maintain a low fire. An infusion of the wood shavings with leafy branch tips of *Hyptis emoryi* are used by the Seri people for curing asthma and other breathing difficulties. A tisane from the leaves is used for treating colds (Felger and Moser 1985; Turner et al. 1995); the seeds are also edible (Mabberley 1987).

Bursera microphylla (torote or elephant tree): a deciduous, pachycaul tree up to 8 m tall, having a trunk diameter of 0.3 m with a papery, exfoliating bark and crooked, rapidly tapering branches. It is a native of south-western Arizona, south-eastern California and north-western Mexico (ZB IIIsr, IIIswr, IIIwr, possibly IVa, i.e. a relatively arid IV), growing in washes, on gravel plains and rocky slopes. In the Sonoran Desert the bark tannin was formerly exported and the gum used for treating venereal diseases; the gum mixed with oil is still used by the Seri people for caulking boats. The wood is pithy and very soft. The dry wood is used for kindling, it burns quickly and is considered by the Seri to be the best wood for smoking out bees. A dark, reddish-brown stain is prepared from the red pulpwood. A belt made of the leafy twigs is worn by the Seri to repel sharks. The sap contains terpines, and the crushed leaves are highly aromatic.

The tree is intolerant of temperatures below -4 °C. The leaves may appear at any time of the year following even a meagre rainfall. The thin sheets of bark admit sufficient light to permit photosynthesis underneath. However, the bark becomes increasingly opaque with age and is shed; exfoliating also prevents the establishment of crustose lichens, which would inhibit photosynthesis. Water is stored in the conductive and parenchymal tissue of the trunk (1.4 kg m^3), lower branches, and wood. The water storage tissues begin to enlarge in the seedling stage, first in the roots and later in the shoots. This water storage serves as a buffer against a year or more without measurable rain (Kearney and Peebles 1951; Felger and Moser 1985; Turner et al. 1995).

Commiphora wightii: a much branched, spiny shrub up to 2 m tall. Distributed from Arabia to north-west India (ZB and ZE II-II, III), it is the source of the oleo-gum-resin known as guggul or Indian bedellium, and is a mixture of 61% resin, 29.3% gum, 6.1% water, 0.6% volatile oil and 3.2% foreign matter. Guggul is largely used for incense, in perfumery and local medicine. The papery bark flakes off from the older parts of the stem exposing a green photosynthetic surface. The oleo-gum-resin occurs in secretory canals within the phloem of the larger veins of the leaf and in the soft bast of the stem. These canals, or ducts, are formed schizogenously in the primary and secondary phloem, and develop vacuoles containing osmophilic substances; the site of gum synthesis has yet to be ascertained (Kumar and Shankar 1982; Bhandari 1990).

CACTACEAE

Carnegiea gigantea, syn. *Cereus giganteus* (saguaro): an arborescent, columnar cacti, with a massive, solitary, succulent, erect and ridged stem up to 15 m or more tall and 0.75 m diameter, and developing as many as 50 arms; it is the largest of the columnar cacti (Fig. 27). A native of the Sonoran Desert (ZB IIIsr, swr), it is nocturnal flowering but with the white blooms persisting through the next day; it is designated the state flower of Arizona. The fleshy red fruit are either eaten fresh or stored in the form of a syrup or preserve. The fermented fruit juice produces an intoxicating beverage. The Papago people prepare a form of butter from the seeds. The dry skeletal wood may be used for building wattle and daub houses.

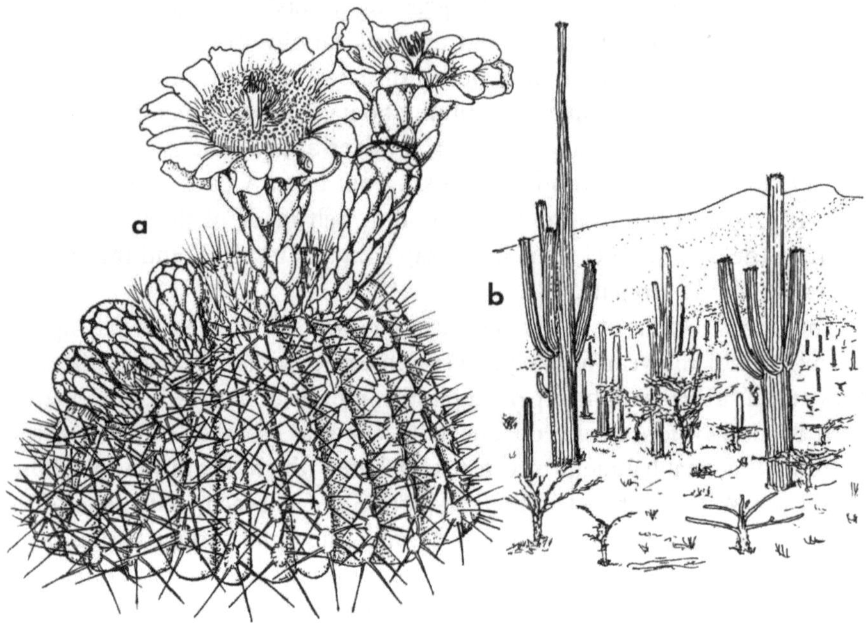

Fig. 27. *Carnegiea gigantea* (saguaro). **a** Tip of a flowering stem. **b** Stand of saguaro growing on a hill slope near Tucson, Arizona. (Adapted from Plates 22 and 23 in Britton and Rose (1920)

Although used in local medicine there are no reports of saguaro being used as an hallucinogen, despite the presence of carnegine, 5-hydroxyphenylethylamine and norcarnegine, plus traces of 3-methoxytyramine and the new arizoniune (a tetrahydroquinoline base), pharmacologically active alkaloids capable of psychoactivity (Kearnsey and Peebles 1951; Havens et al. 1973; Felger and Moser 1985; Schultes and Hofmann 1992).

The diameter of mature stems of saguaro tends to increase towards the northern extent of its range and is associated with an increase in apical temperature and a decrease in mean air temperature. The increase in apical temperature together with the apical pubescence are considered sufficient to confer some degree of frost protection (Nobel 1980). Although the root system of the saguara is little more than 1 m deep it produces numerous subsurface lateral roots which extend for over 30 m and enable the plant to take up water from even the lightest rainfall. As the water is taken up the massive trunk swells, the accordion-like ridges allowing for expansion around a stout but elastic woody skeleton. This skeletal structure consists of a cylindrical framework formed from groups of vascular bundles which anastomise up to middle height and are free above. The water content of the tissues ranges from 75 to 90% of the green weight, being higher towards the apex and lower towards the base. Due to the great capacity for water storage and the slow rate of growth, the saguaro is able to fruit annually, more or less regardless of drought.

Saguaro seeds will not germinate at constant temperatures of 15 °C and lower, and very little (2%) above 35 °C; optimum constant temperature for germination (38%) is 25 °C. Alternating temperatures of 35/25 and 30/20 °C will both give 28% germination in the light, although there is no germination in the dark. The ability to germinate at alternating temperatures of 35/25 °C is of ecological importance since daytime temperatures are often higher than 35 °C.

The seedlings require a 'nurse plant' for survival, since seedlings germinating in the open die from either sunburn or frost. The saguaro favours south-facing, frost-free slopes and well-drained soils for root anchorage. There is little aerial growth initially, the first 10–20 years of growth being spent in developing the root system which, when established, will allow dramatic spurts of growth. The juvenile saguaros are armed with relatively long, downward pointing spines which provide defence against predatory small animals; as the plant matures the spines become erect, pointing outwards (Crosswhite 1980; McGinnies 1981; Gutterman 1993).

Ferocactus cylindraceus, syn. *F. acanthodes* (Californian barrel cactus): a simple, cylindrical and ribbed stem armed with bristle-like spines on the margins of the areoles. It is a native of southern California, Utah and Arizona southwards into Baja California and western Sonora (ZB IIIswr, IIIwr, VIIa). The Seri people prepare a gruel from the seeds. The bitter flowers are rarely eaten; eating the pulp is said to cause headaches and the juice is not potable (Felger and Moser 1985). The rooting system is shallow, with a maximum depth of 15 cm, thus enabling the cactus to respond within 24 h to rainfall. After rain the perennial roots produce ephemeral rain roots. As the soil dries out the two root types undergo reduced respiration. When the soil water potential reaches -1 MPa, respiration of the rain roots is irreversibly inhibited and the roots are unable to rehydrate. For the perennial roots respiration becomes zero at -1.6 MPa; even after the soil dries out the roots fully recover within a day of soil wetting.

The ribbed stem structure permits swelling with water uptake and shrinkage during dry periods. About 6.1% of the water entering the succulent stem is stored. It is this water that permits the nocturnal opening of the stomata for ca. 40 days after the soil water potential has fallen below that of the stem, during which period the plant would be unable to extract moisture from the soil. After 7 months of drought, diurnal stomatal activity was not observed and the stem osmotic pressure had risen to 0.64 MPa, more than double the value measured during wet periods with nocturnal stomata opening. The Californian barrel cactus flowers every year and increases its total biomass by 10% annually. Only 6% of its pool of utilisable water is used for flowering and seed ripening.

It is a CAM species. When the nocturnal stem surface temperature was raised from 8 to 35 °C, the stomata resistance increased four-fold, indicating that cool night temperatures are advantageous for gas exchange. Moreover, the optimum temperature for CO_2 uptake in the dark was only 12.6 °C. Night uptake of CO_2 became maximal for 3 milli-einsteins cm^{-2} photosynthetically active radiation incident during the preceding day, and the minimum number of incident quanta absorbed per CO_2 fixed was 68. The transpiration rates (mass of water transpired per mass of CO_2 fixed) had the relatively low value of 70 for an entire year, consistent with values obtained from other CAM plants (Nobel 1977; Turner et al. 1995).

Lophocereus schottii, syn. *Cereus schottii* (senita): a clump-forming cactus up to 5 m diameter with numerous, long, ascending, deeply ribbed branches up to 7 m long. The odourless, pink flowers open shortly after sunset and wither the following morning. Senita occurs in Arizona, Baja California and Sonora (ZB IIIsr, IIIswr, IIIwr). The fresh, spineless fruit, ca. 2–3 cm diameter, are edible. It may be grown in frost-free areas as an ornamental. The chopped stems are sometimes used in Baja California as a fish poison (Standley 1926; Kearney and Peebles 1951; Havens et al. 1973; Felger and Moser 1985; Turner et al. 1995).

According to Felger and Lowe (1967) the surface to volume ratio of the stem is correlated with the senita's geographical distribution and parallel gradients in the annual, especially summer, rainfall. Those with the thickest stem and thus lowest surface to volume ratio occur to the north, in northern Sonora, northern Baja California and southern Arizona. Those with the more slender branches occur further to the south, near the Gulf of California and along the Mexican coast, where the rainfall is somewhat higher and water losses reduced. As the relative sparse spine cover at the apex of immature stems offers little protection against frost, Nobel (1980) believes this is a factor that limits its northern distribution.

Opuntia: flat-jointed species of *Opuntia* are known as prickly pear (e.g. *O. basilaris*), those with edible fruit are referred to as tuna (e.g. *O. ficus-indica*), while species with cylindrical joints are known as cane cacti, the spinier ones being known as cholla (e.g. *O. fulgida*). Widely cultivated as ornamentals, the genus contains pectin, which is antitussive and antihypercholesterolemic. Clinical studies have shown that *Opuntia* species affect glycaemia, glucose, insulin, serum lipids, beta chlolesterol and glucose tolerance and have a hypoglycaemic action (Winkelman 1991).

Opuntia basilaris (beavertail cactus): forming up to 1-m-diameter and 30-cm-tall clumps of low growing, flat-jointed stems, the joints are usually spineless but with abundant glochids. Distributed through south-western USA and northern Sonora (ZB VIIa), growing on sandy soils it is commonly planted as an ornamental in arid landscapes. The terminal pad is often orientated north-south for maximum absorption of solar radiation during the winter, when the soils are more likely to be wet. The stomata remain open throughout both day and night provided the plant water potential is greater than -0.5 to -0.6 MPa, permitting both C_3 and CAM respiration pathways. Once the plant water potential falls to -0.8 to -1.0 MPa the stomata will open only at night. Provided night temperatures are lower than day temperatures less water will be used for carbon fixation using the CAM photosynthetic pathway than by the daytime fixation by the C_3 pathway. During drought the stomata close, which, together with the impervious cuticle, greatly reduces gas respiration and exchange. With the stomata closed most of the respiratory CO_2 is retained within the joints. The recycling of endogenously produced CO_2 keeps the plants in a moderately active metabolic state, thus enabling a rapid response to slight rainfall. During drought water is first lost from the parenchyma, which has a lower osmotic pressure than the chlorenchyma. It is the high water potential that enables the stomata initially to remain open at night, even when water can no longer be extracted from the soil; this also allows the photosynthetic tissue to remain alive during severe drought (Standley 1924; Turner et al. 1995).

Opuntia ficus-indica (Indian or Barbary fig): a shrubby or arborescent plant up to 5 m tall, often with a cylindrical trunk and bearing elliptic to narrowly obovate, flattened joints up to 50 cm long; leaves early deciduous, spines sometimes present, up to 1.5 cm long. A native of the New World, not known in the wild, it is now widely cultivated and naturalised in the Mediterranean, Red Sea, North and South Africa and Australia (ZB and ZE II–IIIswr, III, III–IV, IV, IV-III). It is cultivated for its edible fruit or as succulent browse for livestock. Several genotypes are recognised, including spineless cultivars; the spiny forms, since they are not browsed, will often naturalise.

CAPPARACEAE

Boscia senegalensis: a small evergreen tree or shrub up to 6 m tall, with leathery, unpleasantly smelling, ovate leaves up to 12 cm long. A Sahelian endemic, it is widely distributed through the Sahel and into north-eastern Africa in areas receiving 100 to 900 mm, mainly summer rainfall [ZB and ZE I–III, II–III, IIIsr, III(tI)]. It usually grows in very dry, barren and often hard soils, but is intolerant of waterlogging. It is extremely drought resistant and is one of the few woody species that has withstood the effects of desertification.

The leaves, which are being used as a famine food, require leaching with several changes of water to remove the possibly toxic, bitter principles before being fit for human consumption. They contain the alkaloids L-stachydrine and hydro-3 stachydrine, which affect the aggregation of thrombocytes and shorten the bleeding time. The leaves are only casually browsed by livestock, who are possibly deterred by the unpleasant smell. The presence in the twigs of glucosinolates, which can hydrolyse to the highly toxic and irritant mustard oils, could affect palatability. The globose fruits, up to 1.2 cm diameter, are also eaten by man and livestock. The fruit, which are more usually used as a famine food, are debittered by boiling in water. The protein content of the debittered fruit has a chemical score of 85, indicating a nutritional value superior to that of the local cereals.

The wood is soft and workable when boiled and, if large enough, may be used in house construction. It burns with much smoke when used for fuel. The leaves, root and bark are used in various local medicinal and veterinary preparations. The thick cuticle and sunken, thick walled stomata with their protective papilllae offer protection against excessive water losses (Osborn 1968; Baumer 1983; Boulos 1983; Booth and Wickens 1988; Salih et al. 1991).

CASUARINACEAE

The genera *Casuarina* and the closely related *Allocasuarina* fix nitrogen by means of root nodules formed by the root hairs being inoculated by the actinorhiza *Frankia*. The nodules do not contain vesicles, suggesting that the infected cells are at a low pO_2 (partial pressure of oxygen). Indeed, the thickened and possibly suberised host cells suggest that the host restricts oxygen sufficiently to make vesicle production unnecessary, also enabling the pO_2 to be sufficiently reduced to permit both synthesis and operation of nitrogenase. Furthermore, the comparatively high level (for actinorhizal plants) of haemoglobin present in the nodules suggests that there are similar requirements to those present in legume nodules for oxygen transport.

Allocasuarina decaisneana, syn. *Casuarina decaisneana* (desert oak): a graceful tree up to 15 m tall with a pendulous habit and vestigial leaves. It grows on deep sandy soils in one of the hottest and driest parts of central Australia with 50–250 mm annual rainfall (ZE II–III, III–I). The presence of groves is regarded by Beard (1969) as an indication of shallow ground water. Cattle may browse the tree when hungry but do not relish it. The main use is for the production of poles for fencing and firewood. The tree may also be planted as a windbreak, and for shade as an ornamental (Hall et al. 1972; Baumer 1983; Lazarides and Hince 1993).

The root hairs of *A. decaisneana*, in common with other species of *Allocasuarina* and *Casuarina*, are the host of the filamentous, nitrogen-fixing actinorhiza *Frankia*. The *Frankia* does not develop vesicles at the hyphal tips, although these may be formed in culture. The failure to do so in symbiosis may be associated with the cell wall material of the host cell wall restricting oxygen diffusion. Many of the *Allocasuarina* and *Casuarina* species nodulate freely, the woody, perennial nodules often forming large masses on the root system. In the sand dunes of the Cape Vert peninsula, Senegal, soil nitrogen around *Casuarina* increased by ca. 60 kg ha^{-1} $year^{-1}$, the fixed nitrogen being added to the soil through the decay of dead roots and leaf litter (Sprent and Sprent 1990).

CHENOPODIACEAE

Agathophora alopecuroides, syn. *Halogeton alopecuroides*: a stout, juicy succulent with cylindrical leaves terminating in a bristle-like rudimentary lamina. A halophyte, widely distributed from North Africa through Arabia to Pakistan (ZB and ZE III, IIIwr, III-IV, IV), it is an important summer forage for goats and camels. The plant is capable of surviving several years of drought by drastically reducing its transpiration area by the leaves becoming globose instead of their normal cylindrical shape (Danin 1983).

Arthrocnemum macrostachyum, syn. *A. glaucum*: an annual herb up to 1 m high, apparently leafless (fused leaf pairs covering the nodes), densely branched, jointed and succulent, and decumbent at the base; it produces long, thread-like, horizontal, adventitious roots when vegetative organs become buried. A mainly coastal halophytic species of the Mediterranean region eastwards to the Indus delta (ZB and ZE III, IIIwr, III–IV, IV), it may be cultivated as a salt tolerant ornamental; it was formerly a source of barilla (Aronson 1985, 1989). Although water losses through transpiration are low, the water content in the shoots may be changed several times during the day. Thick sclereids strengthen the palisade sclerenchyma of the internodes, thus preventing collapse on loss of turgidity. Portions of the fleshy cortex are discarded as a means of reducing the salt content (Kassas and Batanouny 1984; Walter and Breckle 1986).

Atriplex canescens (fourwing saltbush): a C_4, polyploid, evergreen, drought-resistant shrub up to 2m tall and bearing scurfy-pubescent, lanceolate leaves up to 5 cm long and 0.7 cm wide; the deep roots descend to 7 m or more. It is usually dioecious although stress can change the sexual expression, for example, winter cold and competition for water can change pistillate plants into staminate plants. It is widely distributed through western North America (ZB and ZE III, IV, V–IV, VII, VIIa, VII–V, VII–VI) and is salt and alkali tolerant.

It is highly prized as a browse species and is so greatly relished by livestock that its reproduction is often hindered. Too liberal feeding is reputed to cause bloat and scours in cattle. It may sometimes be toxic or cause anaemia in sheep, which is possibly associated with the high salt content. The Seri Indians use an infusion of the leaves as an emetic. The plant may be grown for erosion control (Kearney and Peebles 1951; Havens et al. 1973; Felger and Moser 1985; USDA 1988; Turner et al. 1995). A number of ploidy levels are recognised: di-, tetra-, hexa- and dodecaploid ($2n = 18$), of which the tetraploid is the most widely distributed. Populations of the other ploidy levels form isolated enclaves within the general distribution, reflecting environmental characteristics and morphological expression.

The ecophysiology of *A. canescens* was examined by Senock et al. (1991), who found that the distinct segregation between cytotypes demonstrated an apparent response to ecotypic differentiation and functional significance of differing leaf morphologies in response to leaf temperatures, net photosynthesis, leaf conductances and internal CO_2 concentrations. Mean leaf area increased from diploid to hexaploid and was associated with a corresponding decrease in leaf temperature. Photosynthetic rates of diploids decreased rapidly when leaf temperatures approached 37 °C, and 63% of the variation in photosynthesis was accounted for by increasing leaf temperatures. The increasing internal CO_2 levels in diploids indicated that internal limitations to CO_2 uptake in the mesophyll were responsible for the depressed photosynthesis. However, the photosynthesis of tetraploids and hexaploids was largely unaffected by leaf temperatures. Higher leaf temperatures in diploids were also translated into higher leaf conductances and lower water-use efficiency than the other cytotypes, whereas the higher water-use efficiency in the tetraploids and hexaploids were related to either lower conductances and higher photosynthetic rates, or a combination of reduced conductances and higher photosynthesis. The hypothesis that the various cytotypes are adaptations to soil differences is supported by the decreased conductance and subsequent increased water-use efficiency of the hexaploid, a cytotype that is largely confined to flood plains having clay soils with high moisture-retention properties. It is usually dioecious although stress can change the sexual expression, for example, winter cold and competition for water can change pistillate plants into staminate plants.

Atriplex halimus (silvery orache): a much-branched, up to 3-m-tall, mealy shrub with a creeping rootstock, bearing ovate to elliptic leaves up to 4 cm long. Widely distributed through the Mediterranean region and extending south to East Africa and possibly Malagasy, it hasbeen introduced elsewhere, growing on a wide range of soils in areas receiving 50-200 mm annual rainfall [ZB and ZE II–III, III, III(tI), III–IV, IV]. A salt tolerant species, it is generally a palatable browse, but becomes less palatable as the salt content increases with the salinity of the habitat. It may be cultivated as a fodder plant, as in Arabia. The leaves are eaten by the Bedouins either fresh or boiled. The roots are used for cleaning teeth; it is also a remedy for hydropsy. The plant's ashes are rich in alkaline salts and are used in treating gastric acidity and as a soap. The seeds in small doses are used as an emetic; in large doses they are toxic (Boulos 1983, 1989; Danin 1983; Orev 1992).

In spring new stems develop from buds at the base of the older stems. The number of leaves depends on the moisture available. Both leaves and stem are covered with several layers of vesicular hairs, which on young leaves are filled with saline solution. During summer the vesicular hairs eventually burst and deposit salt on the leaf surface, which may function to decrease absorption of solar radiation. The hairs, despite their unusually high salt content, failed to absorb water when immersed in water, i.e. they do not function as a means of reverse transpiration. The concentration of salts in the vesiculated hairs was considerably higher than that of the leaf sap and xylem exudate and played a significant role in preventing an accumulation of toxic salts in the plant tissues. The bracts enveloping the fruits contain water-soluble germination inhibitors, the leaching of which ensures adequate moisture is available for germination (Oppenheimer 1960; Zohary 1966; Mozafar and Goodin 1970; Danin 1983; Gutterman 1993).

Atriplex hymenelytra (desert holly): an evergreen, dioecious, C_4 shrub up to 1.2 m tall with thick, holly-like, subhastate leaves up to 4.5 cm long, silvery lepidote when young. A native of south-western USA (ZB IIIwr, IVa), growing on rocky slopes and gravel fans, it appears to be the most drought tolerant native *Atriplex* species in North America. The leaves are gathered for Christmas decorations. The plant is browsed by livestock; it is also suitable for growing as an ornamental (Kearney and Peebles 1951; Havens et al. 1973).

Desert holly tolerates moderate salinity (8 dS m^{-1}; see Aronson 1989). The plant is usually winter flowering and, following anthesis, a new crop of leaves is produced and the previous leaves absciss, although there is a brief period during which both old and new leaves are present. The leaves are dimorphic, with relatively large leaves during the rainy season and the development of smaller and more pubescent leaves during the dry season. The leaves also exhibit a number of other adaptations to a hot and arid environment. They are sufficiently steeply angled (ca. 70°) during the summer months to increase reflectance to ca. 60%, thereby reducing midday solar inception, yet obtaining a relatively high inception when the solar angles are less and vapour pressure deficits are at a maximum. The leaves return to a more or less horizontal position during the winter, thereby reducing the overall reflectance to ca. 35%. There is a substantial reduction in the absorptance of incident radiation during the hot season due to a change in leaf moisture and associated dissolved salt contents. During this period the light intensity required for photosynthetic saturation is low coupled with a reduction in the radiation absorbed by the leaves, which consequently results in a greater water-use efficiency. The leaf surface is covered with bladders, which are in the hydrated state in very young leaves, but which collapse as the leaves dry out during the hot summer. The salt content of the bladders, mainly as NaCl, then crystallises on the leaf surface. The high reflectance and steep leaf angle thus provides heat resistance during the hot, dry summer (Mooney et al. 1977).

In field experiments on CO_2 and water vapour exchange, the rates of CO_2 uptake and leaf conductance were substantially higher in March than in July.

Despite the large differences in ambient temperatures for March and July, there was no adaptive acclimatisation response in the temperature dependence of photosynthesis. The photosynthesis to temperature ratios at typical midday leaf temperatures were much higher in March than in July. A winter-active C_4 plant with a Kranz anatomy, *A. hymenelytra,* combines a strong regulation of water loss by stomata control with a low efficiency of the water transport system, yet is unable to prevent depression of plant water potential with increasing transpiration. The high water-use efficiency results from a combination of the C_4 photosynthetic pathway, relatively low leaf conductance, plus maximum growth and photosynthetic activity during the cooler months of the year (Pearcy et al. 1974; Sánchez-Diaz and Mooney 1979; Turner et al. 1995).

Atriplex lampa (zampa): a perennial shrub, a native to the temperate arid region of Argentina, growing where the mainly summer rainfall is 180-200 mm (ZB and ZE III-VIIa, VIIa). It is browsed by sheep, especially during the cold dry season when other fodder is scarce. The wood is used for fuel (Gastó et al. 1985). Although deep-rooted, 60-70% of the absorbing rootlets are concentrated between 10 and 40 cm deep. In the event of the taproot reaching dry or compacted soil, it will either remain stunted or die back from the tip, and will be replaced by strong or numerous lateral ramifications developing near its origin; these increase in vigour and form the main root system (Morello 1951a; Gastó et al. 1985). Trione and Passera (1993) found that repeated conditions of water stress during the growing season reduced growth in all the organs, but the absolute nitrogen content in terms of both free and fixed nitrogen remained unaffected despite the low soil nitrogen. They concluded that the partitioning of assimilates and remobilization of organic nitrogen within the plant would appear crucial to the balance of growth and internal nitrogen economy.

Atriplex lentiformis (quail-brush): a deciduous and usually dioecious, dome-shaped shrub; it is the largest and showiest salt-bush in Arizona, attaining a height of 3.3 m where the water table is high. Its distribution is widely scattered through California, Utah, Nevada, Arizona and Sonora (ZB IIIswr, VIIa). The parched seeds are eaten by the native Americans. The plant is palatable and browsed by cattle. Quail-brush may also be grown for landscaping. It is a halophytic phreatophyte, with the Krantz-type leaf anatomy of a C_4 plant. In a comparison of photosynthesis between coastal and desert populations maximum rates of CO_2 uptake were similar, 15.8 and 16.5 μmol dm^{-2} min^{-1}, respectively. However, the thermal optimum for photosynthesis was 32 and 44 °C respectively. Furthermore, CO_2 uptake below 36 °C was higher in coastal plants whereas above 36 °C it was higher in desert plants. Plants grown at high temperatures showed increased thermal stability of the photosynthetic apparatus, accounting for the greater acclimatisation of desert plants (Kearney and Peebles 1951; Havens et al. 1973; Rea 1991; Turner et al. 1995).

Atriplex mollis: a shrub up to 50 cm tall, much branched from the base and bearing linear, silvery leaves up to 2 cm long. A native of the eastern Mediterranean and North Africa (ZB and ZE III, III-IV, IV), it is a potential fodder plant for use in the reclamation of waterlogged saline areas. The plant reduces its excessive salts by excreting them into inflated vesicular hairs (Kassas and Batanouny 1984; Le Houérou 1985; Walter and Breckle 1986).

Atriplex nummularia (old man saltbush): a predominantly dioecious shrub, up to 3 m tall, bearing succulent, broadly elliptic to ovate or obovate leaves 2–4 cm long. It grows on a wide range of soils in regions of southern arid and semi-arid Australia with 200–500 mm annual rainfall (ZB and ZE III, III-IV, III-IVa, IV-IIIswr) and is regarded as the most suitable species for growing on alkali soils with 200–250 mm rainfall. Arising from a brittle rootstock with a taproot descending to 5 m , it is a densely branched and often spreading bush with long, brittle stems which may root where they touch the ground. A long-lived and highly drought and salt tolerant species, three subspecies are recognized; subsp. ***numularia*** occurs on the heavy flood plain soils of arid and semi-arid central and eastern Australia; subsp. ***omissa*** is widely distributed; and subsp. ***spathlata*** is found in Western Australia from north of Kalgoorlie to the Nullabar Plain and south-western South Australia.

Old man saltbush may be readily trained to provide shelterbelts 2–3 m tall, providing protection down to ground level. It is a useful fodder plant that has been widely introduced into other arid and semi-arid countries of the world. In Australia sheep can exist almost indefinitely on this species, but under heavy grazing pressure it may rapidly disappear. Both productivity and palatability are higher than for most other species of *Atriplex* but it is a fragile species requiring careful management and consequently cannot be recommended for all types of situations. The brittle stem and palatability make it susceptible to trampling and overgrazing. The nutritive value and digestibility is high. However, the palatability may be decreased by the high salt content, so that it is often only browsed in times of drought when other food is unavailable. The species has the ability to accumulate salt in the tissues even when growing in a salt-free soil. When grown in Libya for fuelwood at an initial density of 1000 plants ha^{-1}, the average production during the first 4 years was 2400 kg dry matter (DM) ha^{-1} $year^{-1}$ (Hall et al. 1972; Askew and Mitchell 1978; Le Houérou 1985; Mitchell and Wilcox 1988; Lazarides and Hince 1993).

It does not regenerate readily from seed (Askew and Mitchell 1978; Runciman 1986). The seeds, which are uniform in shape and colour, are like most other species of *Atriplex*, borne between two bracteoles that contain high levels of NaCl (Table 7). When the false fruits are wetted, the salt concentration

Table 7. NaCl concentrations in bracteoles of *Atriplex* species and those required to inhibit germination of naked seeds. (after Beadle 1952)

Species	NaCl in water-satuated bracteoles (M)	NaCl concentration preventing germination (M)
Atriplex holocarpa	0.62	0.8
(syn. *A. spongiosa*)	0.69	0.4
Atriplex lindleyi subsp.	0.90	0.7
inflata syn. *A.*	0.61	0.5
	0.44	0.4
Atriplex nummularia	0.72	0.6
	0.48	0.4
Atriplex semibaccata	0.61	0.4
Atriplex vesicaria	0.48	0.4
	0.55	0.5

in the bracteoles is sufficiently high to prevent germination until there has been sufficient rainfall to leach the salt out. The inhibitory effect of the salt in preventing germination until sufficient rainfall has fallen would appear an adaptive trait to ensure the availability of adequate moisture for seedling development. However, Anderson (1982) points out that this is essentially a proximal interaction between plant and environment, and that there might be a biochemical factor acting as the ultimate control. This idea is supported by the observation that in the related genus *Suaeda*, the dormant seeds showed less gibberellin-like activity than non-dormant seed and that dormancy could be reversed by the application of gibberellic acid.

Another apparently adaptive trait in some species of *Atriplex* is the production of dimorphic seeds, small, hard, black, slow to germinate and possibly with a longer viability, and larger, soft and brown (Table 8). However, the factors influencing the structural development of the seed testa can differ. Thus, in *A. holocarpa* the imposition of water stress during seed set increased the proportion of black seeds but this could be reversed with irrigation. In contrast, *A. hortensis* produced black seeds late in the main flush of fruit set, or early in the season in *A. heterosperma*, while in *A. subspicata* the proportion of black seed increased with day length.

Atriplex polycarpa (desert saltbush): an intricately and rigidly branched, deciduous, C_4 shrub up to 2 m tall and bearing fascicles of linear to oblanceolate leaves up to 1 cm long. According to Danin (1996) rhizosheaths occur on the adventitious roots of buried branches. A native of California, Utah, Nevada, Arizona, Baja California and coastal Sonora, growing in a wide range of climatic and habitat conditions, including moderately saline soils (ZB and ZE II-III, IIIswr, VIIa), it is an important browse plant, especially on moderately

Table 8. Examples of seed variation in the size and colour of *Atriplex* species. (Beadle 1952)

Species	Seed
Atriplex holocarpa	Seeds with every gradation of size and colour
Atriplex lindeyi subsp. *inflata*	Large, light brown seeds and small, dark brown, hard seeds
Atriplex nummularia	Seeds, uniform, brown
Atriplex semibaccata	Large, light brown seeds and small, dark brown, hard seeds
Atriplex vesicaria	Seeds uniform, brown

saline soils. The Seri Indians use the wood for fuel and a solution of the pounded twigs and leaves as a shampoo (Kearney and Peebles 1951; Felger and Moser 1985). It has the Krantz-type leaf anatomy of a C_4 plant. The optimum temperature for germination is 9–15 °C, close to the November temperature, the month of utricle dispersal. It is believed that the bracteoles do not contain any germination inhibitors. Germination is rapid and can occur over a wide range of temperate, from 3 to 33 °C, but is reduced by light and more than 1% NaCl (Sankary and Barbour 1972).

An examination of desert saltbush from various parts of its distribution range has revealed changes in the ratio of dark respiration, as measured by CO_2 release, related to changes in environmental temperatures. The rate of CO_2 release was reduced following a high temperature and increased following a low temperature. Plants from warm areas were found to be less able to adapt to low temperatures than plants from the cooler areas. Acclimatisation of plants from warm areas to high temperatures is such that the respiration rate was not measurable at cool temperatures. It was concluded that the intraspecific differences in acclimatisation appear to be correlated with the environmental temperatures of the areas of origin.

Plant salt tolerance varies between populations and within populations. Salt tolerance is both anatomical and physiological. Salts from the plant tissues are excreted into bladder-like trichomes on the leaves. These crystallise on the leaf surface when the bladders collapse, thereby preventing salt accumulation in the plant tissues. Salt toxicity observed in some plants appears to be due less to the direct osmotic effects of salt than to the deleterious effects of high concentrations of the salt-induced accumulation of amino acids and other intermediate products of protein metabolism (Chatterton et al. 1970; Turner et al. 1995).

Atriplex torreyi (Nevada saltbush): a woody, evergreen, dioecious shrub. Nevada saltbush occurs on the saline playas of the northern Mojave Desert and eastern Great Basin (ZB VIIa). It is a drought fodder for livestock. The fluting of the stem, which also occurs in other *Atriplex* species, has been examined by Groeneveld (1988). It is due to splitting along the lines of weakness created by cork rays of the stem wood and by unequal cambial activity around the bole of the stem. This fluting has certain seasonal advantages. It physiologically isolates the branches of the shrub so that the death or suppression of the canopy due to a reduction in available soil moisture does not affect all the branches equally. This can enable the shrub to reduce transpiration losses and survive a prolonged drought, even as a single branch. The available moisture can then be concentrated into one branch and not dissipated over the entire canopy.

Atriplex vesicaria (bladder saltbush): a dioeceous, brittle-stemmed, dwarf shrub up to 1 m tall with slightly succulent, ovate leaves that are shed during prolonged periods of drought, but following rain new leaves soon appear. The silvery-grey appearance of the up to 2.5-cm-long leaves is due to the presence of salt-secreting, vesicular hairs. The size of the leaves depends on the rainfall. The root system is shallow but widespread, exploiting the sparse rainfall. It is a native of southern arid and semi-arid Australia (ZE II–III, III–II, III–IV, IV-III), sometimes forming pure stands and growing on a wide range of soils and in a variety of habitats. Browsed by livestock, the shrub has a life expectancy of 12–25 years (Hall et al. 1972; Walter and Breckle 1986; Lazarides and Hince 1993).

The leaves contain up to 12% crude protein; however it is relatively unpalatable to livestock due to the high salt concentration - up to 15% in the leaves (Runciman 1986; Mitchell and Wilcox 1988). Because of selective grazing by sheep, a significant proportion of the female plant population may be removed (Anderson 1982). The high salt content of the leaves (20% of the leaf weight during the summer) offers some protection from fire by reducing inflammability. Nevertheless, the species can be killed by fire, probably because there are no buds capable of growing into shoots present beneath the bark above or below ground (Hodgkinson and Griffin 1982). The brittle stems make it liable to damage from trampling.

The seeds are enclosed between two bracteoles, each with a membranous bladder-like appendage, which contains sufficiently high levels of NaCl to inhibit germination. However, the equivalent of 100 mm water is sufficient to leach ca. 95% of the salt from the bracteoles and permit germination (Anderson 1982). Although the plant contains 90% water the salt content is always high, with no clear relationship with the salt content of the soil, although there is a suggestion that the plant salt content may increase with higher soil salt content. In any case the salt content is always higher during a dry period than during the rains. Because of the high salt content the potential osmotic pressure is 4 MPa, with plasmolysis occurring at 5 MPa (Walter and Breckle 1986).

Cornulaca monacantha: a highly variable, spiny subshrub up to 1 m tall. Widely distributed through North Africa to Pakistan (ZB and ZE III, III–IV,

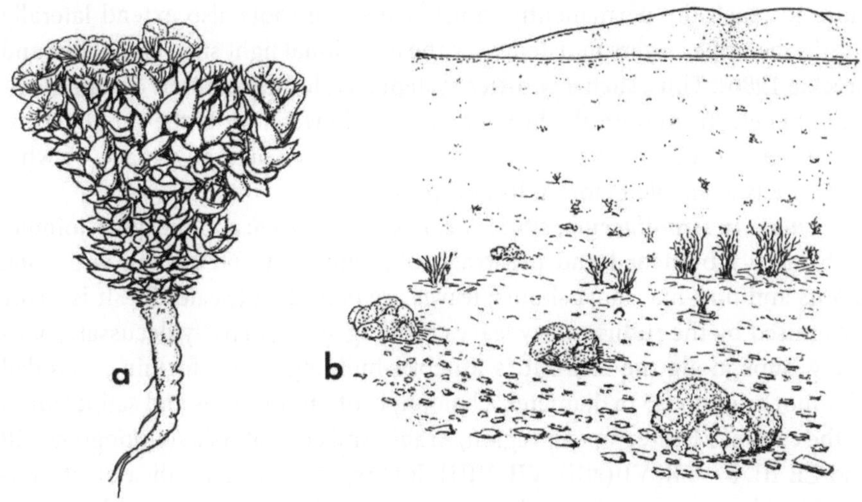

Fig. 28. *Fredolia aretioides* (desert cauliflower). **a** Fruiting plant. **b** Cushion habit of plants growing near Aïn Sefra on the Algeria-Morocco border. (Drawings adapted from Moquin-Tandon and Cosson 1862, plate 11 and photograph in Kew archives)

IV), occurring in a wide range of habitats and highly tolerant of extremes of drought and heat, it is usually one of the last species in the Sahara to succumb to prolonged drought. A decoction of the small spiny leaves is used to treat jaundice (Boulos 1983, 1992). *Cornulaca*, like other members of the Chenopodiaceae, has the xylem and phloem distributed throughout the root and not restricted to the periphery. This allows the root to survive even when the outer layers become dry. The plant's anatomy has been described by Pitot (1956).

Fredolia aretioides, syn. *Anabasis aretioides* (desert cauliflower): a monotypic chenopod forming rounded cushions of densely packed stunted stems and branches, which may be up to 50 cm in diameter and almost as high (Fig. 28). It is a Saharan endemic (ZB III). The dried trunk provides an excellent fuelwood, while the external crust is carefully cut for feeding to camels and sheep. The branches trap sand and as the branches atrophy, the spiny leaves form a peripheral crust. Its cushion habit offers protection against the violence of the wind and the high intensity of solar radiation. The seeds of *Fredolia* germinate extremely rapidly, the embryos unfolding within 10 min of the initial wetting. Ten hours later the roots are anchored in the soil and the spreading cotyledons capable of photosynthesis (Grenot 1974). The plant grows in the shallow depressions where sufficient run-off from a heavy shower may cause the soil to be moistened to a considerable depth. The roots take advantage of this accumulated water by extending downwards for a metre or more to reach the permanently moist layers; the roots also extend laterally within the upper layers moistened by the occasional light shower (Walter and Breckle 1986). Thus, such a two-tier system enables the plant to obtain maximum benefit from both the heavy and light showers, while the rapid germination and seedling establishment is a strategy to take advantage of what moisture is available before it has evaporated.

Halocnemum strobilaceum, syn. *Salicornia strobilacea*: a monotypic, jointed, fleshy, richly branched and prostrate subshrub up to 60 cm tall with long shoots and bud-like short shoots; leaves are paired and scale-like. It is easily recognised by the rudimentary leaves forming small, densely decussate, bud-like groups along the stem. It is a nebka-building shrub forming rounded hummocks up to 1.5 m diameter, a halophyte of salt marshes and saline dunes of the central Mediterranean region, Arabia and central Asia to Mongolia (ZB and ZE III, IV–VII, VII(rIII), VII–VIII). It is regarded as an indicator of non-potable saline water, and eaten by camels during the late autumn, and by sheep in severe winters. The water-soluble salts of the cell sap are a good raw material for domestic soda; it also contains a substance that is toxic to plant pests, especially if applied in a solution of soft soap. The plant may be grown as a salt-tolerant ornamental (Komarov 1960-1977, vol. 6).

Haloxylon ammodendron: an articulate C_4 shrub or small tree up to 2 m tall, bearing opposite, rudimentary leaves (Winter 1981). It is believed to have similar properties and uses to *H. persicum*. It is a native of Central Asia and Mongolia (ZB and ZE VII(rIII), VII-VIII). Extensive afforestation with *H. ammodendron* has been carried out in China to prevent wind erosion and control desertification. Early success was followed by a decline in growth and even deaths. Investigation revealed healthy growth was only possible where the relative water content of the soil was over 50% and that thinning to reduce plant density was necessary for healthy growth (Kebin 1989).

Haloxylon aphyllum (black saxaoul): a small tree up to 6(-10)m tall, with a relatively thick trunk of hard and heavy wood covered by an extremely succulent bark, which apparently serves as a water reservoir (Schimper 1960). Black saxaoul is a thicket-forming halophyte distributed through central Asia and Iran (ZB and ZE VII(rIII), VII-VI). The timber and fuelwood have similar properties to that of *Haloxylon persicum*. The yield of wood is ca. 40 ton ha^{-1}. It also provides a useful winter forage for camels and sheep, as well as being a source of potash and being used for stabilising dunes (Komarov 1960-1977, vol 6; Petrov 1972).

It is noteworthy for the dense mat of root hairs near the root apex and their mucilaginous secretion of pectic substances that cause the sand particles to agglutinate. Such permanent root hairs are believed to remain functional even during prolonged drought (Oppenheimer 1960). The cell sap is saturated with water-soluble sulphates and chlorides, thereby increasing the osmotic pressure of the cell sap and retarding respiration; osmotic pressure may reach as high as 8.8 MPa (Petrov 1972). It is a C_4 species (Winter 1981).

Haloxylon persicum (white saxaoul - owing to its ghostly white appearance during the long dry season): a spartoid, articulate, much-branched shrub or small tree, up to 6 m tall, with very deep roots and with a bole up to 20 cm diameter; bearing triangular, rudimentary leaves 0.5–1.25 mm long (Zohary 1962; Le Houérou 1985). It is a psammophyte of the Middle Eastern deserts, from eastern Egypt and Jordan eastwards to central Asia and Pakistan [ZB III, VII(rIII)].

White saxaoul is an important source of fuelwood, which has a calorific value approaching that of bituminous coal. The wood is durable and heavy, breaks easily but does not splinter. It is used for the production of activated carbon; the ash is used for the extraction of carbohydrates. In the Kara Kum the yield of wood is 25-50 ton ha^{-1}; Zohary (1940) gives the yield of charcoal as 5 ton ha^{-1}. The tree is a useful forage source for camels and sheep, especially during the winter. Extremely drought tolerant, surviving in areas with as little as 30-70 mm annual rainfall, it is suitable for cultivating in areas with upwards of 100 mm annual rainfall for fuel, fodder and shade; it was formerly used by the Bedouins as a detergent. It is also planted for dune stabilisation (Komarov 1960-1977, vol. 6; Petrov 1972; Batanouny 1981; Orev 1992).

Although capable of surviving the crown being covered by sand for a limited (unknown) period of time, the development of adventitious roots are an exception. The plant is also able to survive partial root exposure by deflation to 1 m (Danin 1996). The green cortical tissue from the previous summer, as well as part of the old shoots, dies back in summer and is shed, thereby greatly reducing the surface area for transpiration, although the phreatic water supply to the plant is maintained throughout the summer months. It is a C_4 species (Winter 1981). On a fresh weight basis transpiration reaches as high as 700 mg g^{-1} h^{-1} in January (Walter and Breckle 1986). The species is particularly noted for its ability to maintain the same rate of assimilation over a temperature range of 10-40 °C (Rodin 1981).

Haloxylon salicornicum, syn. *Hammada elegans, H. salicornia*: a shrub with jointed, cylindrical stems and scale-like leaves. Distributed through western North Africa and Arabia (ZB and ZE III, III–IV), it is the dominant species, albeit with less than 5% cover, of very shallow sand of the drainage channels of the Arabian Desert of Egypt. It is browsed by goats and camels. The white sweet fluid excreted by insects during the summer on vigorous plants was once believed to be the Biblical manna; *Tamarix mannifera* or *T. nilotica* are now considered to be the source.

It is physiologically most active during the summer months; new stems elongate in spring and summer, flowers appear in autumn and the fruits ripen at the beginning of winter. The green assimilatory stems become dry in winter and the distal portions that bore the flowers and fruit are shed, thereby reducing the transpiration surface. New stems sprout in spring from buds located at the bases of the previous year's stems. Although the seedlings have leaves ca. 1 cm long, those of the mature plants are minute and scale-like. In the scale axils are buds covered by hairs, which provide a moist microclimate protecting the buds from both heat and aridity (Danin 1983; Kassas and Batanouny 1984). The decline in stomatal conductance in response to increased vapour pressure difference was found to reduce transpiration to about half of the rate that would have been obtained if the stomatal conductance had remained constant. Stomatal conductance was so great as to result in a lower actual transpiration rate at high rather than at low vapour pressure differences (Schultze et al. 1972).

Haloxylon scoparium, syn. *Hammada scoparia, Haloxylon articulatum*: an arido-active dwarf shrub up to 60 cm tall with woody, erect, intricately branched, fleshy, articulate stems, the leaves reduced to minute scales, their photosynthesising function being undertaken by the green bark revealed by the previous year's exfoliating bark. It is distributed from Spain and North Africa to south-western Asia (ZB and ZE III, IIIwr, III–IV, IV, IV–III). Minced fruiting branches mixed with camel's fat or butter are used as a cataplasm for treating snake bites (Boulos 1983; Gibson 1996). It is noteworthy for the dense mat of root hairs near the root apex and their mucilaginous secretion of pectic substances that cause the sand particles to agglutinate. Such permanent root hairs are believed to remain functional even during prolonged drought (Oppenheimer 1960).

The decline in stomatal conductance in response to increased vapour pressure differences was found to reduce transpiration to about half of the rate that would have been obtained if stomatal conductance had remained constant. Stomatal conductance was so great as to result in a lower actual transpiration rate at high rather than at low vapour pressure differences (Schultze et al. 1972).

The photosynthetic organ is the stem and it is composed of a slightly succulent cortex containing two layers rich in chlorophyll and a non-chlorophyllous scleromorphous inner core The temperature dependence of net photosynthesis under conditions of light saturation and maximum air humidity was measured throughout the season for *H. scoparium* growing in the Negev Desert. The optimum temperature for net photosynthesis at the upper temperature compensation point for CO_2 exchange showed a characteristic seasonal variation, with low values in the spring and autumn and high during mid-summer. The shift was exhibited by plants growing under normal soil-water stre ss as well as by irrigated plants. No general correlation was found between the changes in temperature dependence of net photosynthesis, their maximum photosynthetic capacity under experimental conditions, their daily photosynthetic maximum under natural conditions and their rate of dark respiration. The seasonal shift of photosynthetic responses to temperature cannot be explained by changes in temperature sensitivity of the stomata. It may be due to seasonal changes in biochemical and/or biophysical properties. A number of desert and other wild plants, *Reamuria negevensis, Salsola inermis, Seriphidium herba-alba* and *Zygophyllum dumosum,* also showed seasonal shifts of the upper temperature compensation point, with an amplitude of 6-13.7 °C (Lange et al. 1974).

The daily maximum temperature of *H. scoparium* peaked in May and decreased during the second half of the growing season. A correspondence was found between the daily maximum tissue temperature (also the average temperature at the warmest period of the day) and seasonal changes in the optimum temperature of net photosynthesis (OP) and the upper compensation point of CO_2 exchange (CP). This may indicate that shifts in temperature sensitivity of net photosynthesis are adaptations to temperature conditions of the plant. The seasonal changes of OP and CP values corresponded to changes in the daily photoperiod and daily average light intensity. It is also possible that the correlation indicates a casual relationship (Lange et al. 1975).

Kalidium foliatum, syn. *Kalidiopsis foliatum, K.wagenitzii*: a fleshy subshrub up to 75 cm tall, with fleshy, linear leaves 2-6 mm long. A plant of saline habitats in south-western and central Asia [ZB VII(rIII)], it is a poor forage species, being browsed by camels, sheep and goats only in the late autumn and winter (Komarov 1960-1977, vol. 6).

Maireana aphylla, syn. *Kochia aphylla* (cotton bush – from the clusters of white, cotton-like, insect galls): an up to 1.5-m-tall shrub, densely branched with slender, striate, flexible branches that become rigid and spinescent with age. Crowded along the branches, the thick, subterete, 1 to 4-mm-long leaves are soon shed from the short side branches, which remain bare, giving a spiny appearance to the shrub. The inner foliage is maintained well into the dry season. Widely distributed in Australia, especially southern and central regions (ZB and ZE II–III, III, III–II, IV–III, IVa), and highly drought resistant and tolerant of fire, cotton bush occurs on cracking clays and alluvial plains subject to periodic inundation. Not particularly palatable except when young, it is only lightly browsed by livestock who readily eat the young shoots but not the old. Under heavy grazing it is replaced by species of *Sclerolaena*, annual *Atriplex* and other unpalatable annuals (Hall et al. 1972; Jackson and Jacobs 1985; Mitchell and Wilcox 1988; Lazarides and Hince 1993).

Maireana pyramidata, syn. *Kochia pyramidata* (black bluebush): a compact, divaricately branched, up to 2-m-high, greyish-green, long-lived, dioecious shrub with subterete, 2 to 6 mm long leaves, the fruiting bodies with characteristic pyramid-shaped, spongy extensions above the narrow, horizontal wing encircling the short basal tube. In addition to the deep roots there are shallow, deciduous feeder roots that are shed during the dry season. It is a native of south-eastern arid Western Australia (ZE II–IIIswr, III–IV, III-IVswr, IV–III), growing on somewhat saline calcareous red and red-brown soils with the limestone less close to the surface than is the case with *M. sedifolia*.

Long-lived and highly drought resistant as well as producing prolific seed, it is considered suitable for growing as drought forage although heavy browsing by livestock and ringbarking by rabbits may eliminate or greatly reduce stands. The browse is not particularly palatable. The leaves have a crude protein content up to 22% and an in vitro digestibility of ca. 60%; it can be toxic if large quantities are eaten by hungry stock (Hall et al. 1972; Jackson and Jacobs 1985; Mitchell and Wilcox 1988; Lazarides and Hince 1993).

Maireana sedifolia, *Kochia sedifolia* (pearl bluebush): a compact, much-branched, long-lived, bluish-grey, dioecious perennial up to 1.5 m tall, with thick, upright, woody stems forming a dense clump. Very drought resistant, the fleshy, narrowly obovoid, 4-8-mm-long leaves are densely clothed in white hairs and are borne on side-shoots that are rapidly shed during dry periods. There is a deep root system to 3–4 m as well as shallow, deciduous, feeder roots that are shed during the dry season. A native of southern Australia (ZB and ZE II–III, III-IV, IVa), it is the major shrub of the Nullabar Plain, sometimes forming pure stands, although restricted to calcareous soils with limestone close to the surface where the rainfall can penetrate. It is thus able to withstand longer periods of drought than *Atriplex vesicaria*. The shrub has a reputed life expectancy of 30-60 years. The foliage has a crude protein content of 20% and 10% salt, but is not particularly palatable and is usually only browsed when fresh water is available. It is capable of surviving complete defoliation by stock; it may become toxic to stock following rainfall. Regeneration from seed is poor. It is a useful plant for erosion prevention when planted less than 3 m apart.

Pearl bluebush is a facultative halophyte with both a low osmotic pressure of the cell sap and a low chloride content (Hall et al. 1972; Walter and Breckle 1986; Mitchell and Wilcox 1988; Lazarides and Hince 1993). Both *Maireana pyramidata* and *M. sedifolia* have been introduced into other arid and semi-arid countries of the world and are regarded in Israel as being outstanding amenity plants (Aronson 1985).

Rhagodia baccata (sea berry or coastal saltbush): an up to 2-m-tall shrub with a large taproot and numerous fine lateral roots in the upper 10 cm of the soil, bearing numerous lateral shoots with small, semi-succulent, isolateral leaves up to 4 cm long. It is an endemic of south-western Australia (ZB IVa and the relatively humid IVh) and a useful browse species. The cuticle is thin, with simple stomata. There is an extensive palisade mesophyll and a centrally arranged, water-bearing tissue with large, spherical cells containing calcium oxalate crystals (Fig. 29). The young leaves are covered with soon deciduous, vesicular hairs. Normally the leaves on the lateral shoots have an average

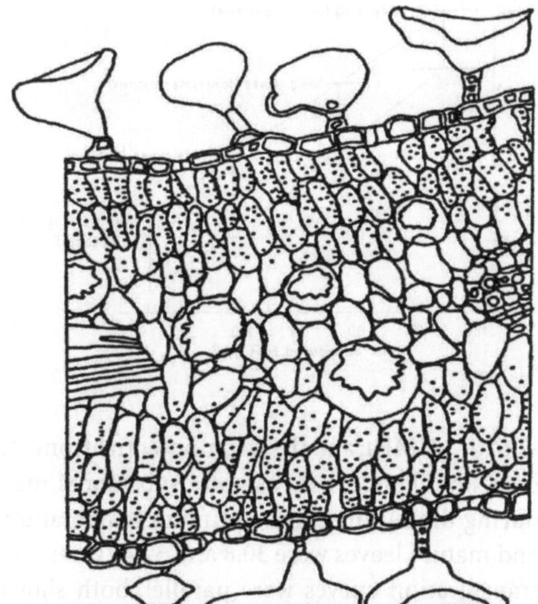

Fig. 29. Portion of transverse section through leaf of *Rhagodia baccata*. Note epidermal hairs with terminal, balloon-shaped and flattened vesicles, simple stomata, extensive pallisade mesophyll and centrally arranged, water-bearing mesophill with large, spherical cells containing calcium oxolate crystals. (Hellmuth 1968)

surface area of 0.45 cm^2, but following rain in excess of 25 mm, quite large leaves with a surface area of 0.85 cm^2 rapidly develop. The leaves have a fish-like smell due to the presence of trimethylamine, one of the hydrolysis products of choline, a constituent of lecithin.

It is extremely drought resistant, but, like other species of *Rhagodia*, it may be damaged in drought years, although recovering after the first rain. No heat damage was observed in the field. Germination is rapid, occurring within ca. 5 days from soaking, with 98% germination, unaffected by light or dark. Vegetative propagation from cuttings was also satisfactory, following establishment in vermiculite. The shrub is in a state of water equilibrium throughout the year. In winter the stomata are open during the day and at ca. 14.00 h the transpiration curve shows a maximum of almost 70 mg dm^{-2} min^{-1}. In summer, with a six-fold increase in evaporation, there is a water deficit in the early morning and the stomata close; between 10.00 and 17.00 h transpiration falls to almost zero, rising again to almost 30 mg dm^{-2} min^{-1} as the stomata open towards evening. There appears to be a direct relationship between water deficit and the progressive loss of indumentum from very young tomentose leaves to the glabrous mature leaves (Fig. 30).

Because of the small water deficit the water and osmotic potential exhibit only very small daily variations. In summer the osmotic potential falls from

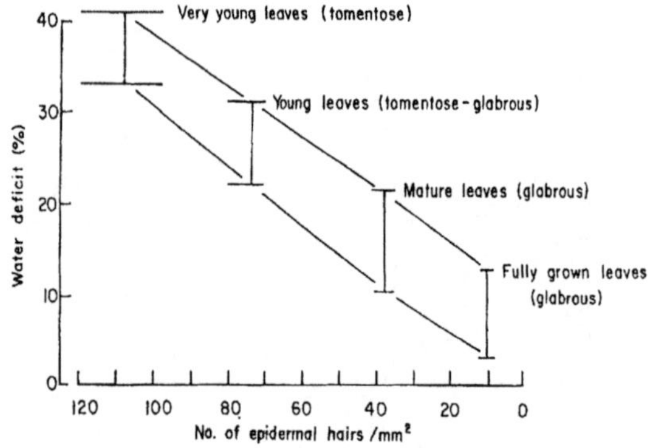

Fig. 30. Relationship between water deficit in *Rhagodia baccata* and number of epidermal hairs on leaves of various ages. *Vertical lines* indicate seasonal (summer and winter) range in water deficit; *horizontal lines* indicate range in the number of hairs counted (average of 14 samples for each age group (Hellmuth1968)

-2.6 to -3.7 MPa, and the water potential from -1.6 to -2.85 MPa. As turgescence does not change, new osmotic substances must have formed in the cell sap during the summer. The optimum temperatures for photosynthesis in young and mature leaves were 30.8 and 37.4 °C respectively. The photosynthetic and transpiration curves were parallel, both showing one winter peak and two peaks in summer, when a nil point was reached around noon (Hellmuth 1968; Walter and Breckle 1986).

Salicornia europaea, syn. *S. herbacea* (glasswort or marsh samphire): a glaucous or grass-green, prostrate to erect, segmented, succulent, annual herb; an aggregate species. With a Mediterranean and Euro-Siberian distribution [ZB and ZE III, III–IV, IV, IV–VII, VII(rIII)], it is an obligate halophyte, often a pioneer of dried-up, saline marshes. The plant may be cooked and eaten in soups or as a vegetable. It exhibits a neotene growth pattern, with provisional flowering in the juvenile stage and starting regrowth when new rains come in time. This can lead to the plant carrying flowers and ripe seeds simultaneously. The flowers occur in groups of three, the central flower producing one large seed and each lateral a small seed. After soaking for 59 days in distilled water the large seeds gave 90% germination, the small seeds 50%.; the small seeds also appear to be less salt tolerant than the large seeds (Launert 1981; von Willert et al. 1992; Gutterman 1993).

Sarcobatus vermiculatus (greasewood): a monotypic shrubby chenopod up to 2.4 m tall with fleshy, narrow, winter-deciduous leaves, evergreen in the warmer areas of its distribution, such as the Mojave Desert. The species is widely distributed, from Alberta to North Dakota and south to California eastwards to western Texas [ZB and ZE III, VII, VIIa, VII(rIII), VII–VI]. A nebka-building halophyte, it has been found in soils with a near-surface sodium percentage of up to 50% and in culture solutions to tolerate a concentration of 1 M NaCl (-4.4 MPa); Smith and Nowak (1990).

The hard, yellow wood is used for fuel and native hand tools. The plant is a possible source of shellac. The young leaves and shoots are browsed by sheep and cattle in winter and spring, but if eaten to excess can cause bloat and perhaps oxalate poisoning. The plant may be cooked and eaten in soups or as a vegetable (Kearney and Peebles 1951; Mabberley 1987). The taproot may descend to 5.7 m and the roots bear rhizosheaths, albeit not as tightly packed as those found in the Gramineae. Stems buried by drifting sand are able to produce adventitious roots (Danin 1996).

CISTACEAE

Helianthemum ledifolium: a rather variable, tomentose, annual herb up to 60 cm tall, bearing elliptic leaves up to 5 cm long. It is a herb of loess soils, distributed through southern Europe to the Middle East [ZB and ZE III, IV, IV-III, IV-VII, VII(rIII)]. It is a host to the much valued subterranean ascomycetous truffle *Terfezia leonis*. The truffles are prized for their flavour and are gathered wild in the Negev and sold in the local markets.

Helianthemum lippii, syn. *H. sessiliflorum*: an evergreen shrublet up to 25 cm tall, intricately branched from the base, bearing small leaves with strongly revolute margins, the branches becoming spinescent after flowering. A plant of sandy soils, it is widely distributed through North Africa eastwards to western Pakistan [ZB and ZE II–III, III, III–IV, IV, VII(rIII)]. It is an important browse plant, contributing ca. 9 kg ha^{-1} $year^{-1}$ towards the diet of domestic livestock. Protecting from livestock during the growing season resulted in the accumulation of total available carbohydrates in all organs, soluble sugars in the reproductive organs and crude protein in the aerial parts, while dry season protection increased the crude protein and soluble nitrogen in the vegetative organs. The plant is the host to the truffle *Terfezia* spp., which are regarded by the Arabs as a great delicacy (Townsend and Guest 1980). The inrolling of the leaves protects the stomata from exposure to the atmosphere and thereby is an effective means of reducing transpiration.

COMBRETACEAE

Conocarpus lancifolius (damas): an evergreen tree up to 30 m tall, with a spreading crown and bearing lanceolate leaves up to 12 cm long. The tree is endemic to northern Somalia and has been introduced to neighbouring countries. Damas tolerates a hot, dry atmosphere and a long, dry season with less than 400 mm annual rainfall [ZB III, III(tI)]. It occurs on the alluvial terraces of seasonal streams and also near the coast where springs emerge or the water table is high; a phreatophyte, it requires a watertable within 8 m of the surface. Regeneration in the wild from seed is good. Although salt tolerant, coastal trees are damaged by salt spray carried by the wind.

Accessible branches of *Conocarpus* are browsed by goats and game animals; in times of drought the tree is lopped to provide fodder for stock. An alcoholic liquor may be extracted from the heavily perfumed flowers. Grown as a street tree in Mogadishu and Khartoum, the tree may also be grown for windbreaks, soil stabilisation and for the reclamation of limestone quarries. The crushed leaves are used for poisoning fish. The gum exudate is used in human and veterinary medicine for the treatment of chest and bowel complaints.

The timber is light coloured, hard grained, medium-heavy density, strong, with long life in salt water. The young branches are brittle, while trunks of mature trees tend to split. The wood was formerly used for dhow construction. The poles are used for house construction, carpentry, fencing posts, carving and turnery. The wood also makes good firewood and excellent charcoal (Little 1983; NAS 1983; IBPGR/RBG 1984; Booth and Wickens 1988; Dahir 1985).

COMPOSITAE

Ambrosia deltoidea, syn. *Franseria deltoidea* (burrobush or triangle-leaf bursage): a drought-deciduous shrub up to 60 cm tall, with a taproot to 1.8 m and bearing narrow, resin-coated, triangular leaves up to 3 cm long It is a native of Arizona, Sonora and Chihuahua (ZB IIIswr). Freshly cut stems may be used to deter rabbits. The plant has a potential; as an ornamental (Turner et al. 1995; Gibson 1996). The photosynthetic pattern was examined by Szarek and Woodhouse (1976, 1977). Summer and winter gas exchange were measured at near optimum soil water conditions. Under such conditions the diurnal gas exchange patterns were characterised by maximum photosynthetic rates and stomatal conductance early in the day and progressively decreased thereafter. The seasonal photosynthetic pattern followed seasonal variation in plant water potential and is characterised by a marked seasonal variation in plant water potential, becoming leafless for approximately 4 months when the plant water potential fell below -5.5 MPa. The species maintained a relatively high water use efficiency throughout the year despite the high evaporation gradient. Following the rainy season the lateral roots of *A. deltoidea* develop short rootlets 1.2–5.0 cm long, which enable the plant to absorb moisture rapidly from the soil. They cease activity when the soil dries (McGinnies 1981).

Artemisia monosperma (sand or single-seed wormwood): a thick-stemmed, up to 70-cm-tall, semi-deciduous shrub. A nebka-building species, widely distributed through north-eastern Africa and Arabia (ZB and ZE III, III-IV), the stems may be used for fuel and building huts. The species recovers well after coppicing. Seeds require long days for germination at 10-25 °C, but this effect disappears above 30 °C. Light, irrespective of temperature, is essential for germination. The mucilaginous achenes require only a light covering of approximately 2 mm sand, sufficient to provide moisture but not to exclude light; the seeds will not germinate if deeply covered and become desiccated and die if completely exposed. The achenes even increase their sensitivity to light after 6–8 days' imbibition in the dark. Given adequate moisture the achenes germinate in 10 days (Gutterman 1993).

Heterophyllous, in winter the new stems, bearing large leaves, sprout from the buds at the base of last year's stems. Towards summer smaller leaves are produced at the distal end of the new stems and in the axils of the winter leaves; the leaf dry weight per unit length is thereby reduced by 76%. Flowering is in the autumn, when the winter leaves are already dry. The deep taproot is protected from desiccation by a thick corky bark. This is an important property for survival in areas where winds may expose a metre or more of the root. Stems covered by a thin layer of sand will develop adventitious roots as well as new stems that grow above the surface, however, if too deeply buried the plant will die. A maximum water saturation deficit of 34% has been recorded and a maximum summer osmotic value of 2.3 MPa, with a transpiration rate of 476 mg g^{-1} h^{-1} for March and 492 mg g^{-1} h^{-1} for November (Oppenheimer 1960; Zohary 1962; Danin 1983, 1986; Gutterman 1993).
Chrysothamnus nauseosus (rubber rabbit-bush): a highly variable, evil-smelling, deciduous, C_3, composite shrub, up to 2 m tall with roots descending to at least 2 m. Numerous, slender, erect, flexible stems are produced annually from a woody base. It is widely distributed in North America (ZB and ZE II–III, III, IV, IV–III, VII, VII–V). Its browse potential is negligible. The flowers yield a yellow dye. The inner bark is chewed as a masticatory by the Piaute people. It may be grown for landscaping. A source of the readily vulcanising chrysil rubber, it also has a potential as a crop for hydrocarbons, insecticides and fungicides. The two gridelic acid derivatives (18-hydroxy grindelic acid and 18-succinyloxy grindelic acid) isolated from the plant have shown significant antifeeding properties against the third instar larvae of the Colorado potato beetle, *Leptinotarsa decemlineata* (Kearney and Peebles 1951; Saunders 1976; Timmermann and Hoffmann 1985; USDA 1988).

Compared with other C_3 species, *Chrysothamnus* exhibits high rates of net photosynthesis. On both a leaf area and dry weight basis *C. nauseous* is similar to the C_4 desert shrub *Atriplex canescens*, although the RuBP Case (ribulose 1,5-bisphoshate carboxylase oxygenase) and intercellular CO_2 contents are similar to those found in most C_3 crop species. The net photosynthesis is not light saturated at light levels equivalent to full sun, possibly due to the presence of the tomentose indumentum on the leaf surface. The indumentum is believed to shade the chloroplasts, reduce heat load and lower leaf temperature. The high rates of net photosynthesis are accompanied by high stomatal conductance values for well-watered plants; the reaction under stress was not examined. Due to the capability of maintaining high net photosynthesis and high stomatal conductance under non-stress conditions, the potential exists for high rates of dry matter accumulation per unit of biomass. The high rubber and resin contents indicate that the isoprenoid biosynthetic pathway is quite active, although the controlling functions are not known (Weber et al. 1985).

Encelia farinosa (brittle bush): a low branching, drought-deciduous, cushion shrub up to 2 m tall and bearing subglabrate to densely white-farinose, lanceolate to ovate leaves up to 10 cm long. Abundant annual rootlets are produced when the soil is moist and are presumably shed when the soil dries out. It is distributed from Nevada to California and Arizona to Sonora and Sinaloa (ZB and ZE II–III, IIIsr, IIIswr, IIIwr, VIIa). It is reported to be browsed by the American bighorn, *Ovis canadensis*. The shrub is often sufficiently abundant to be used as fuel. During the dry season the leaves and stems exude a dark brown, viscous sap from every wound. This 'gum' was chewed by the native Americans and may also be used as an arrow varnish, adhesive and for treating sores; when melted the gum is smeared on the body to relieve body pains. The gum was also used by the Spanish as an incense in the churches of Baja California. There is no exudate during the rainy season (Standley 1926; Kearney and Peebles 1951; McGinnies 1981; Felger and Moser 1985; Mabberely 1987).

During the rainy season the leaves are large, soft and hydromorphic but not very hairy. As the dry season advances the new leaves become progressively more mesophytic, smaller and increasingly hairy, eventually becoming very small, xerophytic and very hairy. The air-filled dry-season hairs may reflect up to 70% of incoming solar radiation between 400 and 700 nm, i.e. in the photosynthetically active range. The leaves are ultimately shed, leaving small, hairy leaf primordia, thereby dramatically reducing transpiration. Absorption of the leaves varied by 10% between plants collected at different sites and the absorptivity declined steadily as the season advanced and the average maximum ambient temperature increased. Together with a concomitant reduction in leaf size during the season, this attenuation of the absorption of solar energy yielded a more favourable energy balance and thus reduced leaf temperature as the environmental conditions harshen. There was also a strong correlation between increased pubescence and decreased amount of precipitation received during the current season prior to sampling; pubescence also increased as irradiance increases. However, leaf pubescence did not affect water vapour diffusion to the extent where it could be considered to be of ecophysiological importance (Walter and Breckle 1986; von Willert et al. 1992; Turner et al. 1995).

The plant contains encecalin, which is toxic to the milkweed bug, *Oncopeltus fasciatus*, and acts as an antifeedant to the bollworm, *Heliothis zea*. The leaves contain a non self-inhibitory toxin identified as 3-acetyl-6-methoxybenzaldehyde, which has an allelophic effect on other plants when released from the decomposing fallen leaves. By inhibiting the growth of annuals, it thereby ensures for itself the available moisture within the immediate vicinity (Harborne 1988).

Grindelia camporum (gumweed): a resinous, evergreen, herbaceous, perennial, 0.5 to 1.5-m-tall shrub. A native of California (ZB IV), it is xerophytic and salt tolerant, the more sparsely branched diploid populations occur in the Sierra Nevada foothills and the north coast ranges, while the tetraploid populations occurring in the saline flats, disturbed fields and roadsides of the Central Valley region of California (Timmermann and Hoffmann 1985). Economically important, gumweed is currently undergoing investigation at the University of Arizona as a potential source of biocrude and as a resin substitute for the wood gum rosins currently being used for naval stores (Timmermann and Hoffmann 1988). A decoction of the leaves and young inflorescences was used by the Californian native Americans as a mild stomachic, taken to purify the blood, as well as to relieve throat and lung troubles (Saunders 1976). Gumweed produces large quantities of terpenoid resin that coat the stem, leaves and involucre. The production and accumulation of the resin appears to be associated with the presence of multicellular resin ducts which occur in the leaf mesophyll and stem cortex, and sessile multicellular resin glands present in shallow pits or depressions on the surface of the stem, leaves and phylaries. These resin glands are most abundant on the involucres of the flower heads, less dense on the leaves and virtually absent on the stem (McLaughlin 1986; Timmermann and Hoffmann 1988).

Timmermann and Hoffmann (1988) noted that the prevailing theories regarding the ecophysical significance of these hydrophobic, non-volatile plant resins are:

1. The resin renders the cuticle less permeable to moisture and thereby reduces cuticular evapotranspiration.
2. By increasing the reflectance of the leaves, the penetration of solar radiation is reduced and internal temperatures decreased.
3. They contain plant defensive substances.

Although such theories are not mutually exclusive and may even act simultaneously, the authors note that the third hypothesis agrees with their own observations and that similar diterpines isolated from other Compositae have known antifeedant activities.

Gutierrezia sarothrae (broom snakeweed): an unpalatable and aggressive bushy, suffructescent composite up to 40 cm high with semi-erect branches and linear and slightly resinous, 2 to 4-cm-long leaves with inrolled margins. It is widely distributed through western USA, extending to Baja California and Chihuahua to Nuevo León (ZB and ZE II–III, III, VII), generally present throughout the short-grass prairie, especially on shallow soils overlying calcrete (caliche). It apparently does not occur on saline or alkaline soils. Broom snakeweed is regarded as an indicator of overgrazing, especially by sheep. The plant may be grazed by horses during the winter and by sheep when growth starts in early spring before alternative fodder is available, but is ignored later when more palatable species are available. The species contains saponins and may be toxic to cattle, sheep and goats causing either death or abortion if eaten to excess. In New Mexico a decoction of the plant is used as an emmenagogue and as a remedy for gastric disorders.

Broom snakeweed, with a shallow but extensive root system, is able to utilise relatively small amounts of rainfall, as is evident from its highly fluctuating predawn xylem water potential ranging from -0.5 to -3.5 MPa. It produces abundant seed, requiring 6 months' maturation at room temperature followed by constant temperature between 15 and 25 °C with an 8-h light period for high germination. The high temperature required for germination prohibits germination during warm winter periods which could be followed by frost mortalities. The seeds remain viable for several years. Although broom snakeweed has no allelopathic effect on *Sporobolus airoides*, it did afffect *Bouteloua curtipendula* and *B. gracilis*, which may account for its invasive ability, although populations tend to be cyclic and relatively short-lived (Standley 1926; USDA 1988; Briedé 1990, McDaniel et al. 1991; Hatch and Pluhar 1993).

Low water-use efficiency and a high degree of drought tolerance are the main physiological defences of broom snakeweed to water stress. Stomatal conductivity of broom snakegrass was more reduced by mild water stress with a soil water potential of -0.2 MPa than were canopy development and photosynthesis. Canopy development, however, was most affected by moderate to severe soil water deficit (.1 MPa), followed by photosynthesis; transpiration and leaf conductance were least affected. Under severe water stress water loss was controlled more through reduced canopy development than by stomatal closure, while photosynthesis was limited more by mesophyll conductance than by stomatal conductance. Water use-efficiency was not affected by initial water stress, but as the soil water deficit developed water-use efficiency declined in response to non-stomatal limitation to photosynthesis and the less sensitive stomata to severe water deficit (Wan et al. 1993a).

Soil water content in the upper 20–30 cm of the soil profile is the single most important factor determining mortality. As the soil water deficit developed root proliferation decreased and, with increasing soil water potential, root growth was favoured over shoot growth. The expansion of the photosynthetic area was more sensitive than stem growth to water deficit. Carbon allocation within the shoot was more sensitive to water deficit than the carbon allocation between root and shoot. Plants died when the soil water potential reached -7.5 MPa (Wan et al. 1993b).

Nassauvia glomerulosa: an evergreen, low-lying composite subshrub. The dominant subshrub of the Patagonian Desert (ZE VIIa-IX), it is regarded as a fairly good sheep browse. Like a number of other xerophytic members of the genus, the hairs contain crystals of calcium oxalate, presumably a method by which the plant neutralises any excessive oxalic acid present (Ragonese 1987).

Parthenium argentatum (guayule): a facultatively apomictic, up to 1-m-tall shrub, much branched from the base and bearing pinnatifid leaves, with a much branched root system descending to 6 m or more. It occurs on the outwash fans and slopes of the calcareous soils of south-western Texas and north-central Mexico, at altitudes between 700 and 2000 m, in areas with an average annual rainfall, although erratic, of 250–373 mm, falling mainly in the late spring to early autumn (ZB and ZE VII, VII(rIII), VII–V), with temperatures varying from -23 to 49 °C.

Guayule is the source of guayule rubber. In the USA commercial extraction began in 1888, but the industry was crippled by the depression of the 1930s and revitalised in 1941 by the Japanese invasion of Southeast Asia and the loss of hevea rubber. Industrial exploitation for guayule rubber began in Zacatecas as early as 1892-1893 and shortly after in San Louis Potosí it proved unsuccessful due to poor processing equipment and the absence of a market, but made some recovery fo llowing the importation of US equipment in ca. 1904 (Standley 1926). There were even experimental plots in Tanzania, then German East Africa, in the early 1900s, but the trials proved to be uneconomic (Anonymous 1906).

The rubber is located mainly in the woody parts, with approximately two thirds in the stem and branches and the remainder in the roots; there is no rubber deposited in the leaves. The chief site of rubber synthesis is the bark, which may contain up to 75–85% of the total rubber weight. Rubber content of the stems is upwards of 20% of dry weight, depending on the season, genotype and age of the plants. Most of the rubber is contained in single cells located in the vascular rays of the stem phloem and xylem with some in the living parenchyma of the stems and roots, not in ducts as in *Hevea.* Rubber synthesis begins in the late autumn and winter, when night temperatures reach 4-7 °C. The relative absence of breeding barriers makes improvements through hybridisation with other species of *Parthenium* a possibility (Goss et al. 1984; Youngner et al. 1986).

The roots produce a specific self-inhibiting growth toxin (E)-cinnamic acid, a simple aromatic compound, the allelopathic effect of which is to ensure that roots of adjacent guayule plants do not intermingle, thereby eliminating competition for water and nutrients. However, it does not appear to affect other species. Since cinnamic acid does not last long in the soil it has to be presumed that the toxin must be continuously produced by the root in order to exert any allelopathic effect. Although toxic to *Parthenium* growth at a concentration of 0.0001%, tomato seedlings were only affected at a concentration of 0.01% (Harborne 1988).

Rhanterium epapposum: an intricately branched, somewhat spinescent subshrub bearing few leaves. It is distributed through Arabia and Iraq [ZB III, VII(rIII)]. Because of its absolute dominance in some areas Walter and Breckle (1986) consider this to be an indication of its poor value as forage; however, Batanouny (1981) considers it to be a valuable browse. The plant dies back almost to the base during drought, with new shoots produced after rain. Sand accumulates around its base, trapped by the numerous branches.

Seriphidium, syn. *Artemisia* subgenus *Seriphidium*: differs from *Artemisia* (syn. *Artemisia* subgenus *Artemisia*) chiefly in having sessile, ovoid or ellipsoid capitulae with all the two to ten florets hermaphrodite and in paniculate inflorescences or compound capitulae, while the inflorescences of *Artemisia* are paniculate, spicate, racemose or corymb-racemose with solitary or clustered capitulae. with heterogamous florets with pappus and capitulae receptacle scales absent (Ling 1991, 1992).

Seriphidium herba-alba, syn. *Artemisia herba-alba*: an aromatic, shrubby perennial. It is distributed through south-western Europe, Asia Minor and north-western Africa (ZB and ZE IV, IV–VII). Despite being aromatic it is browsed by livestock in some areas and in other areas, apparently without any change in floral composition, it is ignored. This is suggestive of the possible presence of chemical races (M. Kernick 1993. pers. comm.). The various parts have a number of medicinal uses; it is widely used in Iraq folk medicine for treating diabetes mellitus. An essential oil distilled from the plant is regarded as a good antiseptic and insecticide. It is also used as a parasiticide in veterinary medicine (Boulos 1983). Al-Khazraji et al. (1993) have shown that the oral administration of 0.39 g kg^{-1} body weight of an aqueous solution of the leaves or bark significantly reduces the blood sugar. The stems are also used by the Bedouin for making huts. The white woolly insect stem galls may be used for tinder.

Towards the desert limits of its distribution the root system is extremely shallow, with the primary root forming branches at 2 to 5-cm intervals, while with higher rainfall the roots may descend to 40 cm or more, penetrating into the vertical fissures of the underlying limestone and ramifying only in the rock. Heterophyllous, the plant achieves a reduction in dry season transpiration by shedding its large winter leaves, which are replaced in summer by smaller leaves with fewer lobes and densely covered with white hairs. In drought years the summer leaves will be retained and the winter leaves will fail to develop. After several years with insufficient rain some sections of the plant may even die (Oppenheimer 1960; Danin 1983).

During the dry season the plant has a very low water content, with a negative hydrostatic pressure in the xylem as low as -16 MPa and an extreme osmotic potential in the leaves of -9 MPa; a diurnal water stress does not decrease greatly during the night. Yet even under these conditions *S. herba-*

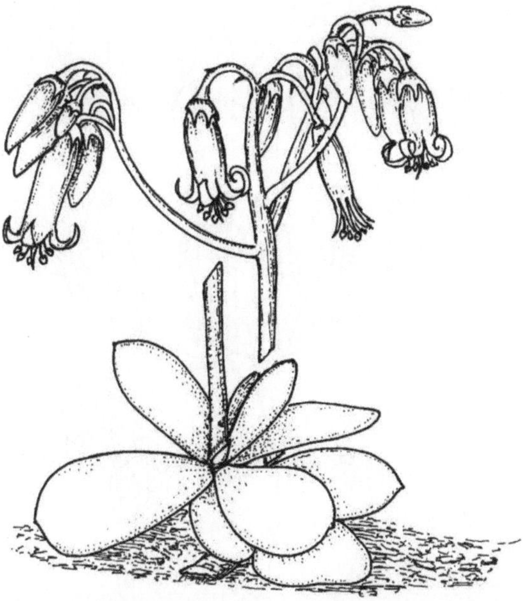

Fig. 31. *Cotyledon orbiculata.* Habit showing a basal rosette of succulent leaves. (Adapted from plate in Kew archives of a cultivated plant from South Africa at Kew in 1933)

alba remains photosynthetically active for a few hours each day throughout the entire dry period (Kappen et al. 1972; Kassas and Batanouny 1984). Friedman et al. (1977) have demonstrated that the shoots contain volatile and water-soluble substances which strongly inhibit the germination of *Helianthemum ledifolium* and *Stipa capensis.* The allelopathic effect is less effective on *Zygophyllum dumosum* and had no effect on *Medicago laciniata.*

CONVOLVULACEAE

Convolvulus lanatus: a low, spiny shrub up to 35 cm or more tall, with shoots woolly when young, becoming glabrescent, bearing linear-spathulate to linear leaves. A drought resistant psammophyte, distributed from Egypt to Saudi Arabia (ZB III, IV), it is a favourite forage for camels. The taproot is adapted for resisting desiccation by a thick corky bark which protects the vascular system even when 2 metres have been exposed by deflation (Danin 1983).

Cressa cretica: a prostrate to erect subshrub up to 25 cm tall arising from a woody rootstock, with leaves ovate to lanceolate and up to 6 mm long. It is distributed from the Mediterranean eastwards to India, tropical Africa, Australia and North America [ZB and ZE II–III, III, III(tI), III–IV, IV–V]. The crushed, dried leaves are taken with sugar as an energetic purge in cases of jaundice (Boulos 1983). *Cressa cretica* is a halophyte excreting excessive salt by means of glands in the leaves.

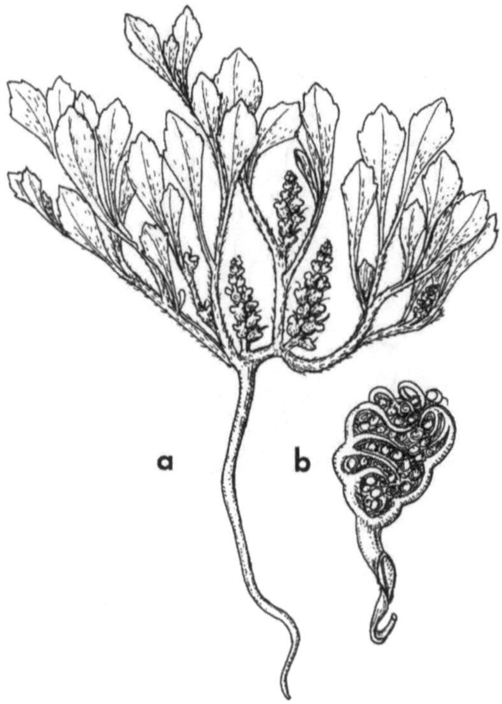

Fig. 32. *Anastasia hierochuntica.* **a** Habit. **b** Detatched, dry and shrivelled portion of fruiting plant which acts as a tumble weed. (Adapted from plate 166 by Laura Ripley in Townsend and Guest 1980)

CRASSULACEAE

Cotyledon orbiculata: an erect, succulent, woody perennial up to 60 cm or more tall bearing broad, succulent leaves (Fig. 31). A native of South Africa (ZE III-II), it is widely used in South African medicine. The leaf contains cotyledontoxin, possibly in sufficient quantity to cause cotyledonosis, which is characterised by spasmodic contractions of the muscles, torticollis and other neuromuscular symptoms in sheep, goats, horses, cattle, dogs and poultry. The plant is cultivated as an ornamental (Watt and Breyer-Brandwijk 1962).

CRUCIFERAE

Anastatica hierochuntica (rose of Jericho, resurection plant or Mary's hand): a monotypic annual herb which sheds its leaves as the seed matures Fig. 32). It is a Saharo-Arabian tumble weed (ZB III), often occurring in pure stands due to the species-specific allelopathic effect which, together with an autotoxic effect, ensures a satisfactory spacing for establishment without excessive competition. An infusion of the dried plant reduces pain and facilitates childbirth; it is also used as an emmenagogue, for epilepsy and colds (Boulos 1983).

The hygrochastic woody branches fold inwards around the pods as the plant dries and open out again when moistened. The small, 1 to 2-mm-long seeds are released when the pod separates along a dehiscence zone about 1.5 h after wetting; the seeds then adhere to the soil by their mucilaginous outer layer; the seeds are also dispersed by surface wash. Under suitable conditions, germination takes place within 6 h. The newly germinated seedling can withstand dehydration for a week provided the rootlets are no more than 4–6 mm long, resuming growth again within 6 h after wetting; they do not survive if the rootlets are longer (Mabberley 1987; Boulos 1989; Hegazy et al. 1990; Gutterman 1993).

Zilla spinosa: a dichotomously branched, blue-green suffructescent herb up to 1.5 m tall, with stiff spinescent branches bearing leaves up to 20 mm long. Distributed through North Africa to Arabia (ZB and ZE III, III-IV), it is an excellent source of camel fodder and may be used in the treatment of kidney stones (Townsend and Guest 1980; Boulos 1983; Danin 1983). The plant sheds its leaves during the dry season in order to reduce water losses (Kassas and Batanouny 1984).

CUCURBITACEAE

Cucurbita foetidissima (buffalo gourd): a coarse, often evil-smelling, usually prostrate, perennial vine from a long-lived, tuberous root up to 2 m long and 15 cm diameter. The aerial growth, which dies back annually, extends over a radius of 7.5 m or more. It is distributed from Nebraska and Missouri to California, Arizona, Texas and Mexico (ZB and ZE IV, VII, VIIa, VII-V). The buffalo gourd has a good potential as a commercial crop suitable for growing on desert sand dunes. The roots contain up to 65% dry weight of a high quality starch. The striped gourds, ca. 10 cm diameter, were eaten by native Americans, either cooked or dried for winter use; the seeds too may be eaten. Saponins are present in stem, gourds and root, and were used for washing clothes (Kearney and Peebles 1951; Saunders 1976; Pasternak et al. 1986).

Marah gilensis (wild cucumber): possessing a large, tuber-like root from which rapidly growing, succulent, annual shoots are produced very early in the year, sometimes even in February or March. However, the shoots are susceptible to frost unless protected by the shade of neighbouring trees, and may die back two or three times before becoming established. It is distributed from Arizona to south-western New Mexico (ZB VIIa). The somewhat fleshy and spiny fruits are bitter, possibly due the presence of glucosides (Kearney and Peebles 1951; Crosswhite 1986).

Elaeagnaceae

Elaeagnus angustifolia (Russian olive or Trebizond date): a deciduous small tree, 6–7 m high, whose roots are either shallow or deep-rooting, depending on the depth of the water table. A native of temperate Asia [ZB VII(rIII)], it is widely cultivated as an ornamental and naturalised in southern Europe. The Russian olive has edible, sour-sweet fruits which can be fermented for an alcoholic drink with a wine-like flavour and the headiness of vodka; 10 kg fruit yields 12-13 l alcohol with a fusel oil content of less than 1.175%. The mealy pulp can be fried or used in pastries. The seeds also produce an oil. The tasty flowers are the original source of 'honey sugar' and a source of bee food; the oil, which is extracted by enfleurage, is used in the perfumery and cosmetic industries. In China it is used for sand fixation along the railway lines (Kebin and Kaiguo 1989).

The wood is extremely hard and used for posts, beams, domestic carpentry, musical instruments and fuel. The bark and foliage are excellent for tanning. The leaves, which in autumn contain 150–250 mg per 100 g vitamin C, are used as fodder for camels and sheep. The tree may be grown for dune stabilisation, forest reclamation or as an ornamental (Komarov 1960-1977, Vol. 15). Extensive afforestation with *E. angustifolia* has been carried out in China to prevent wind erosion and control desertification. Early success was followed by a decline in growth and even deaths. Investigation revealed healthy growth was only possible where the relative water content of the soil was over 50% and that thinning to reduce plant density was necessary for healthy growth (Kebin 1989). The roots produce nitrogen-fixing nodules, the result of an epidermal infection by the filamentous streptomycete *Frankia* penetrating between the epidermal cells. In experimental cultures nodule formation occurred at a pH of 6 but not at 5 (Gardner 1958; Sprent and Sprent 1990). The tree can reproduce vegetatively by both root and shoot suckers, as well as from seed.

ERICACEAE

Arbutus andrachne: an evergreen, large bush or small tree up to 12 m tall, the short trunk with numerous divergent and contorted branches forming a broad, irregular crown, with leaves thickly coriaceous, broadly elliptic and 4–10 cm long. It is a species of the eastern Mediterranean (ZB and ZE IV, IV–VI). The globose fruits are edible; the compact wood is used for turnery It is a true evergreen tree, with shoots and leaves unreduced and physiologically active throughout the year. The summer transpiration rates are low, not exceeding 500 mg g^{-1} h^{-1} fresh weight; summer osmotic pressures are also low - 19.6 atm (Zohary 1962).

Euphorbiaceae

Euphorbia stenoclada (famata botribotrika): a wax-coated, globose, bushy shrub or tree up to 6 m tall with a scaly bark and a flat-topped crown of olive-green, succulent branches bearing caducous leaves and secondary twigs armed with paired, modified, spinose branchlets. Endemic to south-western Malagasy (ZE II–III), it has been introduced to East Africa as an ornamental (White 1983; Rabesandratana 1984); it grows on calcareous sands. Resin from the stem is used to caulk pirogues. Branch apices are eaten by zebus.

Table 9. Chemical composition (%) of *Schinziophyton rautanenii* fruit. (After Booth and Wickens 1988).

	Pulp	Kernel	Seed cake
Moisture	8.4–16.6	4.1–6.5	12
Crude protein	6.8–9.4	22.0–28.8	36.5–60.5
Fat	0.5–1.6	45.0–59.6	1.1–7
Carbohydrates	65.4–76.9	4.6–5	ND
Reducing sugars	ND	ND	5.4–5.8
Sucrose	29.8	ND	ND
Glucose	0.2	ND	ND
Fructose	0.5	ND	ND
Fibre	2.5–3	2.5–2.7	7.7–8.2
Ash	4.6–5.6	3.7–4.1	8.4–9

ND, no data

Schinziophyton rautanenii, syn. *Ricinodendron rautanenii* (manketti or mongongo): a dioecious, deciduous, pachycaul tree up to 24 m tall, with a simple trunk up to 7 m long and 1 m diameter or three to four clustered stems, with a closed, spreading crown and digitately compound leaves; the 25 to 37-mm-long, 19 to 25-mm-diameter fruit has a hard, 2–3 mm thick, difficult to crack shell an d one or occasionally two seeds surrounded by a fleshy, dry and spongy pulp. A drought tolerant species, often occurring as large groves on dune crests, it is widely distributed through the open woodlands of southern Africa (ZE III-II), occurring in areas receiving 150–1000 mm summer rainfall.

Provided there has been no rain the fruit remain edible where they fall for up to 8 months and are eventually destroyed by insects. An important local food source, the fruits are eaten fresh, dried or cooked; pleasant and astringent, especially when raw, they may also be fermented into an alcoholic beverage. The kernels are also edible, either raw or cooked. The kernel yields 30–40% of an edible, quick-drying, fixed oil which, in the past, has been used in margarine. However, the oil reported to deteriorate rapidly on storage or heating due to the high elaeostearic acid content, although its stability against oxidation is aided by its high gamma-tocopherol levels. There are also suggestions of a tenuous link with goitre or interference with iodine. See Table 9 for a chemical analyses of the fruits.

The foliage is browsed by cattle and game. The wood is pale yellow and very soft with a wavy grain and coarse texture; it needs to be rapidly dried after cutting to prevent discoloration by staining fungi. The wood weighs 560 kg m^{-3} green and 240 kg m^{-3} air dried and offers a reasonable substitute for balsa (*Ochroma lagopus* from South America). It is used for crates, pattern making, insulating material, floats, dart boards, furniture backing, etc. and has a potential use for paper and board. The seed oil has a potential in the paint, varnish, linoleum and oil cloth, soft soap and cosmetic industries. Propagated from truncheons, it may be grown for shade, windbreaks, dune stabilisation and live fencing. Methods have been found to break seed dormancy but seedling survival is reported to be poor (Story 1958; Watt and Breyer-Brandwijk 1962; Booth and Wickens 1988).

FOUQUIERIACEAE

Fouquieria columnaris, syn. *Idria columnaris* (boojum – from a character in Lewis Carroll's '*Hunting of the Snark*'): a bizarre tree with a distinctive, tapering, columniform trunk up to 23 m tall, bearing a sparse number of slender, leafy, horizontal branches, which form a bristling armour to the stem (Fig. 2). The stem is occasionally branched, with vertical or contorted branches. The distribution is more or less confined to central Baja California with an outlier in Sonora (ZB IIIwr). The boojum is cultivated as an ornamental in desert landscapes.

The succulent stem consists of a thin periderm, a thick cortex region with an anastomising xylem forming a lattice cylinder surrounding massive water storing parenchymal tissue. During drought, when turgor is reduced, the central or lateral stems may collapse and bend downwards. The primary leaves are borne singly at the apices of the main stems or side branches, and as the apical meristems continue to grow and the stems and branches elongate, the primary leaves and developing spines are left behind as lateral leaves and spurs. The spurs are formed on the newly developing stems and side branches as pseudo-petioles, i.e. extensions of the stem cortex that are attached to the petioles for their entire length. When the leaf is shed the petiole separates from the pseudo-petiole, leaving a sharp, pointed spine. The leaf is never replaced. The secondary leaves are borne in fascicles in the axils of the pseudo-petioles. The boojum may produce more than one crop of leaves a year, depending on soil moisture and water balance, appearing shortly after the onset of the winter rains, and in some years during the summer rains (Humphrey 1974; Henrickson 1972, 1977; Turner et al. 1995).

Fouquieria macdougalii (MacDougal ocotillo): a small, drought-deciduous tree or shrub up to 11 m tall, with one or several, 1 to 2-m-tall and 5 40-cm diameter trunks, a smooth, greenish-bronze bark that peels off in paper-like sheets, and divergent branches. The single primary leaves are borne on new long shoots during the rainy season, the pseudo-petioles, as in *Fourquieria splendens,* developing into spines. The smaller, secondary leaves are borne after the rains in fascicles of two to five on short shoots in the axils of the spines; their petioles do not form spines. The species grows under 155–400 mm annual rainfall in central Sonora and 400–900 mm in the Sierra Madra foothills (ZE II-III). The bark is used as a soap substitute. The wood contains 62.5-67.7% water (Shreve 1933; Henrickson 1972; Mason and Mason 1987; Turner et al. 1995).

Fouquieria splendens (ocotilla): a deciduous, thorny shrub up to 10 m tall with numerous (6–30, ra rely 100), long, unbranched, whip-like stems fanning out at ground level and arising from a short rootstock up to 40 cm diameter; the root system is shallow but laterally extensive. At the onset of the dry season and cessation of growth, the stout pseudo-petioles split along a horizontal plane for its entire length. The upper portion and the attached lamina fall, leaving a stout, sharp spine. The stems bear short-lived primary leaves, the 'petioles' of which develop into spines and bear fascicles of 10–20 secondary leaves in their axils and constitute the major photosynthetic organ. It is distributed from south-eastern California and northern Mexico to western Texas (ZB and ZE II–III, IIIswr).

The wood may be used for hut frames and roofing lathes. Cuttings will root readily and provide a good vermin-proof, living hedge for poultry, etc. The spines offer protection against browsing, so much so that populations may increase in heavily grazed lands. The flowers and capsules are eaten by the Coahuila people. The wax coating the stem is reputed to be a good belt dressing. The Apache people bath in a root decoction to relieve fatigue; the powdered roots are applied to painful swellings. It is recommended for soil stabilisation and amenity planting.

The waxy leaves drop as soon as the soil dries and new leaves quickly develop following a good rain, except in winter when temperatures are low. Depending upon rainfall distribution, the shrub is capable of producing three or four sets of leaves a year. Alternating temperatures following a summer shower of 40/20 °C with a daily exposure or 12-16 h at 40 °C are required for maximum germination (Kearney and Peebles 1951; Henrickson 1972; Havens et al. 1973; McGinnies 1981; Felger and Moser 1985; Gutterman 1993; Turner et al. 1995).

FRANKENIACEAE

Frankenia palmeri: a dwarf, evergreen, salt tolerant, woody shrub up to 30 cm tall bearing dense fascicles of greyish, linear leaves 2–4 mm long; branches rooting at the nodes. The species is more or less confined to the Sonoran Desert with outliers in southern California and northern Baja California (ZB IIIswr, IIIwr), often dominant on maritime, saline and alkaline soils. The Seri Indians regard the dry wood as a valued source of fuel; they also use the brushwood for shelters and roofing. The resin obtained from the base of the plant is used as an adhesive. A tisane from the root is taken as a remedy for colds. Stem cuttings root readily and may be used for live fences (Felger and

Moser 1985; Turner et al. 1995). The shrub is highly salt tolerant, surviving irrigation water containing 3% salt. The salt is taken up by the roots and is excreted by salt glands on the leaves, thereby maintaining a homeostatic balance in saline environments. It also appears to depend on frequent maritime dew for its survival (Turner et al. 1995).

JUGLANDACEAE

Juglans regia (walnut): a robust, deciduous tree up to 30 m tall bearing ovate leaves up to 15 cm long and 10 cm wide. Widely cultivated and generally assumed to be native in Turkey, northern Iran, Transcaucasia and central Asia [ZB and ZE VII(rIII), VII–IV], the walnut is used in classical Chinese formulations for treating breast cancer. The leaves contain the antitumour compound betulin plus pyrogallol tannins and juglone, which has pesticidal and weedicidal attributes. The leaves and the powdered hull are used as an astringent and as a depurative in the treatment of syphilis. Branches and the skin of unripe fruit contain tannin and glucosides with antitumour properties. The outer fleshy part of the fruit is rich in vitamin C; it also produces a yellow dye. The dried fruit, when dry-pressed, yield a valuable oil suitable for the paint and soap industries; cold-pressing yields a light, edible oil which is used to flavour food. The dry, ripe seeds contain alpha-hydrojuglone-4-beta-D-glucoside, juglone and juglanin. It is used in Chinese traditional medicine to treat coughing and bronchitis (Duke and Ayensu 1985; Huang 1993).

The naphthoquinone juglone, which is also present in the North American ***J. nigra*** (black walnut), has an antagonistic effect on both herb and woody vegetation beneath its canopy, the toxic effect being due to leaching from the leaves, stems, branches and roots of a bound toxin, the 4-glucoside of 1,4, 5-trihydroxynaphthalene, which then undergoes hydrolysis and oxidation in the soil to release juglone. In addition to its allelopathic effect, juglone is an important inhibitor of seed germination, concentrations of 0.002% being sufficient to completely inhibit the germination of lettuce seed (Harborne 1988).

LABIATAE

Coridothymus capitatus, syn. *Thymus capitatus*: a rigid, much-branched, aromatic, dwarf shrub forming low, dome-shaped, 5 to 25-cm-high bushes. Widespread in the Mediterranean region (ZB and ZE III–IV, IV, IV–III), it is a saxatile, growing on shallow soils above soft calcareous rocks and sandstones. The leaves contain thymol and carvacrol and may be used for flavouring food. Cold infusions of the flowering branches are also used medicinally as a tonic, for treating coughs; a hot infusion taken internally is reputed to be useful in treating some skin diseases. The plant is valued as a bee food. Adventitious roots develop on buried stems. It has no taproot; the numerous fine rootlets obtain sufficient moisture by penetrating the soft rocks. Heterophyllous, the larger leaves of the winter rainy season are replaced in summer by smaller leaves (Oppenheimer 1960, Zohary 1962, Boulos 1983, Danin 1983, 1996).

Hyptis emoryi (desert lavender): a facultatively deciduous, evergreen, multi-stemmed shrub with slender, moderately flexible stems up to 4 m long, bearing small, canescent leaves and fragrant blue flowers. It is a species of rocky slopes and dry washes of the Sonoran Desert (ZB IIIsr, IIIswr, IIIwr). The flowers are a valuable bee food. The seeds, like those of the annual *Salvia columbariae*, are used as a mucilaginous food known as 'chia'. The plant may be moderately browsed by livestock. The hard wood may be used by the Seri people for making implements.

Deseert lavender is a drought-avoiding phreatophyte with low stomatal conductance. The small, thick and pubescent 'sun' leaves with a high ratio of internal to external leaf area are produced during the summer and autumn when soil moisture is generally low. Large, thin and sparsely pubescent 'shade' leaves with a low ratio of internal to external leaf area are produced during the spring when soil moisture is readily available, thus permitting maximum photosynthesis and absorption of solar radiation (Felger and Moser 1985; Turner et al. 1995).

Salvia fruticosa, syn. *S. triloba*: a strongly aromatic, glandular-pubescent, up to 1-m-tall shrub bearing grey-green, simple to trilobed, elliptic to ovate-oblong leaves. It is a native of the eastern Mediterranean (ZB IV) cultivated in North Africa for its medicinal properties, an infusion of the leaves and young shoots being taken for influenza, coughs and rheumatic pains; it is also used in the preparation of a beverage. It is considered to have a potential as a garden ornamental. The plant is heterophyllous, bearing smaller summer leaves (Zohary 1962; Boulos 1983).

Satureja thymbra: an erect, much-branched, up to 50-cm-tall, aromatic shrub with narrowly obovate to elliptic, 7 to 18-mm-long, thick, sessile leaves dotted with protuberant glands. Adventitious roots develop on buried stems. It is a plant of the eastern Mediterranean (ZB IV). The leaves contain thymol and carvacrol and may be used for flavouring food. Heterophyllous, the winter leaves generally subtend condensed growths with clustered, reduced leaves. Maximum and minimum transpiration rates of 696 and 320 mg g^{-1} h^{-1} have been recorded and are explained as being due to competition between individual plants for soil moisture and this is reflected in the noticeable transpiration differences between species and individuals; the same explanation applies for *Sarcopoterium spinosum** (Oppenheimer 1960; Zohary 1962; Danin 1996).

LAURACEAE

Laurus noblis (bay laurel or sweet bay): a dioecious, evergreen shrub or tree 2-20 m tall, bearing gland-dotted, narrowly oblong-lanceolate, 5 to 10-cm-long leaves. It is widely distributed through the Mediterranean region (ZB and ZE III-IV, IV, IV-VII) and cultivated and sometimes naturalised elsewhere. The aromatic leaves are used to flavour food and are an essential ingredient of *bouquet garni*. It is a source of a relatively non-toxic and non-irritant essential oil (*folio lauri*) that is used to flavour a number of alcoholic drinks as well as acting as a bactericide and fungicide. The leaves also contain a fixed oil. Since early antiquity a wreath of laurel leaves has been a symbol of victory. The aromatic fruit contain a fixed oil, oil of bay, that is used in the perfume industry for making toilet soap and was formerly used pharmaceutically (*oleum lauri*) in the preparation of several ointments, as a narcotic and stimulant, it is also used in veterinary medicine. Both oils contain methyl eugenol, which has been found to be narcotic and sedative in mice. The fruits are also used in a specific remedy for alopecia. The wood is used for smoking meat and cheese. Bay laurel is widely grown throughout the world as an ornamental (Hill 1952, Zohary 1962, Chiej 1984, Duk e and Ayensu 1985; Mabberley 1987; Lawless 1992).

It is a true evergreen tree, with shoots and leaves unreduced and physiologically active throughout the year. The summer transpiration rate is low - not exceeding 500 mg g^{-1} h^{-1} fresh weight. Summer osmotic pressure of the cell sap is also low - 2.72 mPa (Zohary 1962). The sclerophyllous leaves possess a highly viscous protoplasm and are able to suffer a water saturation deficit of 60-65% without damage. The leaves can even withstand a sublethal water content of 44–49% dry weight, the mature leaves of which contain less water than dry matter even at saturation. Under water stress cuticular respiration is very low or even nil (Oppenheimer 1960).

LEGUMINOSAE subfamily **CAESALPINIOIDEAE**

Cordeauxia edulis (jicid - approved Somali spelling; also spelt yeheb, and variations thereon): an unarmed, multistemmed, much branched, evergreen shrub up to 3 m tall, with a deep taproot down to 3 m and small, secondary rhizomes near the surface; nodules occur on the younger roots. The young shoots and leaves are covered with numerous, scale-like, red or purplish-red glands. Two varieties are recognised by the Somalians, moqley and suley, the latter having fewer and larger leaves and less sweet and smaller fruits. Jicid is a monotypic legume and an endemic of south-eastern Somalia and the Ogaden of Ethiopia [ZB III(tI)]. Formerly abundant and locally abundant, it grows on compacted, fine to coarse or gritty sand and loamy sand, tolerating slightly saline soils (0.934 dS m^{-1}) in areas receiving 85–200 mm annual rainfall. Intolerant of waterlogging, the bush can also survive 2 years without rain and, when under stress, the leaves inroll, but under extreme drought stress they will eventually fall.

The fruits are easily shelled and the kernels eaten fresh, roasted or boiled; they are delicious and certainly have a potential as a dessert nut if plantations

Table 10. Chemical composition (%) of *Cordeauxia edulis* seeds. (After Booth and Wickens 1988)

Moisture	7.8–16.9
Crude protein	10.8–15.9
Fat	9.9–13.4
Starch	31.4–41.5
Sugar	12.4–24.9
Reducing sugar	2.5–5.3
Sucrose	11.6–21.6
Ash	2.2–3.8

could be established. It is one of the staple foods of the area, so much so that the 'nuts' are often harvested prematurely in order to prevent other collectors benefiting. Yield is ca. 5 kg per tree. See Table 10 for the chemical composition of the seeds. The seeds contain trypsin inhibitors in similar proportions to those present in *Lablab purpureus* but no toxic phytochemagglutinins or lectins are present; alkaloids and glucosides are also absent.

The leaves may be used to prepare a tisane; the liquor remaining after boiling the seeds is sweet and also drunk. The foliage is browsed by livestock. The wood may be used for fuel. The pulverised leaves produce a magenta dye, which is soluble in boiling water; ca. 200 g of leaves is sufficient to dye calico 10 m long and 90 cm wide. Dye extracted under alkaline conditions (1% ammoniac solution, or similar) tends towards violet while under neutral to slightly acidic conditions the colour is less pronounced. The dye, which is from the glandular hairs, contains a red or magenta naphthazarine-derived pigment known as cordeauxiaquinone, and is the only naphthaquinone found in the Leguminosae. The leaves contain 0.7–0.8% of this mordant red dye, which can form fast and insoluble combinations with many metals. The dye also stains the teeth and bones of the browsing goats bright orange or red, suggesting a possible histological application for the dye. The pigment deposited on the bones is reported by Gutale and Ahmed (1984) as possibly directly or indirectly stimulating the haemopoetic tissue to produce erythrocytes.

The fruit take ca. 10–14 days to mature after rain; should the rain fail, fertilised ovaries will not develop until the next rain. The seeds germinate rapidly, followed by slow top growth. Care must be taken with nursery production not to break the taproot, otherwise the seedling will die. Young shrubs flower and fruit after 3–4 years. Jicid may also be propagated from stem cuttings (IBPGR/RBG 1984; Wickens and Storey 1984; Booth and Wickens 1988; Aronson et al. 1990).

Lemuropisum edule (tara): an unarmed, multistemmed, spreading shrub up to 6 m tall with a dense crown of tangled branches, the short lateral branches with branchlets sometimes spine-like or sometimes reduced to simple nodes bearing microphyllous, semi-persistent, pinnate leaves. It is endemic to south-western Malagasy (ZE II–III), growing on rocky limestone or sandy piedmonts. Ripe nuts are usually eaten raw although roasting could improve the digestibility. The foliage is browsed by goats. The shrub has a potential for cultivation as a food source for arid lands and also for use as windbreak and live hedge (Willing 1989; Wickens 1995b).

Parkinsonia microphylla, syn. *Cercidium microphyllum* (foothill paloverde): a deciduous, stiffly branched, large shrub or small tree up to 8 m tall with pinnate leaves and spinescent tips to the leaf y branches. The young bark is yellow-green and photosynthesising. It is a native of south-western California, southern and western Arizona, Baja California and Sonora (ZB and ZE II–III, IIIsr, IIIswr, IIIwr). The wood is soft and brittle and burns quickly, giving off an unpleasant odour and leaving little ash. The tree is browsed by *Ovis canadensis* (American bighorn); the branches are also lopped and fed to horses, mules and donkeys. Although the pods are fairly palatable they are generally not eaten by livestock except during periods of drought. The sweet, green, young seeds may be eaten raw, when older they are either cooked or ground to a meal. The flowers too may be eaten; they are also reported to be a good bee food (Kearney and Peebles 1951; Coyle and Roberts 1975; Felger and Moser 1985; Rea 1991).

Tylosema esculentum, syn. *Bauhinia esculenta* (marama bean): a prostrate and trailing, tuberous perennial, with stems extending over a 6 m radius and sometimes scrambling over neighbouring shrubs, but dying back during the winter months (Fig. 33). It is endemic to the Kalahari of southern Africa (ZE III-II). With a patchy distribution, it occurs in the grasslands, bushy grasslands or wooded grasslands of southern Africa, in areas of 100–900 mm summer rainfall. It grows on sandy and limestone, including dolomite, soils, but not on soils developed over granite or basalt.

There are conflicting reports about whether the foliage is palatable and browsed by cattle. The young, green legumes and stems are boiled as a vegetable or used in soups. The ripe beans are a staple food of the Bushmen and Khoi-Khoi people and are eaten either roasted or boiled. The dried, roasted beans have a pleasant, slightly coffee-like taste, although bitter forms are known. The beans are considered to have a potential as a dessert nut, as a substitute for almonds or possibly as a coffee substitute. The dry seed weight (36-43%) and protein content (30–39%) is, respectively, about twice and roughly comparable with that of soya bean (*Glycine max*; 37–39%); like most legume proteins, it is rich in lysine (5%) but poor in methionine (0.7%). The

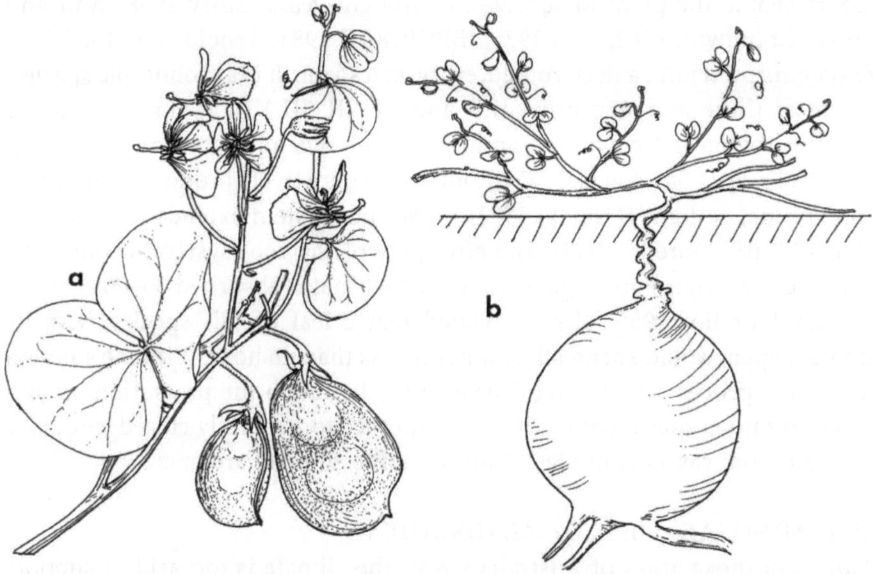

Fig. 33. *Tylosema esculentum* from South Africa. **a** Flowering and fruting branch showing large, leathery leaves. **b** Habit showing tuberous rootstock.[Adapted from drawing by Cythna Letty 1959, plate 1311 and habit photograph in Kew Herbarium (Story 5160)]

seed oil is edible, has a pleasant nutty odour and pleasant flavour; it is also a good source of the nutritionally important linoleic acid.

The San prefer to eat the very young tubers of ca. 1 kg raw, and they may also be eaten baked, roasted or boiled; the astringent taste of the older tubers is removed by boiling in milk. The pounded tubers, even when old and fibrous, are used as an emergency source of water. The tubers are eaten by game, while the beans may be used to fatten pigs. The tuber contains 9% crude protein on a dry weight basis.

To date, attempts to cultivate the marama for the bean have not been very successful. Plants do not flower until the third or fourth year and the yield of beans varies from plant to plant; some selection is obviously desirable. The seeds are hard and germination is improved by scarification. They should be sown in situ; transplanting inhibits resprouting while the drying of the roots inhibits further regrowth. Marama bean may also be grown for ground cover or as an ornamental.

The woody, oval to almost circular pods are up to 6 cm long and 4 cm in diameter and contain one to six spherical seeds. Tubers of plants grown from seed weighed 13 kg after 10 years; weights up to 250 and 300 kg have been recorded from older plants. Containing up to 90% by weight of water, the tubers enable the plant to survive the drought years (Story 1958; Watt and Breyer-Brandwijk 1962; NAS 1979; IBPGR/RBG 1984; Arnold et al. 1985).

Zuccagnia punctata: a deep rooted, evergreen shrub. It is a monotypic species, native of Chile and Argentina (ZB and ZE III, III-VIIa, VIIa). It contains nordihydroguayaretic acid, an antioxidant for oils and fats, and is also the source of a dye (Gastó et al. 1985). Root suckers are freely produced, especially on exposure to light (Morello 1951a). The root system extends over an area of 14–32 m^2 (Morello 1955). The efficiency of the root system of this non-phreatic xerophyte can be gauged by the fact that it absorbed 30–70% of the rainfall. Morello (1955) also calculated that a leaf of this species, with its stomata open, would spend all its water in less than an hour. With the stomata closed the process would take 7 days and 9 h, due to the protection offered by the resinous coating of the leaves. With the stomata fully closed, cuticular transpiration was 77 times less than when the stomata are open.

LEGUMINOSAE subfamily **MIMOSOIDEAE**

Acacia: in those areas of Australia where the climate is too arid to support *Eucalyptus* as the dominant genus, it is replaced by species of *Acacia*, of which the most important is *A. aneura**. These Australian acacias are evergreen, bearing phyllodes (entire, leaf-like and often flattened organs, without pinnae or leaflets) or rarely with transitional leaves, belonging to subgenus *Heterophyllum* (syn. subgenus *Phyllodinae*), with stipular spines usually absent (sections *Pulchellae* and *Botrycephales*) or often only present on juvenile plants (section *Phyllodineae*), or vestigial as in *A. victoriae** (Pedley 1981; Maslin 1988). As a general rule, thin, cylindrical phyllodes predominate in arid areas, while in more mesic areas the phyllodes tend to be relatively broad and flat. Both forms may occur in the same habitat or even on the same bush in relation to their position to sun or shade (Walter and Breckle 1986). While subgenus *Heterophyllum* is mainly confined to Australia, in Africa, Asia and the Americas subgenera *Acacia* and *Aculiferum* predominate. These are mainly deciduous trees, shrubs or climbers bearing bipinnate leaves; in subgenus *Acacia* they are armed with stipular spines in subgenus *Aculiferum* they either bear prickles or are unarmed (Mabberley 1987).

In Africa the dry season diurnal course of transpiration and stomatal conductance of *A. ehrenbergiana*, together with other Sahelian-Sudanian acacias, *A. erythrocalyx*, *A. laeta*, *A. nilotica* subsp. *tomentosa**, *A. senegal**, *A. seyal*, *A. tortilis* subsp. *raddiana** and *Faidherbia albida** were examined by Ullmann (1985). Although no species-specific differences in behaviour were observed, a correlation was found between types of stomatal performance patterns and leaf age. Mature leaves showed a steep, post-sunrise increase in leaf diffusion conductance, an early morning pronounced peak, followed by a continuous decrease during the remainder of the day. Typically this decrease was interrupted in the afternoon by a minor peak associated with an intermittent rise in relative humidity. The diurnal course of transpiration was dome-shaped, attaining a noon maximum associated with the rapid rise in air temperature and a decrease in relative humidity. Senescent leaves that survived the dry season experienced a midday depression of stomatal conductance and leaf water loss, while newly produced leaves exhibited a high conductance from early morning to late afternoon. There was no association with daily maximum leaf diffusion conductance and leaf age; however, there was an association with pre-dawn water potential. The phreatophytic habit of many of these species is clearly an advantage over the need to respond to seasonal rainfall or changes in air humidity. This is evident in the unrestricted dry season gas exchange of trees able to tap ground water resources against the response to drought in trees without access to ground water.

Acacia aneura (mulga): a shrubby tree, up to 14 m tall, and bearing narrow, flat or cylindrical phyllodes on the terminal 10–30 cm of the branches. It is a rather variable species with a number of ecotypes. The root system is extensive, with flat, radially extending roots for anchorage, the others much branched and descending at an angle in search of water. Mulga probably has a life span of at least 100 years (Walter and Breckle 1986). Widespread in Australia, growing in areas receiving 200–380 mm annual rainfall (ZE II–III, III–II, III–IV, IV–III), *A. aneura* is usually found in monospecific stands or patterned groves, the effectiveness of the root system resulting in an extremely poor ground cover or bare ground. Four growth habits are recognised in Australia:

1. 'Low mulga': small shrubs or undershrubs with short branches, the result of constant browsing by sheep and cattle. A useful fodder source during drought.
2. 'Whipstick mulga': immature shrubs with upright branches and bearing leaves only at the branch tips. The browse is made accessible to stock by chaining or bulldozing.
3, 'Umbrella mulga': mature, very leafy shrubs with upright branches. Responds well to lopping which should be carried out to leave low horizontal branches and short upright branches. 175–200 lopped trees ha^{-1} are regarded as giving the best combination of grass and browse for dry season buffer feed. The palatability is highly variable.
4. 'High mulga': stands of old, bare stemmed trees bearing leaves only on the upper branches. The leaves are usually palatable but inaccessible to stock. Furthermore, the trees are intolerant of lopping so that the forage potential cannot be utilised without killing the trees. However, 'high mulga' trees usually produce large quantities of seed.

Where mulga has been introduced into Israel and North Africa it usually takes the form of a tangled shrub 3–4 m high.

Although of only moderate nutritive value, its abundance makes it the principal stand-by browse for sheep during prolonged periods of drought, daily ration of 1.4 kg leaf providing sufficient maintenance for a sheep when supplemented with sulphur and phosphorus. Palatability is variable, some being quite unpalatable with palatability improving with maturity, The phyllodes of immature mulga are often covered in a bitter tasting resin that makes them unpalatable to livestock. An ether extract of the leaves shows that the resin amounts to 4–15% of the dry matter and, in general, the amount increases with the dryness of the habitat (Walter and Breckle 1986).

The pods contain several edible seeds 3 mm in diameter. Their high protein content (23.3%) suggest a bread-making potential; it is also possible that the food industry might find uses for the protein isolates. The ground seeds are eaten by the Aborigines of South Australia. The large, succulent galls, known as mulga apples, produced by the tree are reputed to be good thirst quenchers. The lerp (a honey dew produced by the honey ant, *Austrotatachardia acaciae*) and gum are also eaten (Tanaka 1976; Brand and Cherikoff 1985; Devitt 1992; Aboriginal Communities of the Northern Territory 1993; Lazarides and Hince 1993).

Mulga wood, which is claimed to be one of the hardest in the world, is heavy and durable, with dark heartwood and narrow, pale sapwood. It is used for carvings, fencing, pit props and fuel, although it burns quickly and leaves no cinders (Hall et al. 1972; NAS 1979; Mitchell and Wilcox 1988). In Libya, fuelwood from a plantation with an initial density of 1000 trees ha^{-1} produced an average yield of 400 kg DM ha^{-1} $year^{-1}$. This compared unfavourably with other Australian *Acacia* species in the trials, such as *A. ligulata, A. salicina* and *A. saligna* which produced 5000 kg DM ha^{-1} $year^{-1}$, *A. victoriae* 950 kg and *A. cyclops* 900 kg. Rather surprising mulga has never produced convincing results in the Mediterranean arid region (Le Houérou 1985).

In addition to providing food, fodder and fuel, mulga may be grown for soil stabilisation, amenity planting and bee food (Maconochie 1985; Lazarides and Hince 1993). The Aborigines use the smoke from the leaves for postnatal therapy and for its sudorific activity in the belief that sweating will improve a sick person's condition (Barr et al. 1990; Aboriginal Communities of the Northern Territory 1993).

Although tolerant of drought mulga is intolerant of salt (Walter and Breckle 1986). It is susceptible to fire but will readily resprout if stems are cut off at ground level (Hodgkinson and Griffin 1982). Seed germination is slow due to the hardness of the testa, but is enhanced following fire (Maconochie 1982); profuse germination occurs following 75–100 mm rain falling in a few showers close to the warm season (Baumer 1983). Optimum temperature for germination lies between 20 and 30 °C and is favoured by the high CO_2 content of the layers of humus and leaf litter surrounding the seed. The seedling taproot growth can attain 2–3 m growth with an aerial growth of only 8-10 cm (Walter and Breckle 1986).

A study of the water economy was carried out near Alice Springs with a mean annual rainfall of ca. 250 mm, of which 75% falls during the summer months. The rainfall is irregular with frequent dry periods in excess of 3 months. Fully turgid plants had an osmotic potential of -2.3 MPa, and -2.85 MPa when completely flaccid. They exhibited a high osmolarity, which appeared not to change with season or soil water status. The phyllodes were able to withstand a tissue water potential of -13 MPa without any damage, indicating that *A. aneura* must not infrequently suffer from an acute shortage of water (Slatyer 1960, 1961). Both summer and winter rains, however scanty, are essential for the survival of this species, with vegetative growth and flowering occurring at any time of the year provided rain has fallen.

Acacia caven, syn. *A. cavenia*: an invasive, rather variable, nitrogen-fixing, spiny, deciduous shrub or small tree 2–6 m high. The distribution extends from the Mediterranean region of Chile eastwards to Argentina, western Uruguay, Paraguay, Bolivia and southern Brazil; the northernmost extent in Chile occurs in the southern part of the Atacama (ZB and ZE II–VII, III–IV, III–VIIa, IV). The extremely hard and compact, reddish to yellow wood of *A. caven* is used for turnery (wagon wheels), general carpentry, firewood and high quality charcoal (Hoffmann 1978; Rodríguez et al. 1983). The foliage is browsed by livestock. The flowers are the source of an essential oil that is used in perfumery; it also yields a gum and a dye (Gastó et al. 1985). It is a source of tannin and is used medicinally as an astringent (Zin and Weiss 1980). An aqueous extract of the seed is used as a coffee substitute and, mixed with *Zea mays*, for child diuresis; it is also used with *Peumus boldus* and *Triticum aestium* for bladder diseases (Martín 1983).

In Chile the species is of interest on account of its reversed phenology, in that it loses its leaves at the start of the rainy season and produces new leaves at the beginning of the dry season. A similar behaviour has been observed in Africa for *Faidherbia albida**. However, whereas this behaviour is consistent throughout the *Faidherbia*'s distribution range, including areas with a bimodal rainfall (Wickens 1969), *A. caven* behaves normally elsewhere in its distribution range, shedding its leaves during the winter dry season. Aronson et al. (1994) hypothesise that the reversed phenology in the Mediterranean zone of Chile is the result of a relict behavioural rhythm acquired at its centre of origin in the subtropical, winter-dry climate of the Chaco prior to its migration westwards into Chile. The presence of cyanogenic glycosides was investigated by Aronson (1990) in five populations growing in Chile. He found a good correlation between the level of accumulation and mean annual temperature. Thus, these preliminary results lend support to the theory put forward by Jones (1972, 1981) that the ability to accumulate cyanogenenic compounds may be a response to cold climates.

Acacia craspedocarpa (hop mulga): a hemi-spherical, up to 4-m-tall shrubby tree bearing phyllodes (Fig. 23). It is a native of western Australia (ZE III–IV). The browse is of low nutritive value and is reputed to be unpalatable to sheep. The flattish phyllodes are vertically orientated. They are brownish at first due to the numerous epidermal hairs, and become bright green in colour as the hairs disappear, eventually changing to yellowish green in summer, when a yellow pigment accumulates in the vacuoles and masks the green of the chloroplasts. The mature phyllodes are 2 cm long, 1 cm wide and 0.9 mm thick, with sunken stomata and a cuticle which thickens during late summer to 40 μm. Despite a remarkable reduction in late summer transpiration rate without stomatal closure, the photosynthetic rate was hardly affected.

Owing to the orientation of the phyllodes the winter and summer photosynthetic curves exhibit a steeply rising and falling plateau. The transpiration curve shows the same daily pattern and is dependent on stomatal movements. Photosynthesis is markedly dependent on the temperature, attaining an optimum at 34–37 °C. Thus, the photosynthetic rate is less in moist winters with temperatures of 12 °C than in early summer. The temperature maximum for photosynthesis is 52.8 °C. By late summer transpiration had fallen to 26% of its early summer rate; photosynthesis is, however, only 86% of the early rate, thus demonstrating that water use efficiency is greater in the late summer. Although on summer afternoons the phyllodes had a water deficit, the stomata did not close.

Although the phyllodes overheated to 6 °C above ambient temperature when the transpiration was low, the heat resistance limit of 54.5 °C was never reached. The daily values of osmotic potential in winter were -1.9 to -2 MPa, in early summer -2.5 to -3.6 MPa, and in high summer -4.1 to -5.1 MPa. These fluctuations are largely due to a fall in water content, although when the phyllodes turn yellow this is accompanied by a change in the chemical composition of the cell sap (Hellmuth 1969, 1971; Walter and Breckle 1986; Mitchell and Wilcox 1988).

Acacia georginae (Georgina gidyea – named after the Georgina River): a gnarled, sometimes multistemmed tree or shrub up to 5 m tall, with the main stem branching near the ground. The more or less globular crown is usually dense, with greyish, leathery phyllodes with the odour of rotting onions. It is slow-growing and reputedly long-lived. It is endemic to the Georgina River Basin of Northern Territory and Queensland, growing in areas receiving 220-280 mm, predominantly spring rainfall (ZE II–III), preferring calcareous soils and less common on deep sandy soils.

The wood is used by the Aborigines for throwing sticks; it is also a favoured fuel. The tree may be used for shade, shelter and as a windbreak. It is generally regarded as a useful source of drought fodder despite its poor to moderate nutritive value and poor digestibility, although in some areas the leaves and pods are regarded as highly toxic, especially when young. That some *A. georginae* trees are toxic and others non-toxic is due to differences in their capacity to take up fluoride and convert it into fluoroacetate; the poisonous principle is probably as monofluroroacetic acid, of which the seeds contain more than the pods and the pods more than the leaves. Up to 4.5 kg whole pods fed with bran and chaff has been found to be lethal to cattle and 0.9 - 6.1 kg lethal to sheep (Hall et al. 1972; Askew and Mitchell 1978; Everist 1981; Baumer 1983). According to Dayton (1948; Jackson and Jacobs 1985) toxicity is due to the cattle browsing *A. georginae* and *Eremophila maculata** (spotted berrigan). The acacia pods are harmless, but contain an enzyme that liberates prussic acid from a glucoside present in the berrigan browse. Thus, while both species may be grazed separately with impunity, they are toxic when grazed together. The Georgina gidyea seeds are large and although it has a hard seed coat, its margin fractures easily and the seeds are readily desiccated (Maconochie 1982).

Acacia nilotica subspp. ***nilotica*** and ***tomentosa***: trees up to 14 m tall. They can withstand 5-6 months of inundation annually, although the former subspecies is considered to be slightly less tolerant. On dry sites the leaves may be shed shortly after the flood waters have receded, retaining the leaves for much longer on more favourable sites and in some places may even remain evergreen. It is widely distributed along permanent and seasonal waterways and pools of Sahelian and Sudanian Africa (ZB and ZE II-III, III), and has been introduced elsewhere.

The timber is hard, heavy and durable. It is widely used for boat building, railway sleepers, beams, water wheels, telegraph poles, pit props, agricultural implements, etc. It is also used for fuelwood and charcoal, although the latter tends to emit sparks when burning. Both the pods and bark are used for tanning. The tannin content of the pods is 25–33.8%, which compares favourably with myrobalan (*Terminalia chebula*), divi-divi (*Caesalpinia coriaria*) and wattle bark (*Acacia* spp.). Despite the high tannin content the trees are browsed by stock. The bark, leaves and fruit are used in traditional medicines (Osborn 1968; Sahni 1968; Baumer 1983; Boulos 1983).

The respiratory pathway of the root system under anaerobic conditions following seasonal inundation requires investigation. Under such conditions, the glycolyctic pathway, in the absence of sufficient oxygen, gives rise to acetaldehyde via pyruvate. The acetaldehyde, in the presence of high induced levels of alcohol dehydrogenase, is converted into ethanol, which is highly toxic to the plant cells if allowed to accumulate in any quantity (Harborne 1988). The adaptation of the *A. nilotica* roots to such ethanol toxicity is not known.

Acacia senegal (gum acacia): a deciduous shrub or tree up to 15 m tall. The species is widely distributed through tropical and southern Africa through southern Arabia to India in areas with 100–800 mm summer rainfall on light soils and up to 2280 mm on heavy soils [ZB and ZE I–III, II–III, III(tI)]. The trees are often either the result of coppice regrowth arising from previous gum gardens cleared during the cultivation cycle or from seedlings planted during the last years of a cultivation cycle. The ensuing fallow land is known as a gum garden. Traditionally such gum gardens were for 14 or more years when, fertility restored, the land was cleared for 4 years under cultivation before again reverting to fallow. The ever-increasing demand for cultivated land has resulted in the fallow period being shortened or even eliminated, accompanied by a reduction in tree population. As a result the demand for gum arabic now exceeds availability.

Gum acacia is an extremely important species in the economy of the Sahel, being suitable for growing in both forestry and agroforestry plantations. There are now a number of large scale projects throughout the Sahel involved, with varying degrees of success, in the planting of *A. senegal* as part of the battle against desertification as well as for increasing the production of gum arabic. Gum from *A. senegal* and some 18 other species was, until the recent tightening of the specifications, a major item of commerce. This group of species, which included *A. seyal* (gum talha from West and East Africa), *A. xanthophloea* (from East Africa) and *A. karroo* (from southern Africa), was traded on the international markets. Now only gum arabic, from *A. senegal*, is permitted for the food trade, the remainder being for industrial use only. Nevertheless, acacia gum 'from *A. senegal* and other African species' is still official in the *British Pharmacopoeia* (1993) for use as a bulk-forming laxative and pharmaceutical aid. Similarly, while the US specification for the use of acacia gums in the food trade limits use to that from *A. senegal* (gum arabic), the pharmaceutical specification permits the use of gum talha from *A. seyal*, the rational being that its use would be under medical supervision.

Gum arabic was traditionally defined as 'the gummy exudate from *A. senegal* or *its related species*', embracing a number of species that were not remotely taxonomically related, despite the fact that the Test Article, evaluated as toxicologically safe as a food additive, refers solely to gum from *A. senegal*. It is the calcium salt of arabin, a complex polysaccharide built of glucuronic acid, arabinose, rhamnose and galactose units (Sharp 1983).

The increasing international pressure towards tighter trade specifications and labelling regulations, identity and purity has led to the *Revised Specification* (FAO 1990; WHO 1990a, b) where gum arabic is defined as originating from *A. senegal* or *closely related species,* with a specific optical rotation range of -26° to -34° and a Kjeldahl nitrogen content of 0.27–0.39%. This has limited the designation of gum arabic to members of the subgenus *Aculeiferum* which, in addition to *A. senegal*, includes *A. laeta, A. mellifera, A. polyacantha*, etc., although gum is not available commercially from any of these closely related species. The gums from the tannin-containing, positive optical rotation species such as *A. karroo, A. nilotica, A. seyal* and *A. xanthophloea* are now excluded for use in food and consequently attract a lower price (Anderson 1993). Furthermore, the presence of tannin is considered carcinogenic. Exudates from species other than *A. senegal* occur as small tears and driblets; their collection is consequently extremely time-consuming. This, together with the low price, currently at US$ 1000/t, compared with US$ 5000/t for gum arabic, is likely to kill the export trade in other acacia gums.

Within the European Union, foodstuffs grade gum arabic (E414) is subject to a rigorous specification of identity and purity. It is used in confectionery, foodstuffs and pharmaceutical applications, including use as a foam stabiliser and clouding agent in beer. Poorer grades are used industrially in foundry sands, printing inks, lithography, etc. (Anderson 1985). Originally gum was obtained from natural exudations following damage to the tree, tapping being introduced some 70 years ago. The tapping season begins after the leaves have fallen. The bark is peeled back from one or two branches on a tree to expose a length of ca. 75 cm. The gum nodules then form in 3–8 weeks, exuding from an exposed abscission scar. Tapping ceases during the mid-winter cold spell and restarts in March when it becomes warmer.

The wood is used locally for building purposes, fence posts, firewood and charcoal. The root bark from the lateral roots produces a strong cordage suitable for ropes used in wells. The tree also provides a useful browse for all livestock (Sahni 1968; Baumer 1983; IBPGR/RBG 1984; Booth and Wickens 1988).

Acacia senegal has a deep tap root and extensive lateral roots, especially within the upper 20–30 cms of the soil, thereby enabling the tree to take advantage of even the lightest shower as well as deeper sources of moisture. The laterals extend over a radius of 25 m, with the ends turning sharply downwards (Fig. 34). The root systems of adjacent trees overlap with no evidence of allelopathy. Seeds germinate within a few days of receiving satisfactory moisture conditions, 50% immediately, the remainder about 2 weeks later. However, bruchid damage can be so severe that there can be some years with few or no viable seed (Hunting Technical Services 1964).

The physiology of gum production, a prerequisite for selection and management for high yields, is still imperfectly understood. It is currently under investigation at the Université P. Sabatier, Toulouse. Despite the similarity in the structure of the arabinogalactans of the cell walls and gum arabic, suggesting an arabinogalactan as a precursor of gum arabic (Anderson and Dea 1968), gummosis is now believed to be directly related to starch metabolism within *A. senegal*. However, gum formation may be accompanied by some cell wall restructuring in response to cell adaptation (Joseleau and Ullmann 1985).

Fig. 34. Excavated root system of *Acacia senegal* (gum arabic) growing in low sand dunes at Ed Dumokia, Kordofan Province, Sudan. Tree is approximately 30 yeras old. Excavation of taproot was discontinued at 2.3 m as being too dangerous to continue, root was then 18 cm in diameter. Lateral roots occur mainly 20–30 cm below surface and extend for a radius of 25 m, their extremities urning abruptly downwards. Excavation was by the late P. Vidal-Hall, Gum Research Officer, El Obeid. (Composite photograph by the author)

Histological investigations by Mouret (1985) have revealed a pectic trend in the tissues of both tapped and untapped trees. This tendency was also found in the walls of the phloem parenchyma of seedlings, i.e. before any possibility of gum exudation. These findings suggest an early preformation of gum in the tissues and the possibility of a genetic potential for gummosis. Gummosis appears to be induced by distress signals, the possible agents being drought stress, natural or artificial wounds, parasite attacks, etc.

The first conspicuous phase of gummosis is a thickening and chemical modification of the cell walls in the phloem parenchyma, eventually followed by a breakdown of the cell walls to form lacunae. An intermediate stage involves the modification of the phloem and cortical parenchyma followed by the sclerenchyma and fibres in the vicinity of the point of exudation. The starch grains within the affected tissues appear to be progressively assimilated into the amorphous gum mass, an observation that agrees with the biochemical findings of Joseleau and Ullmann (1985) cited above (Mouret 1985).

Acacia tortilis: a deciduous tree up to 20 m tall but usually smaller. Widely distributed through the arid and semi-arid regions of tropical Africa and eastwards through Arabia into India, four subspecies are currently recognised, of which only subsp. ***tortilis*** and subsp. ***raddiana*** (syn. *A. raddiana*) occur in the Sahel and Sahara (ZB and ZE II–III, III). A drought tolerant species with the taproot descending 35 m or more to the water table, it grows on a wide range of soils, including fossil dunes, but avoids freshly blown sand. It is heat tolerant, withstanding temperatures of 45–50 °C dur ing the harsh, dry season.

The leaves, young branches and fallen pods are browsed by livestock. The red heartwood has a high calorific value and is much valued as a source of fuel and charcoal. The fibrous bark contains tannin; it is also used for cordage. It produces a gum which is inferior to that of *A. senegal*. The flexible roots may be used for hut frames; they are also the source of a vermifuge. The leaves and bark may also be used in local medicines (Sahni 1968; Baumer 1983; IBPGR/RBG 1984; Boulos 1989).

Acacia victoriae, syn. *A. senetis* (gondabluey or prickly wattle): a multistemmed, bushy shrub up to 10 m tall, bearing flat, curved phyllodes up to 3 cm long and 5 mm wide and usually with a pair of straight spines at the base. Widely distributed through the mainland states of Australia, it usually occurs on sandy and saline soils in areas receiving 200–600 mm annual rainfall (ZE II–III, III–IV, IV–III); it may withstand a few days of frost. The phyllodes fall at the onset of severe moisture stress and the older branches rarely shoot again. Nevertheless, it has excellent natural regeneration in favourable years with good growth early in the season and may often form thickets and almost pure stands.

It is readily browsed despite the spines, mainly when flowering, but has only moderate palatability. Because of its abundance it is considered an important fodder source for cattle and sheep during less severe droughts. The seeds provide a source of protein for cattle. The leaves, green and dry seeds are also eaten by the Aborigines. It is an ornamental species that may be grown for hedges as it withstands clipping, but unfortunately it is short-lived (10-15 years). It is a suitable species for low shelterbelts, shade and windbreaks, and also for coastal sand dune fixation (Hall et al. 1972; Askew and Mitchell 1978; Baumer 1983; Mitchell and Wilcox 1988; Devitt 1992; Thomson 1992; Lazarides and Hince 1993).

Acacia willardiana (Willard acacia or palo blanco): a slender, unarmed tree up to 9 m tall, with white, exfoliating bark and drooping branches bearing narrow phyllodes up to 30 cm long and 4 mm wide. The species is endemic to Sonora growing on rocky sites in areas receiving summer rains (ZB IIIswr). The hard wood is used for implements and building poles. The tree is sometimes grown for landscaping. It is the only New World species with phyllodes. The species is among a small, anomalous group of uncertain taxonomic affinity within subgenus *Aculeiferum* where the phyllodes are dorsiventrally flattened. The phyllodic petioles bear up to four pairs of soon deciduous pinnae which appear on the first shoots of each season's growth, after which nearly all of the tree's photosynthetic functions are carried out by the phyllodes; the phyllodes are not noticeably sclerophyllus (Shreve and Wiggins 1964; Pedley 1975; Felger and Mason 1985; Turner et al. 1995).

Faidherbia albida, syn. *Acacia albida*: a deciduous tree up to 30 m high and armed with spinose stipules. Widely distributed through the savanna regions of tropical Africa and extending into the Lebanon [ZB and ZE I–III, II–III, III(tI), IV], it is essentially a riverine species although it can also occur on well drained soils of the interfluves. The tree is of considerable benefit to both grazier and cultivator. The leaves and pods provide an excellent dry season fodder. The dung from the browsing livestock plus litter and possible benefits from nitrogen fixation are freshly available to the cultivator at the start of the rainy season and not subjected to a long period of desiccation. A three- to four-fold increase in cereal yields can be obtained from crops grown beneath the tree canopy compared with crops growing outside the canopy. The timber may be used for furniture, joinery, boxes, boat building, pestles and mortars, etc. The ground and sieved pods may be eaten with sour milk. The gum, bark and pods may be used medicinally. The tree may be grown for agroforestry and live hedges (Wickens 1969).

The species has the unusual characteristic of shedding the leaves at the start of the rainy season and producing new leaves and flowering at the start of the dry season, a pattern that is also followed in areas with a bimodal rainfall. Observations in Darfur Province, Sudan (Wickens 1969) have shown that while trees on the higher ground with a low water table will shed their leaves earlier than trees on the riverine soils with a high water table, the appearance of new leaves and flowering occurs spontaneously, irrespective of water table. On the alluvial soils of riverine terraces, changes in the level of the water table may mean that sections of the root system, apart from the subsurface laterals, will have to withstand from 2 to 5 months of waterlogging annually.

As yet this unseasonal behaviour awaits a satisfactory explanation. It is possible that the unseasonal behaviour is linked with anaerobic root respiration, although it is curious why *Acacia nilotica* subsp. *nilotica**, which undergoes even more severe waterlogging, does not behave in a similar manner. Lebrun (1968) postulated that leaf fall was stimulated by the rise in the water table and anaerobic conditions whereas Fig. 15 shows new leaf occurring under maximum anaerobic conditions. Similar observations on phenology and changes in the level of the water table are required from other parts of the distribution before any conclusions may be drawn.

Germination is epigeal. During the first 6 months aerial growth ranged from 5 to 15 cm, while the seedling taproots were 70 cm or more long. By 18 months, aerial growth may have attained 27 cm, with considerable thickening of the taproot and the development of extensive laterals within the zone of moist soil. Vigorous coppice growth can occur from the stumps of felled trees. Root suckers may be produced from both the distal and proximal portions of severed roots.

Prosopis cineraria: a facultatively deciduous tree. It is one of the underexploited multipurpose trees of the hot deserts of Asia, distributed from Arabia to Pakistan (ZB and ZE II–III, III). The wood is hard but turns and polishes readily. The poor form of the timber from unimproved trees makes it unsuitable for construction purposes, so that it is largely used for fence posts and implement handles. The wood has a high calorific value (1720 J kg^{-1}) and is used for firewood and barbecue charcoal; it is also used for pulp wood and paper. The tapped branches will produce amber gum which is soluble in water. The tannin content appears to be unexploited. The leaves and pods provide a nutritious fodder for livestock. The sweet unripe pods are used in curries and pickles, and the powdered bark mixed with flour may be eaten in times of scarcity (Leakey and Last 1980; Bhandari 1990). The tree may also be grown for soil stabilisation and nitrogen fixation in agroforestry systems (Mann and Shankarnarayan 1980).

The roots may penetrate as deep as 36 m in search of water. However, the lateral roots are few and at the seedling stage tend to be superficial so that establishment can be difficult. Transpiration is high, and in one Indian study in an area receiving 370 mm annual rainfall transpiration amounted to 220 mm water $ha^{-1}year^{-1}$. The tree tolerates saline conditions and may grow in salt concentrations up to pH 9.8 (Leakey and Last 1980).

Prosopis tamarugo (tamarugo): an evergreen, spiny tree up to 18 m high, highly drought tolerant and also with a high degree of salt tolerance. It possesses a double root system. The taproot, which may bifurcate, extends for several metres down to the water table, the waters of which originated high up in the Andes. At 50–100 cm below the surface, a second, richly ramified system of fine roots develops, extending laterally to a diameter equalling that of the crown (Stienen 1985). Tamarugo is a north Chilean endemic of the Pampa del Tamarugal (ZB III).

The tree is of considerable local economic importance. The wood is extremely durable, although difficult to work, and can be used for heavy construction, tool and knife handles, tubes, barrels, sleepers, furniture, turnery and parquet floors. It yields high quality charcoal as well as firewood of high calorific value; in the past it has been overexploited for buildings and fuel for the smelting of silver. The fruits may be fed to sheep and cattle and the leaves provide a useful browse. The dry matter digestibility is only 45–50%, sufficient for the maintenance requirements of sheep with free access to it. However, stock feeding on tamarugo do require mineral supplements containing cobalt (Co), iron (Fe) and magnesium (Mg) plus vitamin A (Göhl 1981; Stienen 1985).

The studies of Sudzuki (1969) were purported to show that at night the leaves of *P. tamarugo* took up large quantities of water from the camanchaca, or fog, and conducted the water to the roots. The dense mass of fine roots, or root felt, then excreted water so that the soil became thoroughly wetted to a depth of about 1 m, while the soil above and below remained dry. This water was then used by the tree during the day. It has since been proved that the water requirements of *P. tamarugo* are met by the groundwater and that the leaves have no anatomical feature that could possibly permit the required large uptake of water. The thick mass of fibrous roots and the accumulation of water in the soil with an osmotic potential of -1.5 MPa are less easy to explain. The mean osmotic pressure of the cell sap of the leaves was -2.13 MPa. The nightly change in direction of the vascular flow of water from the shoot to the roots observed by Sudzuki may indeed occur temporarily during the spring, when the water potential of the leaves at night rises to -1 MPa, and that in the roots is -1.5 MPa. Normally, however, the flow of water is from the roots to the leaves. It has been speculated by Mooney et al. (1980) that the wetting of the soil in the rhizosphere might arise when, owing to the lower night transpiration, some of the water absorbed by the taproot is excreted by the felt of fine roots to act as a reserve for use during the day when transpiration increases. Walter and Breckle (1986) suggest that this root felt may have been formed when the water table was at that level, and that the water table has since fallen due to extensive reforestation and extraction. It is perhaps significant that the present tamarugo stands grow in the lower parts of the terrain, where the groundwater is closest to the surface.

The soils on which *P. tamarugo* grows are thickly encrusted with salt. The roots of germinating seed can only penetrate this crust and reach the groundwater in exceptionally rainy years, when the salt crust is sufficiently dissolved to permit the roots to descend to the water table. Reforestation practices with tamarugo require the digging of a pit through the salt crust. Stienen (1985) reported that in periods of high humidity, due to more than normal fog, the leaves respond by increasing in size. The stomata are mainly open at night, but never entirely open. When the soil moisture is such that the humidity beneath the crown reaches 95%, the stomata will then open during the day.

LEGUMINOSAE subfamily **PAPILIONOIDEAE**

Alhagi graecorum, syn. *A. maurorum sensu auct.*: a bushy, deep-rooted, thorny subshrub up to 1 m tall, with branches bearing axillary spines, and with roots descending to 15 m. Widely distributed through the eastern Mediterranean from Greece and Tripolitania, and eastwards to Arabia and Iran (ZB and ZE III, III–IV, IV, IV–III, IV–VII, VII–IV), it is often confused with the more slender ***A. maurorum*** whose distribution extends from Turkey through the Middle East to Afghanistan and central Asia (Meikle 1977).

It provides valuable forage for camels and young sheep and goats. The dry plant is used for fuel. The tangled spiny twigs are used in summer as air-cooling window-screens, the rapid evaporation from dripping water onto the twigs providing a pleasant stream of cool air. The plant is a source of manna and may also be grown as a garden ornamental. It is regarded as an indicator of fresh to saline water. Heterophyllus, the surface area of the summer leaves is finally no more than 2% that of the winter leaves, which disappear in July; the thorns increase in inverse proportion (Oppenheimer 1960; Townsend and Guest 1974; Danin 1983).

Astragalus microcephalus (gum tragacanth – a general name for a number of species): a spiny cushion shrub up to 1.5 m tall, bearing pinnate leaves with deciduous leaflets and persistent, spinose rachis up to 4 cm long. Distributed from Anatolia and Caucasia to Iran [ZB and ZE VII(rIII), VII–IV]. It is reported by Gentry (1957) to be an excellent source of gum tragacanth in Iran, the gum being obtained by tapping the rootstock some 5 cm below the soil surface. It is not cultivated. The gum is widely used in pharmaceutical preparations, ice cream, liquors, confectionery, salad dressings, as a glaze or filler (sizing) for the textile industry, etc. (Towsend and Guest 1974).

In the gum tragacanths the pith and medullary rays of the central cylinder are metamorphosed into undifferentiated material forming the central gum cylinder. The gum cylinder is contained by a woody cylinder, with annual growth and with numerous conspicuous medullary rays radiating out from the central cylinder, and apparently functions as a nutrient and moisture storage organ. The cylinders are layered with gum during the dry summer season, the gum being contained in the cylinder under pressure. The gum cylinder of the various gum-producing species is relatively very large; near the root apex it is nearly half the diameter of the root, i.e. 8-10 mm diameter. The yield of gum increases if rain falls before tapping, but decreases if it rains during tapping. Most of the exudation occurs at night, when the air humidity is higher than during the day and thus favours the less rapid drying of the exudate. During the day the exudation is minimal, partly due to loss of moisture through transpiration; the high day temperatures also increase the drying rate and aid the sealing of the wound. As transpiration ceases the exudation pressure appears to increase (Gentry 1957).

Caragana jubata: a shrub up to 5 m high, branched from the base of the main stem and with the densely leafy branches studded with numerous, slender, 6 to 7-cm-long, spinescent rhaches of former leaves, which presumably function as a protective sheath. It is distributed through eastern Russia. Mongolia and north-western Tibet [ZB VII(rIII)]. The shrub is grown as an ornamental (Komarov 1960-1977, vol 11).

Caragana microphylla: a deep-rooted, thicket-forming, psammophytic shrub to 1 m tall. A native of eastern Siberia and eastern Mongolia and the Kalgan and Shensi provinces of China [ZB VII, VII(rIII)]. It has been cultivated in the former Soviet Union for fodder since 1861; it is also planted for stabilising sands (Petrov 1972). The stipules are usually persistent as recurved woody spines 3–10 mm long; the leaflets, like other species in this genus, are soon

Table 11. Chemical composition (%) of *Geoffroea decorticans* seeds and pods. (After Becker 1983)

	Total solids	Total nitrogen	Non-protein nitrogen	Protein	Fat	Fibre	Ash	Sugar
Pericarp	96.9	0.8	0.06	4.4	1.2	17.2	2.4	48.4
Seed	96.5	4.7	0.1	28.8	48.9	ND	ND	7.7
Whole pods	93.1	0.8	ND	5.2	3.1	20.0	1.9	36.8

	Hemi-cellulose	Mono-saccharides	Sucrose	Raffinose	Stachose	Total
Pericarp	32.7	+	44	Trace	0	44+
Seed	10.5	+	5.6	1.6	0.4	7.6
Whole pods	26.1	+	33.1	2.9	0.1	36.1

ND, no data
+, trace, not quantified

shed in order to reduce transpiration, the leaf rachi then deflex to form protective sheaths to the young green shoots throughout the hot dry summer and are shed during the winter months.

The species has a high transpiration rate (ca. 1 g water g^{-1} h^{-1}) and remains vigorous even when the surrounding herbaceous layer is severely degraded under grazing pressure. The mean water potential of the roots is low at -1.09 MPa, so that the shrub is able to obtain water from deeper soil layers than the associated shallow-rooted grass *Leymus chinense* (syn. *Aneurolepidium chinense*) with its high water potential of -0.31MPa (Song 1990).

Geoffroea decorticans (chañar): a deciduous tree up to 12 m tall bearing pinnate leaves and oval to globular fruit up to 3 cm long. It is widely distributed in Peru, Chile, Bolivia, Argentina, Paraguay and Uruguay (ZB and ZE II–VII, III, III–IV, III–VIIa, V-VII) and salt and frost tolerant. The attractive yellow wood is easier to cut than that of *Prosopis chilensis* and is used for carving, turnery, pit props and tool handles as well as for fuel. The specific gravity of the wood is 0.58-0.70. The bark is the source of a champagne-coloured textile dye and is also used in the herbal treatment of respiratory complaints. The leaves are eaten by sheep and rabbits, and also used for green manure. Goats avidly eat the branches, leaves and bark. The leaves are also reputed to be an anthelmintic. The pericarp of the fruit contains 48.4% sugar, predominantly sucrose, and the seed 29.4% protein; low levels of trypsin inhibitor and cyanide are also present. The poor nutritional quality of the pods is attributed to a high seed haemagglutinin content; see Table 11 for the chemical composition of the seeds and pods. The fruit is stored and fed to pigs. A refreshing alcoholic beverage, *aloja*, may be prepared from the fermented fruit, which may be distilled to give the potent spirit *aguardiente*. A black, treacle-like syrup, *arôpe*, is prepared by first warming and then boiling 5 kg fruit in 10 l water until reduced by half; the mass is then churned, chopped and the syrup decanted after settling. *Arôpe* is reputed to have medicinal properties as an antiasthmatic and chest decongestant. The species has a potential as a source of sweetners and as a fermentation substrate and possibly seed oil if trees with high yielding seeds can be found. The species may also be used for land stabilisation and erosion control (Soriano 1972; Becker 1983; Wickens 1988; Aronson 1989).

According to Morello (1951b) and Becker (1983) root suckers may be produced either spontaneously or from damaged roots, so that the species may become invasive if not controlled. Becker (1983) also reported nodulation with *Rhizobium* (mesquite) sp. and *R.* (tesota) sp.. Young stems and shoots are green, the smooth stem eventually turning to bark. From 15 to 20 years of age large sheets of the bark start to peel, revealing a smooth, green, photosynthesising underbark (Aronson 1990).

Halimodendron halodendron: a monotypic, 30 to 40-cm-tall shrub with long and short shoots, bearing spinose stipules and spine-tipped leaf rachi which persist after the leaflets have been shed. A plant of the salt steppes of the southern former Soviet Union to Siberia, central and eastern Asia [ZB and ZE VI–VII, VIIa, VII(rIII), VII–IV], it is grown as an ornamental. The short shoots appear in the axils of the previous year's spiny leaf rachi. (Mabberley 1987).

Olneya tesota (ironwood): a monotypic, evergreen tree up to 9 m tall, with a short trunk up to 45 cm diameter, the branches are armed with pairs of hard, sharp spines in the axils of the pinnate leaves. Nitrogen-fixing nodules have been observed by Felker and Clark (1981). The tree is distributed from southern Arizona, south-eastern California, Baja California and Sonora (ZB and ZE II–III, IIIsr, IIIswr). Ironwood is regarded as an indicator of almost frost-free sites as its young branches are sensitive to frost; it may be used therefore, for indicating sites suitable for citrus cultivation (Mason and Mason 1987). The wood is brittle, hard and remarkably heavy; it is too hard for ordinary woodworking tools and too heavy to float. The wood burns very slowly and with great heat and is consequently used by blacksmiths. The wood was used by native Americans for arrowheads and tool handles. Native Americans may eat the young seeds raw or when nearly mature and parched; the seeds contain the amino acid canavalin, a mild toxin which is also present in *Canavalia ensiformis* (jack bean). The fallen flowers are eaten by goats (Kearney and Peebles 1951; Coyle and Roberts 1975; Felger and Moser 1985; Rea 1991).

The photosynthetic patterns of ironwood and the drought-deciduous *Ambrosia deltoidea** have been examined by Szarek and Woodhouse (1976, 1977). Summer and winter gas exchange was measured at near optimum soil water conditions. Under such conditions the diurnal gas exchange patterns were characterised by maximum photosynthetic rates and stomatal conductance early in the day and which progressively decreased thereafter. The effect of

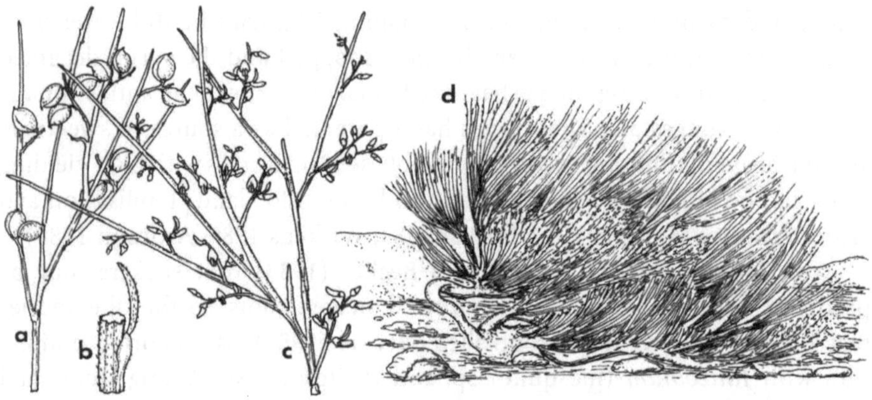

Fig. 35. *Retama raetum* (white broom). **a** Fruiting branch. **b** Soon deciduous leaf on young stem. **c** Leafless flowering branch; **d** Habit on a wind-swept peak in Tenerife, Canary Islands. (Adapted from Jaffri and El-Gadi 1980, Fig. 10) and plate in Kew Herbarium)

plant water potential on gas exchange was both direct and indirect. Decreasing plant water potential indirectly affected $^{14}CO_2$ assimilation by stomatal closure, and the effects were similar for both species. However, the direct effect of decreasing plant water potential were dissimilar. The *Ambrosia* had a higher water potential and maximum photosynthesis, but was more sensitive to water stress that the ironwood.

The seasonal photosynthetic pattern followed seasonal variation in plant water potential. *Olneya tesota* was photosynthetically active throughout the year, but with maximum photosynthetic rates of 12 mg CO_2 dm^{-2} h^{-1} in spring and summer. The tree maintained a favourable water balance throughout the year due to its deep roots, with the plant water potential seldom falling below -3.3 MPa. The extremely hard seeds are scarified by the abrasive action of the sand and gravel carried by the valley storm waters, ensuring germination when water is readily available (McGinnies 1981).

Retama raetam (white broom): an erect, soon deciduous, spartoid shrub up to 2 m tall (Fig. 35). It is a nebka-forming shrub, widely distributed through western Asia and Arabia (ZB III, IV). The fruits and flowers are an important fodder source; the germination of the hard seeds is greatly improved after having passed the seeds through the digestive tract of browsing animals. The stems produce a charcoal of a high calorific value. The plant has a number of medicinal applications as a febrifuge, purgative, vermifuge and abortifacient. It is also used for the cut-flower industry (Boulos 1983; Gutterman 1983; Danin 1996).

During the winter months the new green assimilatory stems are densely hirsute and sparsely covered with minute, scale-like leaves. Towards summer the leaves fall, the stems harden and many are shed; most of the hairs are also shed, remaining only in the furrows where the stomata are located. All are strategies for reducing transpiration losses. Following the winter rains, new twigs bearing small leaves and flowers develop, the number of flowers produced being dependent upon the available water. The possession of a taproot down to more than 10 m and with lateral roots extending over an area of 40 m^{-2} permits an annual water consumption of 1500–3500 kg. At the end of the dry season its transpiration rate can sometimes drop to ca.100 mg g^{-1} fresh weight h^{-1}. The maximum winter osmotic value was 1.7 MPa, with higher values during the summer months (Oppenheimer 1960; Zohary 1962; Danin 1983).

Robinia pseudoacacia (black locust or false acacia): a deciduous tree, 7-8 m high, with thorny stipules and a shallow root system. A native of eastern and central North America, introduced and now naturalised in Eurasia, it can be reproduced from seed, stump and root suckers. Unfortunately, it has little resistance to frost. Its ability to grow on poor soils has led to its use as a pioneer plant for eroded areas and strip-mine soils; it may also be grown for windbreaks, hedges, shade and as an ornamental.

The North American Indians eat the dried pod as a vegetable and also the cooked seed. The bark, which is sweet like liquorice, leaves and roots are toxic due to the presence of the toxalbumin robin, especially in the young shoots, which precludes their use as a fodder source for livestock, although the floral nectar is a useful early season bee fodder. Mitogenic lectins, which stimulate mitosis in leukocytes, have also been isolated.

Black locust has been found unsuitable in the USA for planting with conifers because it overtopped the latter, although understorey planting in native pine plantations or in mixtures of rapidly growing hardwoods may be useful, the other species benefiting from the nitrogen fixed by the mycorrhizal nodules of the black locust. The black locust nodulates and grows best at high levels of phosphate and medium levels of available nitrogen.

In the Loire valley, southern France, coppiced trees yield 6 t ha^{-1} $year^{-1}$. The timber is yellow, very heavy, very hard, strong and very durable and also rot-proof; it is used for construction, furniture and fuel. The resistance to rot is on account of the presence in the wood of 4% of an isomer of dihydroquercetin or dihydrorobinetin (taxifolin), which inhibits the growth of wood-destroying fungi. It also yields a 3-hydroxyflavonone which is readily convertible to the flavonol robinetin. A yellow crystalline pigment has also been isolated from the wood consisting of penta-acetyl and pentamethoxy compounds (Watt and Breyer-Brandwijk 1962; Crane et al. 1984; Blackwell 1990; Sprent and Sprent 1990).

In autumn the RNA content in the bark cells of *R. pseudoacacia* increased prior to and during the induction of freezing tolerance and paralleled and preceded a similar increase in soluble protein; the DNA content remained constant. Both the RNA and protein levels remained high until early summer, when the cells were in their least hardy state, and then fell rapidly to their former levels. These changes were closely associated with changes in tissue freezing tolerance. Information regarding the number of species involved and the role of RNA in cold acclimatisation is still poorly understood, while the accumulation of soluble protein in cold-acclimating tissues, while not universal, is generally accepted as an important step in the development of freezing tolerance (Guy 1990).

A useful feature of the species grown for protective afforestation in the semi-desert is the variability of the root systems. This enables the roots to make full use of the soil and soil moisture available. Their heat resistance allows them to withstand tissue heating up to 45–50 °C. They are also salt resistant and can withstand seasonal soil water salinities of up to 2–3%, the most salt tolerant species being *Tamarix*, which can tolerate up to 15–20% salinity. Their growth and development is characterised by rapid initial growth, especially of the root system. They enter the reproductive phase early, after which their growth slows down; they also tend to be short-lived (Petrov 1979).

MYOPORACEAE

Eremophila maculata, syn. *Stenochilus maculatus* (spotted fuchsia bush): a low, dense bush up to 1 m tall, bearing dark green, lanceolate leaves up to 4 cm long and 6 mm wide. Widely distributed in Australia (ZE II–III, III–II, III–IV, IV–III), it usually grows in saline environments and also in alkaline soils of flood plains and valley floors. The plant contains up to 9% of the cynogenetic glucoside prunasin. Under normal conditions, due to the absence of any hydrolysing enzymes, it may be safely browsed without any ill effect. However, cases of severe poisoning can occur when browsed with *Acacia georgina**, whose pods contain the hydrolysing enzyme, especially if eaten in large quantities by hungry stock, 30 g of leaf being sufficient to kill a sheep. The browse contains up to 21% crude protein and has an in vitro digestibility of 74%. The tree is a source of bee food as well as fibre and medicine for the Aborigine (Jackson and Jacobs 1985; Mitchell and Wilcox 1988; Lazarides and Hince 1993).

MYRTACEAE

Eucalyptus: the Australian shrublands are dominated by species of mallee eucalypts. These are species of *Eucalyptus* where the two swellings produced between the angles of the cotyledons grow down over the hypocotyl and give rise to a swollen rootstock or lignotuber composed of accessory strands of meristematic tissue capable of producing numerous stems (lignotillers). The lignotubers are protected from lethal temperatures by being underground. It appears that lignotuber development is accentuated in soils of low fertility, especially where phosphorous is limiting, suggesting a wider adaptive function than a regenerative base following fire. Because of the regular, but not necessarily annual, shedding of mallee bark during the main summer growing season (December to April), the remaining periderm is quite thin and offers little protection against fire. It is the shed bark, some of which may be caught up in the branches, that constitutes a fuel hazard.

Seed from most eucalypts is shed gradually over 2–4 years. Fire greatly accelerates the process, often resulting in a massive seed fall following an intense fire. Most of the aerial 'seed-bank' retained in the capsules may be protected from fire by: (1) insufficient fuel to produce crown fires; (2) capsules being located beneath the leaves so that even when a crown fire develops the heat flux received by the capsule is small; and (3) insulation of the seed within the capsule. However, when a high intensity crown fire does develop, especially in whipstick mallee, the crown is often totally consumed by the fire and it is unlikely that many seeds could survive. Provided there is a sufficient interval between periodic fires for seed set, periodic fires may be required to sustain genetic diversity by facilitating seedling recruitment.

Mallee eucalypts can be susceptible to repeated burning, possibly due to an imbalance between photosynthesis and respiration. The mallee relies on rapid refoliation following burning to replace the pre-fire photosynthetic surface area as rapidly as possible and offset any respiratory burden imposed by the lignotuber and roots. Root dysfunction may also be involved in mallee death after repeated burning. The lateral roots act as water storage organs (27–44% moisture) and may form an important water source for rapid refoliation after summer fires, when water stress is at its highest. Radiocarbon dating of the lignotubers of *E. oleosa* suggests a life span of 300–400 years, and it is likely that other semi-arid eucalypts may have a similar longevity. Hence, a fire frequency of about once every 200 years or more may be required to maintain population stability (Rotherham et al. 1975; Noble 1982).

All species of *Eucalyptus* yield a volatile oil known as eucalyptus oil or eucalyptol, the percentage of which varies from species to species. For medical purposes it must contain not less than 70% w/w of the menthane-1:8-diol anhydride cinerol. The commercial possibilities depend on the chemical composition of the eucalyptus according to the following four groups: pinene, cinerol (eucalyptol), phellandrene and peppermint ketone, and miscellaneous. Thymol may be readily prepared from the peppermint ketone piperitone; the peppermint ketone eucalypts are known as 'peppermints'. Eucalyptus oil is colourless or pale yellow, with an aromatic camphoraceous odour and a pungent camphoraceous flavour. It is a recognised household remedy for colds and an ingredient of nasal drops. In India the oil is used as a mosquito and vermin repellent; it is also an important component of such preparations. The oil is toxic if taken internally, causing death by respiratory paralysis.

Eucalyptus camaldulensis (river red gum): a deciduous tree up to 45 m tall with a trunk 2.5 m or more in diameter. The most widely distributed of the Australian eucalypts, it occurs throughout Australia except in the south and south-west and Tasmania, growing in areas receiving from 200 to 1300 mm annual rainfall (ZB and ZE II–III, III, III–II, III–IV, IV–III, IV–V). Provenances vary in salt, waterlogging and calcium carbonate tolerance; of its numerous provenances in Western Australia, a few have been shown to be highly salt tolerant (Sprent and Sprent 1990). Probably the most widely cultivated timber species in the semi-arid tropics and subtropics, it will grow well in poor soils and in areas where there are prolonged dry seasons (provided the roots can reach groundwater) or where periodic waterlogging occurs.

The tree may be grown for reforestation, wind breaks and shade. The transpiration rate is such as to provide a dry road surface when grown as a tree-lined border to a road through marshy areas. It is not suitable for planting in coastal areas exposed to salt-laden winds. Although recommended for planting as a source of bee food, bee keepers in certain areas are warned that when the flowers are infested by the insect *Nysius vinitor*, the pollen flow will cease. Selected trees may be browsed by sheep. The timber is red, hard, heavy, very durable but of varied strength due to the irregularity of the grain. The timber is used for sleepers, flooring, pole timber, fuelwood and charcoal, and also for paper, rayon and particle board manufacture; the bark contains 2.5-11% tannin and the wood 2–14%. The fuel wood has a calorific value of 20.1 MJ kg^{-1} .

The Aborigines use an unstrained decoction of the boiled leaves as a wash for internal pains and drink the strained liquid to relieve congestion of the nose and sinus. The leaves contain 0.3–2% essential oil consisting chiefly of 1,8-cineole and alpha-pinene, with medicinal properties as a decongestant, expectorant and counter-irritant *Eucalyptus camaldulensis* was formerly the main source of a dry, gummy exudate known as eucalyptus gum, eucalyptus kino and red gum, which exudes from the stem following insect injury or slashing. The gum contains 45% of kinotannic acid as well as kino red, a glucoside, catechol and pyrocatechol. It is a standardised pharmaceutical product which is extensively used in lozenges for its astringent and antiseptic properties for coughs, colds and relaxed throat, and also for dysentery.

The allelopathic effect of the complex phenolics present in the leaves is reported to effectively inhibit herbaceous growth on a clay loam, but is totally ineffective on sandy soils (Watt and Breyer-Brandwijk 1962; Hall et al. 1972; NAS 1979; Anderson 1982; Crane et al. 1984; Barr et al. 199 0; BOSTID 1990; Sprent and Sprent 1990; Aboriginal Communities of the Northern Territory 1993; Lazarides and Hince 1993).

NYCTAGINACEAE

Abronia maritima (coastal sand verbena): a prostrate or trailing, much branched, more or less evergreen, perennial herb with viscid, succulent leaves. It forms dense mats on the upper beaches and beach dunes of western Mexico (ZB III). The stout taproot descending to 5 m and up to 4 cm diameter apparently functions as a water storage organ. A tisane of the root bark is taken by the Seri people to aid the expulsion of a torn placenta (Wilson 1972; Felger and Moser 1985).

OLEACEAE

Olea europaea (olive): an evergreen tree up to 15 m tall. The cultivated olive, subsp. ***europaea*** var. ***europaea***, and its wild form, var. ***sylvestris***, are regarded as characteristic trees of the Mediterranean evergreen oak woodland and maquis (ZB IV). It has been introduced into California and other subtropical regions of the New World by Spanish and Portuguese settlers and missionaries, and by Italian migrants to Australia; it has also been established in South Africa since 1925. There is now considerable interest in developing olive cultivation in China, India and elsewhere. It is an ancient cultigen dating back to at least 3700–3500 BC (Renfrew 1973; Zohary 1973; Townsend and Guest 1980; Kiritsakis and Markakis 1991). The var. ***europaea*** is unknown in the wild and its seed does not produce the parental type, but reverts to var. *sylvestris*. There is a divergence of opinion as to whether the cultigen arose from the wild form or from the small-fruited Afro-Asian subsp. ***cuspidata***, syn. *O. africana, O. chrysophylla, O. ferruginea* and *O. cuspidata*. The chromosome numbers ($2n = 46$), wood anatomy, palynology and flavonoid content show remarkable consistency throughout the *O. europaea* complex. Interestingly, the flavonoids clearly separate the complex from other members of the genus (Green and Wickens 1989).

The anatomy and pollen of subsp. ***laperrinii***, syn. *O. laperrinii* from the Saharan mountains have been described in detail by Benichou (1962) and minor differences claimed between this subspecies and other members of the complex. Both anatomical and palynological studies at the Royal Botanic Gardens, Kew, have failed to support these differences. The sclerophyllous leaves of *Olea* contain such xeromorphic features as thick outer walls to the epidermal cells, thick cuticle, high stomatal density and abundant sclerenchyma, characters that are also present in *Pistacia, Quercus* and *Styrax* (Benichou 1962; Metcalfe 1983). The stomata on the undersurface are not sunken and are protected by two guard cells. The peltate scales on the undersurface offer additional protection against evapotranspiration losses. Transpiration rates may be as high as 717 mg g^{-1} h^{-1} under favourable conditions or as low as 208 mg g^{-1} h^{-1} under drier conditions in Dalmatia, while even lower rates have been obtained in Israel. In November summer values of osmotic pressure may reach as high as 4.2 MPa, while the maximum water saturation deficit recorded was 26.9% in September (Zohary 1962)

The fruits increase rapidly in size with the first rains of the autumn and when this growth has ceased they are hand-picked for table olives. They should have a large amount of easily separated flesh and an oil content of not more than 12-20%. The oil-producing olives are picked later, when they are fully ripe and have maximum oil content. The whole fruit should contain 35–70% oil (dry basis), with the pulp often containing more than 75%. The fragrant oil has a delicate flavour and is high in nutrients and health value. It is almost unique among the vegetable oils in that it may by consumed without any refining treatment. The first pressing yields the highest grade virgin oil. Edible olive oil is a blend of virgin oil with refined oil that has had two pressings. Commercial oil is the refined lampante oil produced after the pulp has been pressed with hot water and sulphide following treatment with solvents and repeated refinement. Inferior grades of oil may be used for soap-making, perfumery and in other industries (Manousis and Moore 1988; Kiritsakis and Markakis 1991). The wood is extremely heavy and tough (specific gravity 0.8–1.1) and is used for house beams, furniture, inlays, carvings, minor joinery and carpentry. It is a valuable source of a slow burning fuelwood and charcoal. It is a source of manna. The tree also has a number of medicinal uses (Townsend and Guest 1980; Miller and Morris 1988).

PASSIFLORACEAE

Adenia globosa: a shrub or climber up to 8 m, arising from a succulent, somewhat warty, grey-green, subglobose, half-imbedded, water-storing, lumpy trunk up to 2.5 m diameter from which arise semi-succulent stems bearing caducous leaves with axillary thorns up to 9 cm long, the thorns a morphological adaptation of the tendrils that are present in other species. It is a native of Somalia, Kenya and Tanzania [ZB and ZE I–III, III(tI)]. The stem is used by the Masai for a cattle medicine (Watt and Breyer-Brandwijk 1962).

PEDALIACEAE

Harpagophyton procumbens (grapple plant): a perennial herb with a tuberous taproot, up to 20 cm long and 5 cm diameter, from which radiate horizontal secondary roots, on which one or more enlarged storage roots, weighing up to 4 kg, may be formed. The prostrate, semi-succulent and eventually woody, annual stems radiate for up to 1.5 m, dying back annually. The woody, elliptical fruit case has four rows of curved arms bearing recurved spines (Fig. 36). The grapple plant is a native of Namibia and the Karoo to the Transvaal and Botswana (ZB and ZE III, III–II).

It is reported that seed germination is dependent on the seed being trampled well below the soil surface. The older seeds have a higher germina-

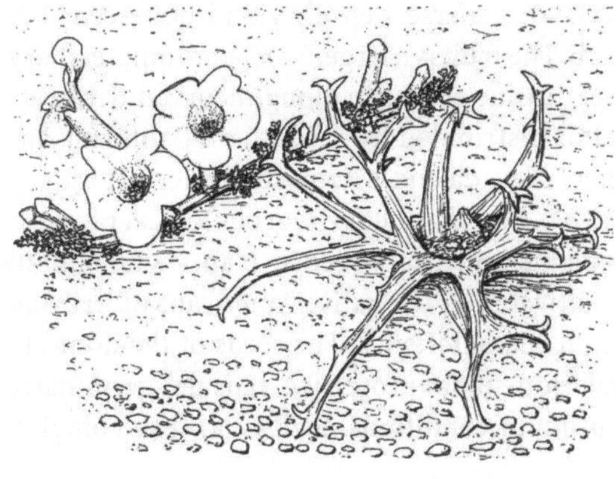

Fig. 36. Habit and fruit of *Harpagophyton procumbens* (grapple plant) growing in southern Africa. (Adapted from a photograph in Kew archives)

tion; sclarification also improves germination. Stem sections treated with rooting hormone may be used for vegetative reproduction.

The plant may be browsed by livestock in times of drought since it is one of the few plants remaining green; gemsbok may even dig up the tubers for nourishment. The older stems tend to form a tangled mass in the stomach, blocking the digestive system, especially of horses. The fearsomely armed fruit can become jammed in the foot or hoof, and also in the mouth, as well as becoming entangled in the coat. The grapple tuber may be used medicinally by Europeans as well as by the indigenous populations. The secondary storage roots are carefully harvested in order to avoid damaging the primary root, which is left to regenerate and maintain populations. The harvested roots are cut, dried and exported mainly to Germany for processing. It is reputed to provide one of the world's most effective herbal remedies for arthritis. The tubers, but apparently not the root, contain the glycosides harpagoside, harpagide and procumbide, as well as a phytosterol mixture containing mainly beta-sitosterol and stigmasterol, together with unsaturated fatty acids, triterpenes, flavonoids and free acids (chlorogenic acid and cinnamic acid). It is claimed that phytochemical and pharmacological studies show that grapple possesses two major properties for use as herbal medicine: (1) the presence of antirheumatic and antiphlogistic properties corresponding to those of pyrazolone derivatives and phenylbutazone, also certain analgesic action but with no diuretic or spasolytic properties; (2) its effect as a bitter, with a bitter value of 6000 or more, equivalent to that from one of the most powerful bitter tonics, that of gentian root (*Gentiana* spp.). It is used for treating rheumatic complaints such as arthritis and spondylosis, rheumatoid arthritis, neuralgia and headaches; it is also used in cases of diabetes, arteriosclerosis, gall bladder, liver and kidney complaints (Watt and Breyer-Brandwijk 1962; Volk 1964; Taylor 1985; Ihlenfeldt 1988; Weiss 1988).

Proboscidea parviflora (devil's claw): a coarse. densely viscid-pubescent, large-leaved, decumbent annual herb. The fruit is a large, 5 to 7-cm-long, leathery, two-valved, imperfectly dehiscent capsule ending in a long, incurved, dehiscent beak twice as long as the body. It is distributed from southern California and northern Mexico through southern Nevada and Arizona to western Texas (ZB and ZE II–III, III, IV–III, VII, VII–V). Two distinctive morphological races are recognised: black-seeded wild plants and white-seeded plants with slightly larger seeds, domesticated and cultivated by the native Americans of Arizona. The latter have better germination characteristics. The seeds of the domesticated race, after being removal from their protective capsule, germinated readily after 10 years. Under optimum conditions devil's claw can complete its life cycle in 8 weeks.

The oily seeds of both the wild and cultivated plants may be eaten as a vegetable, as are also the young pods. The black designs in the baskets made by the Arizona native Americans are woven from the black fibres stripped from the claws of mature pods. The pods may be regarded as a pest since they can become entangled in the wool of grazing sheep. The plant shows potential as an oilseed crop for arid regions. The seeds of the domesticated race contain 35–40% of an edible oil and 23–27% protein, producing estimated yields of 1000 kg oil and 675 kg protein ha^{-1}, which compares favourably with other crops.

The relatively stout stems, which are considered by Shreve and Wiggins (1964) to be extremely large for summer desert annuals, are anchored in the soil by a relatively shallow, delicate and diffuse root system mainly within 10 cm of the surface and with a taproot that seldom exceeds 20 cm long. The leaves too are unusual in that the surface area of the blades of the domesticated race may be up to 750 cm^2, whereas few arid plants have entire leaves with a blade surface area greater than 15 cm^2, i.e. they are more mesophytic than most arid plants. Typically the leaves wilt at midday and regain turgidity by dawn of the next day (Kearney and Peebles 1951; Berry et al. 1981; Rea 1991).

PLANTAGINACEAE

Plantago coronopus (buck's horn plantain): an acaulescent annual or biennial herb with a rosette of narrow leaves up to 15 cm long. Widespread through the Mediterranean region eastwards to Afghanistan (ZB and ZE III, IV, IV–VII), the herb is often cultivated for its palatable leaves. The seed capsules are enclosed in the dry inflorescences which, when wetted, open out and the capsules dehisce; the mucilaginous seeds are released by the raindrops and adhere to the soil. The timing of the rain affects germination; while rain at the beginning of the night may induce germination, poor germination may result if the rain falls during the day. Seeds may survive in the soil for a number of years; a mass germination may be required to break the surface crust, but only a few dominant seedlings will survive to produce seeds (Low 1991; Gutterman 1993).

POLYGONACEAE

Calligonum polygonoides subsp. ***comosum***, syn. *C. comosum*: an early deciduous, glabrous, C_4 shrub up to 2 m tall with articulated and often clustered, white, fragile branches and young, green branchlets. A Saharo-Sindian species (ZB III), it is browsed by livestock. Young branchlets photosynthesise after the leaves have fallen. Provided the seed is upright and not too deeply buried, light penetrates through the micropyle to the embryo and inhibits germination. High temperatures (30 °C) and close contact with water also inhibit germination. In following seasons, as soil movements change the position of the seed, the percentage germination will also change (Boulos 1989; Gutterman 1993; Gibson 1996).

Calligonum polygonoides subsp. **polygonoides**: a rigid, much branched and almost leafless shrub. It is distributed from Syria and Iran to the Indian Desert (ZB and ZE III, IV–VII). The wood is used for hut building and fuel, the charcoal is used by blacksmiths and the branches are browsed by camels. The young stems, which appear soon after the start of the monsoon season, are green and fleshy (Bhandari 1990); its distribution and ecology are discussed by Charen and Sen (1985).

PORTULACACEAE

Portulaca oleracea (common purslane): a succulent annual herb. Widespread throughout the warm-temperate regions of the world, it is one of the eight most common weeds of the world. It is widely used as a pot herb and in local medicine. The New World diploid occurs mainly as a coastal halophyte, while the tetraploids are mainly weeds. The hexaploids are found in the higher latitudes and altitudes. Common purslane is a photoperiodic facultative or quantitative short day-plant. The photoperiodic treatment of the mother plants affects the germinability of the seeds, and is dependent on the final 8 h maturation on the mother plant. The shorter the daylength the faster the germination. Red or far-red light treatments of the mother plant also affect germination, but there is no dark germination, even after 8 days of seed maturation under continuous red light. The results suggest that the influence of light treatments on the mother plant and the effect on seed germination are not regulated via phytochrome but rather through some other pathway (Gutterman 1974, 1993).

RHAMNACEAE

Ziziphus lotus: a deciduous tree up to 7 m tall, but due to browsing camels and goats it usually is a bushy shrub with numerous, intricate, twining and erect, zig-zag branches, usually armed with slender spines. The distribution extends from southern Greece and south-western Europe through North Africa to Israel and Arabia (ZB and ZE III, III–IV, IV). It is often found as a nebka-forming shrub due to wind-blown sand being trapped by the prostrate lower branches. The accumulation of sand is sufficiently slow for the shrub to maintain itself on top of the mound, and for annual layers of leaf litter, etc. to alternate with the sand. The extensive lateral root system of *Z. lotus* extends for 10-15 m, far beyond the limits of the nebka, mainly in the direction of the wind and it is thus able to benefit from any rainfall that falls between the nebkas. The olive-size fruit are eaten by man and livestock, and the leaves are browsed by goats and camels. The fruits are also used in local medicine as a tonic and febrifuge. The dense spiny branches make this a suitable species for live hedges in desert regions (Baumer 1983; Le Houérou 1986; Walter and Breckle 1986; Boulos 1989).

Despite severe atmospheric drought, the transpiration rate of *Z. lotus* does not follow the potential evaporation curve during the forenoon, but remains at a more or less constant rate throughout the day, thus enabling minimum water consumption to be maintained without the danger of a heat death. However, this restriction is insufficient for balanced water relations, so much so that increasing water deficits lead to a deterioration of water potential, resulting in a rapid decrease in photosynthesis shortly before noon. Net photosynthesis reaches the compensation point, or falls below it, and recovers only with the decrease in evapotranspiration in the late afternoon (Kassas and Batanouny 1984).

Ziziphus spina-christi: a shrub or small tree up to 20 m tall, evergreen in wet habitats, and partially deciduous or deciduous in drier sites. It is widely distributed through the drier parts of northern Africa and the Middle East [ZB and ZE II–III, III, III(tI), IV]. The shrubby var. ***microphylla*** often occurs as a nebka-forming species, especially in arable and fallow lands, while var. ***spina-christi*** occurs both as a shrub or a tree. The trees may occur in ring clumps. One such clump examined by the author arose from root suckers after the parent tree had been felled, the suckers arising several metres from the old stump to form a circle (Hunting Technical Services 1968); other rings may possibly originate from stock enclosures constructed with cut fruiting branches that had regenerated in situ. The fruits are edible and rich in vitamin C. The possibility of using *Z. spina-christi* as the root stock for grafting on the large fruited cultivars from India of *Z. mauritiana* to improve fruit production should be investigated. It is browsed by livestock. The branches are often cut to make temporary stock-proof deadwood hedges. The wood is hard and is used for local furniture, tool handles, spear shafts, fence posts and hut building. The wood also provides a good fuel, which burns with an intensive heat, and charcoal. It is a useful plant for stabilising dunes, windbreaks and stock-proof hedges (Sahni 1968; Baumer 1983; Booth and Wickens 1988; Boulos 1989; Orev 1992).

ROSACEAE

Coleogyne ramosissimus (blackbrush): an evergreen, up to 2-m-tall, rounded shrub with short, intricately branched and spinescent branches and bearing small linear leaves. The rounded habit is the result of the terminal twigs dying back and forming spines, with all subsequent regrowth from subterminal buds. During drought intercalary cork is formed between the stem segments, creating multiple stems. It is distributed from south-western Colorado to northern Arizona and south-eastern California (ZB VIIa). It is browsed, especially in winter, by sheep and goats, and to a lesser extent by cattle; it has the great advantage of being able to withstand heavy grazing pressure. It does not resprout readily following fire and reseeds with difficulty. It may be grown for landscaping (Kearnsy and Peebles 1951; Havens et al. 1973; USDA 1988; West 1988).

According to Smith and Nowak (1990), in a comparative study using leaf area index as a function of declining water potential in three shrub species, blackbrush maintained pre-dawn water potentials down to -8 MPa throughout the summer, although over 50% of the leaves were lost. However, the two drought-deciduous shrubs *Grayia spinosa* and *Hymenoclea salsola* became completely deciduous by midsummer, when pre-dawn water potentials fell below -5 to -6 MPa.

SANTALACEAE

Santalum acuminatum (sweet quandong): a small tree or shrub up to 10 m tall, a semiparasite, particularly during the early stages of growth. The grey-green, opposite leaves are up to 6 cm long and 3 cm wide, tapered at each end, and with a short hooked tip when young. The edible fruit are ca. 2 cm diameter, bright red and rattle at maturity. Except for the south-eastern coastal regions, sweet quandong is widely scattered through southern Australia, especially in the mulga communities on the rocky hills and plains (ZB and ZE III–IV, IVa, IV–III). The tree is reputed to be highly resistant to drought, high temperatures and, in an experimental orchard, has survived for 7 years using irrigation water with a conductivity of 4.7 dS m^{-1}. Germination will occur within 2 months of removal of the seed from the ripe fruit, although germination is better from more mature seeds. The optimum temperature for germination is between 16 and 20 °C.

The timber is aromatic and used for carving and cabinet making. The foliage is eagerly browsed by sheep and cattle. The trees may be used for shade. The kernels of the fruit are used for ornaments, necklaces and games; they are also used by the Aborigines to relieve pain. A staple food of the Aborigines as well as a popular fruit of the white settlers, the rather acid, pulpy flesh may be eaten fresh, but is usually dried to improve the flavour and then cooked and eaten in jams, chutneys and pies. The fruit are a good source of carbohydrate (19–23%) and relatively high in protein (1.7–2.8%). The kernel is also edible and contains an edible, inflammable oil (58% fat) which may also be used medicinally. Unfortunately, some tasters regard the kernels as possessing an objectionable aroma, mainly due to the presence of methyl benzoate. The methyl benzoate varies between 32 and 1294 μg g-1, not only between individual trees but also seasonally from the same tree. There is obviously potential for selection for flavour (Brand and Cherikoff 1985, 1988; Maconochie 1985; Mitchell and Wilcox 1988; Rivett et al. 1989; BOSTID 1990; Aboriginal Communities of the Northern Territory 1993; Lazarides and Hince 1993; Wickens 1995b).

The major fatty acid in the oil-rich kernel is santalbic acid (trans-11-octadecen-9-ynoic acid), also known as ximenic acid, a characteristic fatty acid of both the Santalaceae and Olacaceae. The santalbic acid is present in relatively large amounts and is accompanied by small amounts of a second acetylenic fatty acid, stearolic acid (9-octadecynoic acid). Several long-chain acetylene fatty acids have been shown to by physiologically active, inhibiting enzymes involved in some important biological pathways and the suggestion has been made that toxicity problems may arise when an acetylenic acid has a structure analogous to a polyunsaturated fatty acid substrate for essential enzymes. Thus, the consumption of quandong kernels cannot be recommended until current work by the Commonwealth Scientific and Industrial Research Organisation (CSIRO) on the metabolism of santalbic acid has been completed (Rivett et al. 1989).

SAPOTACEAE

Argania spinosa, syn. *A. sideroxylon*: usually occurs as an up to 10-m-tall, multistemmed bushy tree with massive contorted boles, or, when single-stemmed, branching low down with stout, wide-spreading branches. A Moroccan endemic, requiring 100–300 mm annual rainfall (ZE III–IV), it is the dominant tree in south-western Morocco, extending from Safi to Oued Noun, and penetrating inland as far as the plain of the Sous and the lower southern slopes of the High Atlas and the north-western slopes of the Anti-Atlas. It is largely confined to wadis south of Oued Noun, with its southern limit just south of Oued Drar; there are also small enclaves south-west of Rabat and in the Moulouya valley near Berkane. It is tolerant of most soil types except drifting sands and waterlogged soils. With increasing humidity it gives way to *Olea-Pistacia* bushland or *Tetraclinis* forest. Much of its distribution area has been degraded through cultivation or browsing goats.

The seeds germinate readily, the young shoot appearing within 1 month and fruit production in 5–6 years. The mature nuts contain 66–77% oil known

Table 12. Nutritional analyses of the seeds and oilcake (Göhl 1981), and the amino acids (Farines et al. 1984) in the oil of *Argana spinosa*

	DM	CP	EE	NFE	CF	Ash
Seeds, Morocco	94.8	9	71.7	10.4	6.8	2.1
Oilcake, mechanically extracted	73.7	33.4	25	12.7	23.9	5

Amino acids in oil (%)							
Arachidic	Trace	Linoleic	31.8–34.6	Myristic	Trace	Palmitoleic	Trace
Benenic	Trace	Linoleic	Trace	Oleic	45.2–46.9	Penta-decanoic	Trace
Eicosenoic	Trace	Mono-decenoic	Trace	Palmitic	13.5–13.9	Stearic	5.6–5.7
Heptadecanoic	Trace						

DM, dry matter; CP, crude protein; EE, ether extract; NFE, nitrogen free extract; CF, crude fibre

as argun oil. The strong flavoured, edible oil may also be used for making soap. The kernel, after first roasting to eliminate the saponins, is then ground and mixed with tepid water and the oil rises to the surface; it is removed by precipitation. The residues are allowed to settle before use. See Table 12 for nutritional analyses of seeds and oil cake. The greyish-green oil cake contains the slightly toxic, haemolytic saponin sapoarganine, which does not harm ruminants and passes out with the urine, and beef cattle may consume up to 2 kg day^{-1} without harm, although horses will not eat the cake. However, when fed to dairy cattle it can contaminate the milk and cause diarrhoea in children. The fleshy parts of the fruit, containing 10% protein and 7% fibre (dry matter basis), and the leaves are eaten by livestock (Göhl 1981; Baumer 1983; Wickens 1995b).

SCROPHULARIACEAE

Monttea aphylla: a 1.5 to 4 m high shrub with rudimentary, early deciduous leaves. A native of the Chilean Andes, the plant is browsed by sheep and the wood is used for fuel (Cabrera 1955, 1961). The plant is noteworthy for the very thick cuticle characterised by vesicles of fatty substances secreted by the epidermal cells of the terete stems, the very large and prominent external atrium of the stomata and the internal structure consisting of several layers of palisade tissue uninterrupted by sclerenchyma (Fig. 18). The poor regeneration of this species may be attributed to late summer drought following the ripening of the seeds (Oppenheimer 1960; Metcalfe 1983).

SIMMONDSIACEAE

Simmondsia chinensis, syn. *S. californica* (jojoba): a monotypic, tardily drought deciduous to evergreen, dioecious, up to 3-m-tall shrub bearing leathery, vertically orientated, ovate to oblong-elliptical leaves up to 5 cm long; the taproot may descend to 12 m (Fig. 37). The species owes its misleading specific epithet to a misunderstanding as to its country of origin (Bloomfield 1985). Its distributed ranges from California to Arizona southwards into Baja California and Sonora (ZB and ZE II-III, IIIsr, IIIswr, IIIwr); it has been introduced into many arid and semi-arid areas of the world. It is locally common on gravelly or rocky hillsides and valley slopes and, in drier areas along watercourses, grows in non-saline to slightly saline and near neutral to alkaline soils.

The acorn-like fruit mature in the early autumn and shed the oily, chocolate-brown seed on to the ground, where they are avidly eaten by children, native Americans, sheep and goats. The Mexicans prepare a nourishing and savoury beverage by roasting and grinding the kernels, mixing with the yolk of a hard boiled egg and then boiling the paste in water with the addition of milk and sugar. The oil extracted from the boiled seed is used to promote growth in deficient eyebrows and as a hair restorer. The Mexican historian Clavijero (1789) describes the oil as a cure for wounds and a remedy for cancer.

The oil was first analysed by Diguet (1895), who recommended the cultivation of the shrub in the French colonies of North Africa for its edible oil (Douglas 1947; Mirov 1952; Saunders 1976; USDA 1988). The fruit contain 40–60% oil, technically a liquid wax, composed of one molecule of a long-chain alcohol to which one molecule of a fatty acid is attached. It is the only known plant to produce such an acid-alcohol ester in its fruit and this has immense conservation significance in its similarity to that of sperm whale oil. It has considerable economic potential as a lubricant for delicate instruments and as an industrial lubricant for high temperature and pressure processes, and also as a chemical intermediate for products used as wetting agents and extreme pressure additives. The oil can be sulphonated, oxidised and chlorinated for these and other uses. The seed oil can be hydrogenated to form a hard, white, crystalline wax, with a melting point of ca. 70 °C, which ranks in hardness next to carauba wax from *Copernicia prunifera**, and can be used for polishing waxes, carbon paper, impregnated heat-resistant containers and smokeless, long burning candles. The oil reacts with sulphur chloride to form a rubbery compound known as factice, which is used in varnishes, rubber, adhesives and linoleum. It is also a source of straight chain alcohols and acids used in detergents, disinfectants, emulsifiers and bases for creams and ointments. The oil may also be used in shortening and other food products to delay the onset of rancidity. The decorticated seed may be used as a high protein stock feed and in the plastics industry. The growing plant may be used

Fig. 37. *Simmondsia chinensis* (jojoba) from south-western North America. Branch showing thick, leathery leaves jointed at base and male inflorescence – female flowers are usually solitary. (Part of photograph no. 73 12992 in the Kew Herbarium by Suendé dated 8 January 1980)

for browse, soil stabilisation and amenity planting (Douglas 1947; Brooks 1978). Jojoba is currently being grown as a commercial crop, mainly for the cosmetics industry. It will have to be grown far more extensively, with higher yields and at a far lower selling price before it will be able to replace the role of the sperm whale oil in industry.

The life span of the shrub is 100-200 years or more. It prefers neutral to alkaline soils with a high phosphate content and subject to annual drying. The plants are dioecious in a roughly 1:1 ratio, with the male plants tending to survive seedling establishment better than the females. Although honey bees collect pollen from the male flowers they never visit the female flowers as there is no floral attraction or reward in the form of volatiles or nectar in the green apetalous flowers (Gentry 1958; Buchmann 1987).

The preferred annual rainfall is 300 mm plus; with less than 120 mm the shrub is confined to run-on sites. It tolerates high daily temperature fluctuations ranging through 30–40 °C during the morning, as well as high daily extremes of 43–52 °C shade temperature and can tolerate temperatures as low as -9°C without serious damage. However, the seedlings are sensitive to light frost of -3 to -4 °C, while the flowers are destroyed by late frost. Exposure of the growing fruit to high temperatures can cause scorching and loss of seed. Low winter temperatures together with good winter rainfall encourage regrowth. However, light rains often fail to break shoot dormancy because the moisture fails to reach the deep roots. Deep soil moisture is required early in the season for good fruit development, i.e. a Mediterranean-type climate is preferred. In drier environments, perennial growth is limited and die-back common, producing shorter internodes and prolific lateral branching, twiggy and intricately branched bushes, while a wetter climate favours taller bushes with a broom-like habit due to the longer shoots and fewer lateral branches (Gentry 1958). The seeds contain simmondsin, an unusual cyanogenic glucoside, which acts as an antifeedant to most desert rodents except the pocket gopher, *Perognarthus baileyi*, which appears to have evolved a detoxification mechanism (Elliger et al. 1973; Sherbrooke 1976; Harborne 1988).

On germination the taproot descends for 27.5–40 cm before the shoot breaks surface. Adult jojoba may have several taproots which develop by repeated forking from below the crown and descend for 2.5–4 (-10) m. No shallow or subsurface feeder roots have been observed. The leaves persist for 2–3 years, depending on moisture and shade conditions, with growth continuing from one season to another. However, the older leaves lose most of their initial fine pubescence and have a waxy coating (Gentry 1958).

It is a valuable drought and winter browse, especially for sheep and goats. Roundy and Ruyle (1989) observed that greatest herbivory activity was in late spring and summer when annual herbaceous species were no longer available. Grazed shrubs had greater twig growth than ungrazed, resulting in similar net increases in twig length biomass. Greatest twig growth is in spring, and the ability for regrowth from the lower lateral or apical buds after removal of the outer twig tips by browsing allows the jojoba to tolerate herbivory.

Environmental stress placed on jojoba growing coastal and inland desert habitats has been examined by Al-Ani et al. (1972). In the relatively mild environment of coastal habitats, the leaf water potential values were consequently high and the plants showed relatively high metabolic rates. During the summer, although the leaf water potential value was lower than in spring, the higher temperatures were more favourable for metabolic activity. Photosynthetic rates were approximately double those for spring, and dark respiration rates and carbohydrate content were also higher. The plants remained in a relatively water-stressed condition through the autumn, with no apparent growth and low photosynthetic and dark respiration rates. The carbohydrate content was somewhat lower than in summer. During the relatively cooler winter season, there was a fair amount of precipitation. Thus, at the beginning of the growing season the leaf water potential and the photosynthetic and respiration rates increased greatly while the carbohydrate was slightly lower than in autumn. Winter growth continued into the summer, with the growth pattern corresponding with the other physiological responses. With the mild coastal climate active CO_2 uptake continued throughout the year, with the highest metabolic activity during the summer and lowest in autumn; spring and summer diurnal patterns of activity did not show any strong midday depressions.

Jojoba has the greatest part of its distribution in the harsher desert environment, and exhibited distinctly greater physiological responses compared with jojoba growing in the coastal habitat. The main rainy season was during the winter; temperatures showed characteristic diurnal and seasonal fluctuations. During a relatively wet year the leaf water potentials were moderate and the seasonal temperatures mild and, as indicated by the photosynthetic and dark respiration rates, the plants exhibited their highest seasonal metabolic activity and highest total carbohydrate content. During the summer, the leaf water potential values were low and showed greater diurnal ranges than during the spring. The photosynthetic rates were lower than in the spring; the dark respiration rates and carbohydrate levels also declined rapidly. The continuing severe environmental conditions into the autumn brought the leaf water potential values and metabolic activities to their lowest levels.

With a good winter rainfall the leaf water potential noticeably increased. The winter temperatures were relatively moderate so that the shrubs commenced their growth, accompanied by an increase in metabolic activities. The photosynthetic rates increased considerably over the autumn rates, there was also a large increase in respiration rate. The carbohydrate contents declined concomitantly with stem growth and seed and fruit production. The seasonal growth patterns started in the winter were ended in the spring. Spring photosynthesis showed moderate midday depressions, although there were no clear depressions observed during the summer. Metabolic activity continued even at quite low leaf water potentials.

During dry years the plants were under high water stress throughout the year and gave a distinctly different response to that in a wet year. Metabolic activity was limited by the arid conditions. There was no apparent seasonal growth. Leaf water potential levels were low; photosynthetic patterns were low throughout the year. Summer diurnal photosynthetic patterns exhibited higher rates in the morning and dropped sharply during the afternoon before partially recovering. Low dark photosynthetic rates continued throughout the year.

The coastal plants appear more sensitive to drought than the desert populations, with a rapid decline in net photosynthesis and rapid water loss. Al-Ani et al. (1972) conclude that there is a trend towards ecotypic variation between Pacific coast and Arizona upland populations. Furthermore, the desert populations have adapted to the arid environment by possessing distinctive physiological characteristics compared with those from the milder coastal environment. These include the ability to sustain very high rates of metabolic activity during the short growth period when leaf water potentials are high. The plants also maintain low photosynthetic rates and sharply reduced dark respiration rates during the non-growth period, thus permitting a balanced carbohydrate economy. Provenance is clearly an important factor to be considered when selecting seed for introduction.

According to Glat et al. (1981), like many desert plants, the stomata on both surfaces of jojoba are recessed for approximately half the depth of the epidermal layer and surrounded by a protective collar of cuticular material. There were an average of 95 stomata mm^{-2} on the upper surface and 119 on the lower. When subjected to water stress the upper stomata tended to close first. An intricate wax deposit on the lower surface possibly diminished transpiration losses from below; the wax could also aid energy dissipation from the lower leaf surface and thereby provide a cooling-off effect. Total pore area is ca. 0.8% of the total leaf area, which is close to the 1% reported for other species, suggesting that total pore area is not a significant factor influencing drought tolerance. However, a 30% difference in stomatal densities was observed between five different clones; such a significant difference could indicate important effects on gas exchange of these clones.

STYRACACEAE

Styrax officinalis: a spreading, winter deciduous, up to 7-m-tall shrub or tree bearing ovate leaves up to 7 cm long. The var. ***officinalis*** is a native of the eastern Mediterranean (ZB IV). It is a valued winter bee food; its seeds may be used as beads. It is the mastic from the bark of the local endemic var. ***chia*** from the island of Chios that is tapped and is the source of friar's balsam and incense (Zohary 1973). The sclerophyllous leaves contain such xeromorphic features as thick outer walls to the epidermal cells, thick cuticle, high stomatal density and abundant sclerenchyma, characters that are also present in *Pistacia, Olea* and *Quercus* (Metcalfe 1983). The seeds are noteworthy for a 50% germination even after heating to 140 °C, the optimum dry heat pretreatment being not less than 30 °C (Oppenheimer 1960).

TAMARICACEAE

Tamarix aphylla, syn. *T. articulata* (athel or tamarisk): a heavily branched, monoecious, C_3, evergreen tree or shrub which, under favourable conditions, may attain a height of 18 m with boles 1.7 m diameter; the root system may extend downwards for 10 m or more and laterally for 34 m. The small branchlets are articulated, with the leaves reduced to tiny, bluish-grey scales with epidermal salt-excretory glands. These glands, which also occur on other *Tamarix* spp., result in a high salt accumulation beneath the canopy which prevents germination of other species and also competition. Tamarisk is widely distributed on dunes, in wadi beds, loess and salty soils through the arid, subtropical, summer rainfall and Mediterranean climatic regions with 100–500 mm rainfall of northern Africa eastwards to India [ZB and ZE III, IIIsr, IIIwr, III(tI), III–IV, IV–II]. It is cultivated in South Africa, Israel, Jordan, Kuwait, northern Iraq, Australia and North America.

Tamarisk grows on a wide range of soil types, especially sands, including alkaline and saline soils, the latter with a maximum electrical conductivity of 56 dS m^{-1}; light soils and a pH of 5.7 provide conditions for optimum growth. Heat, drought, wind, salt and frost tolerant, the tree is capable of rapid growth, and is very resistant to browsing by goats and camels. Mature trees, however, are liable to blow over and the extensive shade, litter and salt drip inhibit the growth of other plants. Because of the ease by which it can be propagated from root cuttings and rapidity of growth, it is highly regarded in the Negev as a species for dryland reforestation. The abundance of salty litter below the canopy is formed by the nightly drip of salt 'tears' and the continuous shedding of twigs. The salty litter kills the surrounding vegetation, so that when planted as a windbreak an open space must be allowed between the trees and the crops; the absence of other vegetation and the inability of the litter, because of its high salt content, to burn also makes it a useful tree for growing as a fire break. It is readily propagated from both seeds and cuttings and is cultivated for its shade, dune stabilisation and as an ornamental. Despite the high salt content of the foliage the tree is often readily browsed by cattle, sheep and goats.

The trunks of *T. aphylla* are a useful source of plank and pole timber and fuel. The wood is light-coloured, hard, heavy and fibrous with a close and often interlocked grain and having an excellent silver-grained figure; it polishes well, the veneer is soft when steamed and hard when dried. The timber is used for construction purposes, furniture, cabinet work, agricultural implements and turnery. Although considered suitable for the manufacture of particle board, debarking is difficult. The cellulose content is lower than is normally considered satisfactory for pulping.

Tannin and dyes may be obtained from the bark and galls; the flower galls, which are easy to collect, contain 40-55% tannin. Most of the tannins are hydrolysable and predominantly ellagitannin; the ether fraction of the ethyl acetate extract contains gallic, ellagic and dehydrodigallic acids, di-hydrojuglone-5-glucoside and other polyphenolic compounds. In the past the gallic acid was of some importance in photography (Waisel 1960a; Baum 1978; Booth and Wickens 1988; Aronson 1989; BOSTID 1990).

Although capable of germinating within 15 h of a favourable wetting, the seeds rapidly lose their viability. The germinating seeds tolerate quite high salinity but this decreases quite rapidly with the age of the seeds. When subjected to dehydration and a loss of a third of their water content the seedlings lose their vitality (Waisel 1960a). On NaCl-free soils only $CaCO_3$ is excreted. When grown in solutions of increasing NaCl concentration, the excretion rate of the Ca^{2+} remains constant, while that of Na^+ and Cl^- at first increase but decrease when the concentration exceeds 0.3 mol. The growth of *T. aphylla* is inhibited at concentrations of 0.1-0.2 mol NaCl, ceases at 0.3 mol, and slow death commences at 0.4-0.5 mol. With a good water supply the transpiration rate is relatively high, with a midday peak. With water shortage the transpiration curve first exhibits two maxima, but with increasing shortage there is a peak in the morning and the second peak disappears, thereby demonstrating an early reduction in transpiration (Waisel 1961; Walter and Breckle 1986).

Transpiration is strongly dependent on soil moisture. The excreted salt crystals covering the branches are highly hydroscopic and absorb water vapour from the atmosphere. The resulting salt solution considerably reduces the duration of transpiration. Under laboratory conditions water may be reabsorbed by the twigs only under high saturation deficit and continuous spraying. In the field the twigs may become wet almost nightly from fog, but little water is absorbed due to the low water deficit and the short duration of the available moisture. The small amounts absorbed are certainly insufficient to cover any water saturation deficit, but may have some beneficial value when under conditions of sub-lethal water stress.

In coastal areas, the mists condense, a phenomenon whereby fine, aeolian, water vapour collides with a solid body, such as a tree, and coalesces to form larger droplets, which fall to the ground. Where such fogs are frequent, the condensation is more pronounced under the salt-excreting *Tamarix* than under other tree species in the area and, during the long summer months, often moisten the ground below to a depth of 50–70 cm (Waisel 1960b).

During periods of high atmospheric activity the twigs are covered by an alkaline solution, the pH of which fluctuates between 8–8.5 at night and 10.5 during the day. Such a solution produced by the glands acts as an efficient trap for atmospheric CO_2. When the pH falls, much of the previously absorbed CO_2 is gradually released into the immediate atmosphere, producing a concentration of 1000 ppm. As net photosynthesis reaches maximum rates at CO_2 concentrations above 500 ppm, the plant may exploit the surplus during the morning hours for maximum assimilation. The 'salt glands' responsible for the production of the alkaline recreation (removal of those ions that pass unchanged through the plant without being metabolised) may serve a triple function: (1) by removal of excess salts from the tissues, (2) by providing a coating of hydroscopic solutes that moistens the twig surface and shortens the transpiration time; and (3) by providing the plants with an enriched CO_2 microenviroment. The reduced leaf surface area for gas exchange, together with the endogenous rhythm of transpiration, and a 10-h daily opening of the stomata are all factors restricting photosynthesis and growth, yet *T. aphylla* is outstanding in the Negev for its rapid growth, a fact attributable to the high concentration of ambient CO_2 (Waisel 1991). The scale insects *Trabutina mannifera* and *Najacoccus serpentinus*, which live on the shoots, produce honey dew excretions which crystalise to form a sweet, edible manna. The manna is dissolved in warm water, boiled with sugar, egg white and almonds and, on cooling, forms cakes of nougat.

Thymelaeaceae

Thymelaea hirsuta: a dwarf, evergreen, dioecious shrub up to 2 m tall, with erect, spreading or decumbent, much branched stems densely clothed with small (2 to 8 mm), leathery, scale-like, densely imbricate, epistomatic, green leaves. It is a circum-Mediterranean species (ZB and ZE III–IV, IV). A massive and sporadic germination of *Thymelaea* seeds occurs every few years. The seedlings up to 20 cm tall exhibit leaf dimorphism, with lanceolate leaves on the main stem and bluish, relatively large, glabrous and spreading ovate leaves on the branches. Seedlings are soon established, since they are avoided by all livestock, including the goat. The shrub is planted in Cyrenaica for dune stabilisation (Tan 1980; Danin 1983).

The leaves are regarded as an efficient anthelmintic, powerful hydragogue, cathartic and expectorant. A decoction of the leaves is used as a shampoo for dandruff. The powdered leaves may also be used for treating skin inflammation, and the smouldering inner bark for healing wounds (Boulos 1983;

Schmidt and Stavisky 1983). The plant contains ü-sitosterol and stigmasterol, the latter a phytosterol used in the preparation of progesterone, a steroid hormone used in the treatment of recurrent abortion in humans and in oral contraceptives.

A strong rope may be prepared from the braided branches or from the bark, with or without the outer bark. The dried inner bark (bast fibres) may also be used as tinder. Piles of the uprooted shrub may be used for stock pens, windbreaks and for mattresses. The branches may be used for traps, brooms and fuel. The long phloem fibres are comparable with those of softwoods, but have a higher length:width ratio, while the xylem fibres are comparable with those of hardwoods, but with a higher length:width ratio, but lower than that for rice straw. The coastal sand dune populations have the longest bast fibres, followed by those from saline depressions, and the shortest from the inland ridges. However, such a relationship was not observed for the xylem fibres. In 1979 a paper mill was established in Israel for using the bast fibres of *T. hirsuta* to produce a high quality, hand-woven paper. Mitnan paper is used primarily by artists, but also by paper conservators for repairing documents and works of art (El-Ghonemy et al. 1974, Schmidt and Stavisky 1983, Shaltout 1992b).

The low nutrient status of *T. hirsuta* when compared with other Egyptian desert plants has been examined by Shaltout (1992a). It is regarded as a copper-rich species with a tendency to accumulate copper and iron in the stem rather than in the roots or leaves, with the roots to some extent acting as a heavy metal filter. Transpiration losses from the leaves have been reduced by the exposed abaxial (lower) surface being glossy, glabrous and without stomata, while the adaxial (upper) surface is hairy, with stomata and closely appressed to the stem. The profusely branching roots occupy two distinct horizons, one a few centimetres below the surface and able to take advantage of the slightest shower, the other descending to 2 m and thus able to tap the deeper lying soil moisture.

UMBELLIFERAE

Azorella compacta (llareta): a yellowish-green, compact cushion shrub up to 1 m high (Fig. 21). A native of the high Andean belt of south-eastern Peru, south-western Bolivia, north-eastern Chile and north-western Argentina, llareta is of considerable potential economic importance. An infusion of 20 g in 0.25 l water is reputed to reduce the blood sugar and has been used under medical supervision in the treatment of diabetes associated with obesity, a use that warrants further investigation. Other medicinal properties are also reputed; the plant is also possibly anticarcinogenic. In the past the llareta was widely exploited by the mining and railway communities as a source of slow-burning, almost smokeless fuel with a high calorific value. It is now classified as vulnerable, in danger of local extinction and its use as fuel is forbidden. In a seed sample tested at the Royal Botanic Gardens, Kew less than 1% of the seed was potentially viable and germination gave only a 30% success rate (Wickens 1988, 1993, 1995a). It is extremely slow growing and is reputed to live for up to 3000 years (Ralph 1978)!

ZYGOPHYLLACEAE

Guaiacum coulteri: a deciduous tree or shrub, 1–12 m tall with crooked branches bearing crowded, drought deciduous, pinnate leaves. The distribution in Mexico ranges from Sonora to Oaxaca (ZB and ZE II–III, IIIswr). It is a relative of lignum vitae of commerce (*Guaiacum officinale* and *G. sanctum*); the wood is very hard and resinous, strong and durable and is used where timbers of great strength are required. It is also an good fuelwood and was formerly used in steam engines. Resin from the tree mixed with clay is used by the Seri people to make a blue paint, while a tisane prepared from the crushed fruits is used to treat dysentery (Standley 1923; Felger and Moser 1985; Mason and Mason 1987).

Larrea divaricata (creosotebush): a much-branched, evergreen shrub with thick, glutinous, strong-scented, bifoliate leaves. The root system descends to ca. 2 m with laterals predominately in the top 10–25 cm of soil. The rather regular, circular distribution pattern observed in stands of creosote bush is the result of vegetative spread of clones (von Willert et al. 1992). Creosote bush requires coarse textured and freely draining soils, which, according to Lunt et al. (1973), is due to the high oxygen requirement for root growth. The species is also intolerant of salinity. The distribution of subsp. ***tridentata***, syn. *L. tridentata* is from western Texas to southern Utah, Arizona, California and northern Mexico (ZB and ZE II–III, IIIsr, IIIswr, IIIwr, VIIa), while subsp. ***divaricata***, syn. *L. divaricata* occurs in Chile and Argentina (ZE II–VII, III–VIIa).

The strong odour, especially when wet, normally deters stock. Nevertheless, there have been cases of pregnant ewes browsing the shrub and dying. The plant may cause dermatitis to those who are allergic to it. The leaves have been used by the Pima people as an emetic and to poultice sores. The twigs and leaves steeped in boiling water are used as a tonic and also reputed to have antiseptic properties. Sucking a 1 cm length of the peeled stem is used by the native Americans to assuage thirst. The lac formed on the twigs by the parasitic red scale insect *Tachardiella larreae* makes an excellent adhesive for fixing arrowheads and mending pottery (Kearney and Peebles 1951; Saunders 1976; USDA 1988).

The glutinous coating of the leaves (Fig. 24) functions as a barrier to water loss, a solar ultraviolet filter, as well as an insect antifeedant. It consists of a complex mixture of phenolics, saponins, terpenoids and wax esters and accounts for 5–20% of the leaf's dry weight. Over 80% of the coating is composed of phenolic aglycones, of which the main component is the catechol lignan nordihydroguaiaretic acid (NDGA), the remainder being a complex mixture of partially oxygen-methylated flavones and flavonols, plus small quantities of acid and basic components. The NDGA is a powerful antitoxidant as well as exhibiting important antibactericide, fungicide, antiherbivore and allelopathic properties. The latter accounts for the absence of other species in the immediate vicinity of the creosote bush (Oliveto 1972; Mabry et al. 1977; Rhoades 1977; Greenfield and Shelley 1987; Gonzalez-Coloma et al. 1988; Harborne 1988; Meinzer et al. 1988).

According to Went (1979) woody perennials appear to germinate primarily in response to summer rainfall, i.e. in certain favourable years in deserts with a winter rainfall regime. This is certainly true for *L. divaricata* since it also requires high temperatures for germination. Barbour (1968) gives the optimum conditions for germination as exposure to vernalisation, wetting and drying cycles with leaching of the pericarp, and night temperatures of 23 °C, with maximum root growth occurring at 29 °C. Exposure of the dry seed to temperatures of 37–71 °C prior to sowing produced a marked reduction in germination.

Bewlage and Krochko (1982) regard subsp. *tridentata* as probably the most drought tolerant species of North America. Its buds and immature leaves can survive considerable dehydration, but not the mature leaves. Sammis and Gay (1979) examined the evapotranspiration and discovered that the total water loss approximated to the precipitation. The actual transpiration losses were only a small proportion of the total evapotranspiration, since the creostotebush had both a high diffusion resistance and a small leaf area index. The authors conclude that the results confirm that the control of creosotebush or similar desert vegetation in order to reduce evapotranspiration would be ineffective. In addition to seasonal fluctuations in osmotic potential, it also varies diurnally. Changes in the temperature optimum for photosynthesis as plants age during the growing season or under different controlled-environment conditions are often less that 5 °C, but temperature acclimation between 5 and 10 °C has been reported for the creosotebush (Smith and Nowak 1990). Bamberg et al. (1975) summarise the adaptive strategies of the C_3 creosote bush as an evergreen with low photosynthetic rates and the ability to depress transpiration loss during moisture and temperature stress, yet taking advantage of any favourable period to photosynthesise.

Nitraria retusa: a low, spreading, halophytic shrub up to 3 m tall, bearing leathery leaves mainly on arrested, short shoots. Widely distributed through the Sahara to Israel and Arabia (ZB and ZE III, III–IV, IV), it is used by the Bedouin as a source of fuel; its drupes are sweet and edible. The plant may be grown as a salt-tolerant ornamental. Heterophyllous, the large spring leaves are replaced in summer by axillary clusters of two to seven thickish summer leaves, which are gradually reduced in size, with the smallest and innermost remaining until the winter rains (Oppenheimer 1960; Zohary 1973; Danin 1983; Aronson 1989).

Monocotyledons

AGAVACEAE

Agave deserti (desert agave): a sparingly or prolifically suckering rosette plant up to 70 cm tall and 60 cm in diameter with shallow roots; the margins of the thick, rigid leaves are armed with slender-cusped teeth. It is distributed through southern California and south-western Arizona to Baja California and Sonora (ZB IIIswr). All parts of the plant are eaten by native Americans. The distilled caudex of mature plants is a source of an alcoholic beverage known as mescal; the caudex may also be roasted and eaten. Cattle may browse the new flowering shoots. The leaves are a source of fibre. The steroid content of the leaves is erratic to absent, absent in the edible stems, and 0.6-1% in the fruits; the main steroid is hecogenin.

Desert agave is a CAM species, although when soil water is available for long periods the plant can switch to the C_3 synthetic pathway during the day, a ploy that permits greater productivity when nights are short even though net photosynthesis is about the same at 18.3 μmol cm^{-2} 24 h^{-1} as a CAM plant and 21 μmol cm^{-2} 24 h^{-1} when following the C_3 pathway. An extremely low transpiration ratio (mass of water transpired/mass of CO_2 fixed) hs been reported during winter. When the soil is wet, 55% of night-time transpiration is from water stored in the leaves, with recharge taking place during the day. Because of the large amount of water stored per unit of leaf area, the plant can use this stored water and the stomata can continue to open at night long after the soil water has dried up (Gentry 1972; Turner et al. 1995).

Yucca baccata (banana yucca): a shallow rooted, rather variable shrub, up to 5 m tall, acaulescent or with few short procumbent stems. The up to 70-cm-long leaves are often twisted, with broad and coarse marginal fibres. It reaches maturity after 2 or 3 years; provided sufficient moisture is available the female plant will produce fruit every other year. It is a native of south-western USA (ZB VIIa) and a CAM species. Due to the thin leaves and low water storage capacity it is extremely sensitive to water stress.

The succulent banana-shaped fruit are borne in large upright clusters and are eaten raw, or roasted and dried for winter use or ground into a meal. A fermented beverage may be made from the fruits. The cooked flower buds are regarded as a delicacy. The leaf fibre may be used for cordage, matting, cloth and baskets. The leaf sap can be used to produce a waterproof pigment for painting pottery. The rootstock has saponifying properties and, when pounded, may be used as shampoo or as a soapy scrubbing brush; the root starch also acts as a stiffener. The soap, known as amole, has been produced commercially. The root also has laxative properties. The plant is cultivated as an ornamental (Kearney and Peebles 1951; Saunders 1976; Potter-Basso 1991; Turner et al. 1995).

ASPARAGACEAE

Asparagus stipularis (thorny asparagus): an erect or sprawling, rhizomatous, generally much-branched, perennial shrub up to 1 m high with large, fusiform tubers and woody, glaucous-green, persistent stems, with leaves reduced to persistent papery scales ca. 12 mm long and spinose cladodes up to 1 cm long. The species is widely distributed through the Mediterranean region and islands of the Atlantic (ZB and ZE III, III–IV, IV). It is browsed by sheep and goats. A root infusion is taken for pulmonary tuberculosis and the bark as an emetic; the roots and shoots are used in the treatment of syphilis. There are numerous other local medicinal uses. The reduction of the photosynthetic processes by the leaves being replaced by the stem and cladodes plus the rhizomatous storage organs are clear adaptations to an arid environment. (Watt and Breyer-Brandwijk 1962; Boulos 1983).

GRAMINEAE

Aeluropus lagopoides, syn. *A. repens*: a stoloniferous, C_4 perennial, either densely tufted with erect stems up to 30 cm tall or long, prostrate stems arising from a woody rootstock and rooting at the nodes, bearing pungent, subulate-lanceolate leaves up to 50 cm long and 3 mm wide. *Aeluropus lagopoides* is a nebka-building grass of saline communities in North Africa, Sudan, Eritrea and Somalia eastwards through Arabia and central Asia to Mongolia, Pakistan and India [ZB and ZE II–III, III, III–IV, IV, VII(rIII)]. It is an obligate halophyte with salt-secreting glands whereby it is able to excrete excessive salts. One estimate of the daily rate of excretion is 870 $\mu mol\ g^{-1}$ dry weight of leaves. It is regarded as a good fodder plant, especially after winter rainfall, dying back at the end of the rains and producing new growth at the start of the rains. It is a useful sand binder (Bor in Townsend and Guest 1968; Boulos 1983, 1989; Kassas and Batanouny 1984; Fitter and Hay 1989; Sen and Mohammed 1994; Danin 1996).

Agropyron cristatum (crested wheatgrass): a caespitose perennial with culms up to 75 cm tall and leaves convolute. A native of southern, central and eastern Europe (ZB and ZE IV, IV-VI), it provides a nutritious grazing or hay. The success of its introduction and extensive cultivation in North America is attributed to its exceptional resistance to drought and low temperatures; it is also available earlier in the spring than the indigenous North American grasses. In order to conserve moisture during periods of low water availability the rigid leaf-blades are rolled, while during very hot dry periods the grass tends to remain dormant.

Alloteropsis semialata (cockatoo grass): a tufted perennial with culms ascending up to 120 cm and leaf-blades up to 12 mm wide, usually expanded, rarely rolled, tapering to a fine, rigid point (Chippindall 1955). A polymorphic species, *A. semialata* shows an unusual degree of variation in leaf anatomy, some having the Kranz-type with parenchyma sheath anatomy associated with panicoid grasses and others with the non-Kranz type associated with temperate grasses. The two forms cannot be separated morphologically, but the non-Kranz would appear to occupy higher and more mesic habitats (Ellis 1974a,b). Widely distributed through the tropics of the Old World, it is a characteristic grass of the highland sourveld of eastern South Africa (ZE II–III, III–II, III–IV, IV–III). It withstands burning, heavy stocking and climatic extremes, regenerating rapidly. It is grazed by livestock, and also grown as an ornamental (Lazarides and Hince 1993).

Ammophila arenaria subsp. ***arundinaceae***, syn. *A. litoralis* (marram grass): a tufted, perennial grass with stout, scaly, branched rhizomes and stout culms up to 1 m tall, the leaf-blades tightly involute, up to 60 cm long. It is widespread in the Mediterranean and the coasts of western Europe, extending northwards to Norway (ZB and ZE IV, IV–V). It is a valuable sand-binding grass.

Andropogon gerardii, syn. *A. furcatus* (big bluestem): a vigorous, coarse, shortly rhizomatous, perennial bunch grass growing up to 2 m tall. Widely distributed in North America from Maine and Quebec to Saskatchewan and Montana, south to Mexico (ZB and ZE VII, VII–V, VII–VI), it is palatable to all classes of livestock, providing late spring foliage; it can also be made into good quality hay (Hoover et al 1948; Hatch and Pluhar 1993). A drought resistant species, the leaf blades have the stomata predominantly on the abaxial surface. In response to low water potential the blades fold, with adaxial surface inwards. The relatively few adaxial stomata are closed in completely folded leaves, this closure having only a minor effect on total leaf conductance. Energy budget calculations have shown that folding reduces transpiration by 7–13% in water-stressed plants due to reduced leaf temperature and, therefore, leaf-to-air vapour pressure deficit. This reduction is small relative to the direct effect of stomatal closure on transpiration. However, small decreases in transpiration, leaf temperature and incident irradescence associated with leaf folding (or rolling) may decrease the potential for photoinhibition, prolong physiological activity and ensure survival during drought. The folded leaves of *A. gerardii* are typically orientated towards the sun, so that incident irradiation per unit leaf area is identical for both open and folded leaves. However, in rolled leaves, which are essentially cylindrical, the incident irradiation must be corrected for cosine effects since energy flux density on a curved surface is less than for a flat surface of equal width because the incident irradiation is distributed over a larger surface area (Heckathorn and DeLucia 1991).

Aristida purpurea (purple three-awn): a densely tufted, perennial grass with culms up to 80 cm tall. Distributed through south-western USA and northern Mexico (ZB and ZE VII, VII–V), purple three-awn provides useful grazing provided it is eaten before the seed heads are formed. It is of interest since the roots, like those of *Orizopsis hymenoides** and *Stipa comata**, are enclosed in sheaths of sand grains (rhizosheaths) which may serve as repositories for moisture and organic energy sources required for nitrogen fixation. Although there is evidence of associated nitrogen fixation, the uptake of the nitrogen remains to be demonstrated. Such rhizosheaths were first reported by Price (1911) for the Saharan *Stipagrostis pungens**, but again any associated nitrogen fixation has yet to be demonstrated (Wullstein et al. 1979; Hatch and Pluhar 1993).

Astrebla (Mitchell grasses): a genus of only four perennial species, remaining dormant during the winter months of April to September. The tussocks range from 0.1 to 0.5 m diameter and are spaced 0.5–2 m apart, depending on soil, rainfall and age. The rhizomes are well protected by dry, closely packed leaf sheaths. The Mitchell grasses are capable of extremely rapid growth following light rainfall at the start of the rainy season, with a rapid growth of new shoots from the lower three to four nodes of the old stem; vegetative growth is completed within 2 weeks, flowering by 3 weeks and seeds set in a further 10 days and ripe after 6 weeks. Heavy rainfall will result in the production of new basal tillers, which grow and develop more slowly. Mature plants may grow through a wide range of temperatures, from 15 to 42 °C. Drought resistance is partly conferred by a robust root system with strong, deep roots capable, in a good season, of storing large amounts of food reserve. The species are widely distributed through Austalia (ZE II–III, III–II, III–IV, IV–III). Mitchell grasses are susceptible to heavy grazing due to the removal of the lower nodes after light rains, resulting in a depletion of food reserves preventing the production of new shoots and a reduction in the size of tussocks. The grasses have two disadvantages, a highly variable growth rate and the low protein content, less than 3%, of the mature herbage.

Bromus tectorum (cheatgrass or downy brome): a slender, tufted annual up to 60 cm tall. A native of central Europe and the Mediterranean basin extending eastwards to Siberia and China [ZB and ZE III, III–IV, IV, VII, VII(rIII), VII–IV], it has been introduced and naturalised in western Europe, Australia and North America where it is now widely distributed throughout the shadscale saltbush and greasewood communities.

It is palatable when young but matures rapidly and when dry becomes a fire hazard. The sharply pointed and awned, bearded florets can cause injury to the mouths of livestock eating the mature grass or hay. In North America its ability to dominate native perennial grasses and annual forbs is attributed to its ability to germinate under a wide range of temperature and moisture conditions. Studies on the cheatgrass have shown that the roots grow well at temperatures as low as 3 °C, whereas *Agropyron spicatum* grows very slowly below 8-10 °C. The rapid establishment of an extensive fibrous root system gives cheatgrass an early competitive advantage in the utilisation of soil moisture and nutrients and it thus out-competes the native species. Whereas the annual evapotranspiration in the Great Basin is roughly equivalent to the precipitation for most of the vegetation, cheatgrass communities are the exception with evapotranspiration substantially less than the precipitation. The extent of moisture exploitation by cheatgrass is related to plant size. Under competition with other species in both wet and dry years the roots may only descend 15 cm, whereas isolated plants and pure stands may exploit the soil to at least 1 m and have some roots penetrating to 1.5–1.7 m. Although pure stands of cheatgrass can affect the rate of soil moisture depletion, the depth and total water extracted by the end of the growing season is not affected. The extent of subsoil water accumulation under cheatgrass or its contribution to groundwater recharge is not known (USDA 1988; Dobrowolski et al. 1990; Smith and Nowak 1990).

Cutandia dichotoma: an annual with culms up to 30 cm tall and narrow, involute or rolled leaf-blades up to 8 cm long. A grass of sandy soils of the eastern Mediterranean and central Asia [ZB and ZE III, III–IV, IV, VII(rIII), VII–IV]. it contributes ca. 4 kg ha^{-1} $year^{-1}$ to the diet of domestic animals. Its protection from grazing resulted in an insignificant change in total available carbohydrates, a remarkable increase in soluble sugars during the winter and decrease in spring, and a significant increase in crude protein during the winter and spring. Thus, the metabolic activities were seen to be directed towards building up the plant body at the beginning of the growing season, with most of the photosynthate being transformed into structural carbohydrates (El-Ghareeb and Bidak 1992).

Distichlis spicata var. ***stricta***, syn. *D. stricta* (inland or desert saltgrass): a C_4, tufted, coarse, strongly rhizomatous, dioecious perennial up to 45 cm tall, with short, folded, blue- or grey-green leaves. It is distributed from Washington to Saskatchewan south to southern Arizona, Texas and Florida (ZB and ZE III, VII, VIIa, VII–V). Like *Atriplex confertifolia*, the male and female plants tend to segregate into xeric and less xeric microhabitats, respectively (Smith and Nowak 1990). Although of low forage value, it is both drought and salt tolerant. Inland saltgrass may be used to colonise sandy soils; it is also grown as an ornamental. The seeds were formerly eaten by the Kawaiisu (Kearney and Peebles 1951; Doebley 1984; Hatch and Pluhar 1993).

At low light intensity growth rates are significantly reduced by salinity and low temperatures, but are not significantly affected in high light. The capacity for high light intensities to overcome any depressed growth rate due to high salinity cannot be completely explained by rates of net photosynthesis since high salinity results in decreased net photosynthesis under all environmental conditions. This salinity-induced decrease in net photosynthesis is due to stomatal closure, although plants grown under low light and low temperatures show a significant increase in internal leaf reaction to CO_2 exchange. However, salinity and light during growth did not significantly affect the rates of dark photosynthesis. The increase in salinity resulted in generally thicker leaves with lower stomatal densities, but with no significant differences in the ratio of mesophyll cell surface to leaf area. The mechanism by which *Distichlis* tolerates salt appears to be clearly linked to the utilisation of light energy, and the salt-induced leaf succulence is regarded as being of questionable importance to gas exchange at high salt concentrations (Kemp and Cunningham 1981).

Eragrostis eriopoda (woolly butt grass or wire wanderrie grass): a robust, perennial, tussock grass up to 60 cm tall, with thick, woolly leaf bases and hairy roots. The leaf-blades are flat, up to 10 cm long, becoming stiff and pointed and spreading stiffly from the culms. The tussocks live for more than 20 years, tending to break up into apparently separate plants, or alternatively may develop into a ring-shaped tussock. It is a widespread drought and trampling resistant grass of sandy, outwash plains and sandplains of Australia (ZE II–III, III–IV, IV–III) with moderately deep, acidic sandy loam or sands. It is only of moderate value as a fodder plant, being eagerly and selectively eaten by sheep only when green, becoming very fibrous when dry. The crude protein content is ca. 10% when green and 4% when dry. The seeds are edible and eaten by the Aborigines (Brand and Cherikoff 1985; Jackson and Jacobs 1985; Mitchell and Wilcox 1988).

Lasiurus scindicus, syn. *L. hirsutus*: a tufted, perennial grass, the culms up to 90 cm tall and often woody below, from a thick, woody rhizome up to 1 m long. Widely distributed from North Africa and the Horn of Africa through tropical Arabia to north-western India, growing best on alluvial or light brown, sand soils, it is drought resistant and moderately salt tolerant [ZB and ZE III, III(tI)], often forming large mounds in sandy areas. Where abundant, the ground seeds are mixed with flour for human consumption. It is regarded as an excellent desert fodder grass when young, relished by all livestock, but intolerant of heavy grazing; it is also a good sand binder. The root fibres are used for weavers' brushes. The grass may be planted to stabilise sand. It has a typical Kranz-type leaf anatomy with a low CO_2 compensation point showing a C-4-dicarboxylic acid pathway for photosynthetic carbon reduction (Bor in Townsend and Guest 1968; Huber et al. 1973; IBPGR/RBG 1984; Boulos 1989; Skerman and Riveros 1990).

Leptochloa fusca, syn. *Diplachne fusca* (kallar grass or brown beetle grass): an aquatic to semi-aquatic, rhizomatous perennial with culms up to 1.5 m tall with tough, inrolled or flat, linear leaf-blades up to 55 cm long. Widely distributed through tropical and southern Africa across to Southeast Asia and Australia, kallar grass often forms pure stands, tolerant of both saline and alkaline soils, growing well at soil pH 10 (ZB and ZE II–III, III, III–II, III–IV, IV-III). It has received considerable attention in recent years as a potential fodder grass for saline areas (BOSTID 1990; Skerman and Riveros 1990). A significant variation in salinity tolerance and uptake of Na^+, K^+ and Cl^- has been observed by Warwick and Halloran (1991). The Na^+ and Cl^- contents were higher in the leaf sheath than in the blade, indicating the grass's capacity to sequester high levels of Na^+ and Cl^- in the sheath away from the leaf blade as well as maintain a high selectivity for K^+ over Na^+.

Lygeum spartum (sparte or albardine): a somewhat diffuse, stoutly rhizomatous perennial up to 60 cm tall, with stiff, subcircular and folded leaf-blades. A monotypic circum-Mediterranean grass (ZB and ZE III, III-IV, IV), sparta may be used for paper making as well as for mats, sails, cordage, etc. It is also grazed by sheep, goats and camels (Nuttonson 1961; Mabberley 1987). The dead roots are strongly hygroscopic, absorbing 100% of their original weight from moist air (Oppenheimer 1960). The leaves in-roll when the moisture deficit is 8–20%; this results in a 70% reduction in transpiration (Walter and Breckle 1986).

Oryzopsis hymenoides, syn. *Achnatherum hymenoides, Stipa hymenoides* (Indian rice-grass): a slender, tufted perennial up to 60 cm tall, with firm, slender, rolled leaves; the maximum rooting depth is ca.1.5m. Distributed from British Colombia to Manitoba southwards to California, Arizona, Texas and Mexico (ZB and ZE II–III, III, VIIa), Indian rice-grass is a palatable grass; in former times it was cut for hay, and the blackish seeds were widely utilised for food by the native Americans during times of scarcity and even cultivated by the Paiute. The seeds are also an excellent dove food. It is an important grass for range reseeding because of its drought resistance, palatability and capacity to grow and spread by natural reseeding where few other grasses can be established. Seed formation is erratic, with good yields in years of good rainfall. The germination of newly harvested seed is low, due to the hard, moisture-resistant seed coat. Good germination may be obtained following scarification, acid treatment or holding for a year in dry storage. Good germination will also occur if the newly harvested seed is planted in the autumn so that the seed coat can be softened by the changing soil temperatures. The roots are enclosed in sheaths of sand grains (rhizosheath) which may serve as repositories for moisture and organic energy sources required for nitrogen fixation. Although there is evidence of associated nitrogen fixation, the uptake of the nitrogen has yet to be demonstrated (Hoover et al. 1948; Wullstein et al. 1979; Doebley 1984; USDA 1988).

Panicum turgidum: a perennial with stiff culms up to 1.5 m tall, rooting at the lower nodes. The leaves are inrolled, thereby reducing transpiration losses.

It colonises unstable sand dunes, forming loose tussocks up to 1 m in diameter. The extensive root system may descend to 2 m or more and extend laterally for 3.4 m or more, with the root hairs covered in protective sheaths of bound sand (see Sect. 4.2 for further discussion on rhizosheaths). The distribution is from northern Africa to Pakistan, mainly in areas with less than 250 mm annual rainfall (ZB and ZE II-IV, III).

Panicum turgidum is a useful grass for stabilising sand in areas receiving 100-400 mm annual rainfall; however, wind blowing between the tufts may leave the grass standing on columns of root-bound sand up to 2 m high. When these columns eventually fall the culms root at the nodes. Seeds require 20-30 mm rainfall before germinating. The percentage germination of stored seeds from dispersal units dried over calcium chloride before sowing increased below 30 °C but not below 5 °C. The ground grains are eaten, especially in times of famine. A two-fold increase in grain yield was obtained in Middle Eastern compared with Mauritania collections and in the Middle East high yielding selections may be cultivated for grain. It also provides a palatable fodder when young and a useful but less palatable fodder when dry. It is used by the Touregs for thatch and matting. The ashes are added to chewing tobacco and the powdered stems used to dress wounds (Williams and Farias 1972; IBPGR/RBG 1984; Boulos 1989; Skerman and Riveros 1990; Gutterman 1993).

Plectrachne schinzii (oat-eared or feather-topped spinifex): a perennial grass forming dense hummocks, rarely rings, up to 50 cm tall and 1 m diameter, with resinous culms and dense green, fluted and inrolled leaves. It is a grass of western Australia (ZE II-II, III-IV). The fresh growth following burning is readily eaten by livestock; young leaves have a crude protein of 10% and when old 3% and a corresponding in vitro digestibility of 35 and 25%; the seedheads are also eaten. The grass persists well during drought, although it is highly inflammable (Beard 1969; Mitchell and Wilcox 1988; Lazarides and Hince 1993).

Sporobolus rigens: a rhizomatous grass, the rhizomes may extend for over 15 m, growing up to 6 m year^{-1}, and lie between 25 and 35 cm below the surface, producing aerial shoots at 18 to 20-cm intervals; the roots are ca. 40 cm long and may descend to 70-80 cm below the surface. The terminal, very hard bud sheaths act almost as drill heads to penetrate the soil. The species also reproduces extensively by seed (Cabrera 1955; McGinnies et al. 1968). It is widely distributed in the moving dunes of the Argentinean Monte (ZE III–VIIa). In addition to its obvious use for stabilising sand, it is used for basketry and making brooms (Gastó et al. 1985); the leaves may also be woven into a cloth.

Stipa comata (needle-and-thread grass): a deep rooted, erect and leafy bunchgrass, growing up to 1.2 m tall, with a maximum rooting depth of ca. 1.5 m. Widely distributed through the Great Plains of North America (ZB and ZE

VII, VIIa, VII–V, VII–VI), it is a valuable forage grass as it begins growth early in spring when other grasses are still dry, and produces new growth in summer and autumn with the advent of sufficient precipitation. Although palatable early in the season, the awned seeds can cause injury when the grass matures. It may be grown as an ornamental (Hoover et al. 1948; Kearney and Peebles 1951; Havens et al. 1973; USDA 1988; Dobrowolski et al. 1990). The roots of *S. comata* are enclosed in sheaths of sand grains (rhizosheath) which may serve as repositories for moisture and organic energy sources required for nitrogen fixation. Although Wullstein et al. (1979 have demonstrated associated nitrogen fixation, the uptake of the nitrogen has yet to be demonstrated.

Stipa tenacissima (esparto grass or alfa): a robust, tufted perennial with wiry culms up to 2 m high, with leaf blades 15–100 cm long. The roots are relatively shallow, down to ca. 30 cm. A native of the Iberian Peninsula and North Africa (ZB and ZE IV, IV-III), it has been introduced in various parts of the world. In North Africa the grass is particularly abundant on shallow soils following woodcutting and burning without cultivation - it is unable to withstand more than 2–3 years in cultivated land. It is a companion forest species and can only regenerate from seed in the shade; in the open it survives vegetatively by means of its rhizomes. The rhizomes slowly extend radially from the original tuft and give rise to a ring of daughter tufts which, after the death of the mother tuft, become independent of each other.

The leaves contain cellulose fibres, the basis of the important esparto-grass industry. The leaves are plucked and baled for export from Spain and North Africa and made into high-quality paper, including cigarette paper. The leaves may also be used for cordage; ship's cables made from esparto are so light that they float. The leaf epidermis is the source of esparto wax. The grass is mainly grazed by camels. It is sometimes cultivated in gardens as an ornamental (Hill 1952; Nuttonson 1961; Williams 1981; Meikle 1985; Le Houérou 1986).

The 30 to 120-cm-long leaf blades of esparto grass have strong nerves and a waxy epidermis. During moist weather the blades are flat and ribbon-like, but at other times transpiration losses are reduced by the blades becoming tightly inrolled. Further protection is afforded by the thickened epidermal walls reinforced by hypodermal fibres, with the few stomata sunk in protective grooves (Oppenheimer 1960). The leaves remain photosynthetically active for 2 or 3 years, after which they persist for several years as a grey mat of dead and dying leaves through which the young leaves emerge.

Working in Almeria, Spain, the seasonal differences in total water potential and the relative water content of the leaves in relation to soil moisture have been examined by Domingo et al. (1991); the results were consistent for species growing in a semi-arid Mediterranean climate.

Similar observations were also made by Volkens (1887) regarding the thickening and cutinisation during the late summer of the guard cells in the two desert grasses *Sporobolus spicatus* and *Stipagrostis ciliata* (syn. *Aristida ciliata*) so that it appears doubtful if the stomata are capable of any movement. Such an adaptation is considered to be of value during the dry season, when transpiration and photosynthesis are reduced to a minimum (Fahn and Dembo 1964).

Stipagrostis pungens, syn. *Aristida pungens*: a C_4 grass with sclerophyllous, spine-tipped, inrolled leaves with sunken stomata. The root system extends laterally for a radius of 20 m or more. It is a nebka-forming grass, widely distributed through North Africa and the Sahara (ZB III). The grass has been investigated by Harche et al. (1992) as a possible candidate for leaf fibre production similar to that from *Lygeum spartum** and *Stipa tenacissima**, but, although rich in hemicellulose, the lignification is less developed than in *Stipa tenacissima**. The strong, cylindrical, hollow culms are often used to probe wounds (Boulos 1983).

The anatomy of the vegetative parts of *Stipagrostis pungens* has been described in detail by Pitot (1956) and that of the roots by Price (1911). The roots are covered throughout their length by a sandy sheath, which is penetrated by the root hairs, which occur throughout the entire length of the sheath. The piliferous layer of the root cap acts as a mucilage-secreting gland, the action of the mucilage easing the passage of the root tip through the grains of sand. At the same time the sand grains are lubricated by the mucilage and adhere to each other and to the expanding root to form a rhizosheath. The cells of the piliferous layer then develop into root hairs which penetrate the sheath. The root hairs do not secrete mucilage but serve as absorptive organs. The mucilage has high absorptive properties, concentrating any soil moisture around the sheath, from which it may be taken up by the root hairs. Thus, the sandy rhizosheath not only protects the root and root hairs against heat and drought but also increases the absorptive efficiency of the root. Furthermore, recent work by Wullstein et al. (1979) has shown the presence of nitrogenase activity to be associated with the rhizosheaths of a number of American grasses, although the uptake of the fixed nitrogen has yet to be demonstrated. Price (1911) also noted the presence of similar rhizosheaths in herbarium specimens of the perennial grasses *Lygeum spartum** and *Stipagrostis obtusa*, syn. *Aristida obtusa* and in the annual grasses *Bromus tectorum**, *Hordeum murinum* and *Schismus barbartus*, syn. *S. calycinus*, although in the annual grasses the sand sheath was retained by the root hairs and not by a mucilaginous secretion. Rhizosheaths have also been observed by the author on *Panicum turgidum**.

Stipagrostis scoparia, syn. *Aristida scoparia* (triple-awned grass): a tufted and often hummock-forming perennial grass with filiform, pungent leaves. A native of North Africa and south-western Asia (ZB and ZE III, III–IV, IV), the grass usually occurs in pure stands on mobile sands. The species has a more southerly distribution than that of *Ammophila arenaria* subsp. *arundinacea**. The grass is grazed and the stems used for fuel. The seeds will germinate only after several consecutive rainy days and then rapidly establish themselves. Those nodes that are covered by sand will produce both roots and stems. Both have a short life span and the grass requires a continuous sand cover for further development of new roots and stems. (Danin 1983).

Triodia pungens (soft spinifex or porcupine grass): a soft, highly resinous, viscid, cushion-like spinifex with stiff, pungent, inrolled leaves, attaining a height of up to 2 m and 4 m diameter Stolons are produced with alternating long and short internodes. Following rain, adventitious roots form on the short internodes, which then spread through the upper 100 cm of the sandy soil and descend to a depth of 3 m. Widely distributed through western and central Australia (ZB and ZE II–III, III, III–II, III–IV), soft spinifex is a subsistence fodder grass, the young leaves may be eaten by cattle and is probably more palatable than other members of the genus. The whole plant is used by the Aborigines in post-natal therapy and for treating colds and influenza (Aboriginal Communities of the Northern Territory 1993; Lazarides and Hince 1993).

The leaf blades are inrolled (hence the name of porcupine grass) and contain little chlorophyllous tissue, but with several layers of sclerenchyma below the epidermis. The epidermal cells contain silica and secrete a resin that coats the leaf surface. The stomata lie in grooves on both the adaxial and abaxial surfaces. When the leaf blades are rolled up the adaxial stomatal grooves are closed, while the abaxial stomata are closed by the shrinking of the parenchymal cells, thereby reducing water losses to an almost undetectable minimum. Their sclerenchymatous structure thus allows the blades to survive long periods of drought. Under favourable moisture conditions and high temperatures the young *Triodia* leaves can grow up to 3.5 mm day^{-1} (Beard 1969; Jackson and Jacobs 1985; Walter and Breckle 1986).

JUNCACEAE

Juncus maritimus (sea rush): a robust, densely caespitose, very tough, rhizomatous perennial with erect, rigid and longitudinally ridged stems up to 1 m high. A pantropical, mainly coastal species, the rhizome is used in local medicine for insomnia (Boulos 1983). The plant sheds its leaves when they become saturated with undesirable salts (Kassas and Batanouny 1984). *Juncus maritimus*, in contrast to other members of the genus, exhibits little osmotic adjustment over a range of external NaCl concentrations from 0 to 300 mM, since its osmotic potential at zero NaCl can be as low as -1.9 MPa. Furthermore, its low growth rate is unaffected by salinity and solute potential, which is maintained at a high level at low external salt concentrations by high internal potassium concentrations. It apparently manages to survive by a constitutive adaptation to the most severe and unpredictable states of the environment (Fitter and Hay 1987).

PALMAE

Copernicia prunifera, syn. *C. cerifera* (carnauba wax palm): a slow growing, long lived, evergreen, single-stemmed, monoecious fan palm up to 20 m tall and bearing wax-coated leaves. It is endemic to northeastern Brasil, where it occurs in enormous numbers; it is reputed that there are 100 million of these palms growing wild in Piauí and Ceará states of Brazil (Smith 1945; Ferri 1954; Andrade-Lima 1981; Kirmse et al. 1983). The palm grows best in areas of very low humidity, on heavy alluvial soils which are often inundated during the rainy season [ZB I(rIII)].

The leaves are the source of the most important of all vegetable waxes, carnauba wax. The cut leaves are sun-dried to loosen the wax, which is removed by beating; the wax is then melted, cleaned and moulded for export. A leaf ca. 1 m diameter may be expected to yield from 3 to 10 g wax. The amorphous wax is hard, tough, lustrous and edible. Its chemical and physical properties are: melting point 83-86 °C; acid number 3-8; saponification number 72–85; iodine number 8-12; acetyl number 55; specific gravity at 25 °C of 0.990-0.999; and the hardness according to the standards of the American Society for Testing and Materials (ASTDM) is D-5 (McLoud 1970, cited by Johnson 1985). The wax is used for the finest wood and car polishes, and also in the manufacture of candles, paints, high-lustre varnishes, carbon paper, gramophone records, sound films, ointments, cosmetics, etc. (Hill 1952; Mors and Rizzini 1966; Purseglove 1985).

Wax production and processing is still rather primitive. However, efforts are now being made to organise the cultivation of the palm, and large plantations have been established in the states of Piauí and Ceará. The dried roots are used locally as a diuretic and depurative. The timber is used for heavy construction and furniture, the leaves for thatching, cordage, paper and fodder. The fruit are edible but contain only a small amount of pulp, livestock eat the fruit, and chopped leaves when forage is scarce. The seed contains ca. 14% oil and can be roasted and ground to make ersatz coffee. Rot and insect resistant, the durable wood is used for all types of construction purposes. When a tree is felled, the heart is taken for food or fed to livestock. Starch can be extracted from the pith of the upper trunk, which can also be tapped for sap (Robinson 1977; Johnson 1985; Balick 1989).

It is reputed to be slow growing and may reach 75–100 years old, but reports of a greater age are considered to be fallacious. Pyrophytic, the adhering leaf bases help to insulate the trunk from fire. Unlike species of the *Ceroxylon* palms, where the wax acts as an insulator against lower temperatures, the wax covering on the upper leaf surface of *C. prunifera* reduces water losses during the long dry season. The wax is not formed if the humidity is too high. Otherwise the physiology of the carnauba palm does not appear to have been studied (Corner 1966; Andrade-Lima 1981).

Phoenix dactylifera (date palm): a solitary or sparsely clustering palm, frequently solitary due to the removal of the basal suckers. Dioecious, the massive trunk can attain 25 m or more and 50 cm in diameter. It is not known in the wild and probably originated in Arabia (Dransfield 1986). Drought tolerant but requiring adequate supplies of underground water or irrigation, it is susceptible to frost and salinity - palms bordering salt marshes and sea beaches have access to underground s ources of fresh water (Purseglove 1985; Walter and Breckle 1986; Danin 1996).

The date palm is widely cultivated in the semi-arid tropics for its edible fruit. In addition to the commercial importance of the fruit, almost every part of the plant may be utilised in the rural economy. The palm may be tapped for a refreshing drink or made into alcohol, the heart, flower and pollen are eaten, while livestock can be fed on date pulp and the softened or ground seeds as well as the roots of felled palms. The wood, although light and soft, lasts for centuries and is used for building purposes, especially for roofs, rafters, revetments, pipes, etc. The leaf petioles are used for making cages, fences and roofs; the thick basal parts are used for fuel. The leaflets are used for making mats, bags, baskets and cordage; the brown leaf fibres are used for making ropes and as substitute sponges. Charcoal from the date stones formerly found a ready market with silversmiths. The date palm also has a number of medicinal uses (Townsend and Guest 1966–1985, vol 8; Osborn 1968; Boulos 1983, 1989).

Up to 3.3 mg kg^{-1} of the human sex hormone oestrone has been found in the pollen. Heavy cell wall deposits of hemicellulose almost fill the cell lumen of the seed, where they form the major food reserve. Dry weather is required for flowering, fruiting and propagation – the pollen is susceptible to rain. The date palm is a shallow-rooted evergreen with large leaves requiring relatively high rates of daytime transpiration for evaporative cooling of the leaves (Corner 1966; Street and Öpik 1984; Harborne 1988; Gibson 1996).

XANTHORRHOEACEAE

Xanthorrhoea thorntonii (grass tree or black boy – blackened by fire): a slow growing, long lived, pachycaul plant with a thick stem up to 5 m tall and bearing a crown of strongly reflexed leaves which are quadrate-rhomboid in transverse section (Fig. 38). It is endemic to Western Australia (ZE III–IV). Black boy produces an acaroid resin from the bases of the old leaves which was used by the Aborigines as an adhesive to fix spear heads to their shafts. It is now used to varnish or lacquer metals and leather. The wood is used for carving bowls, etc. Cattle browsing on the young shoots of black boy develop a peculiar

Fig. 38. *Xanthorrhoea thorntonii* (grass tree, black boy) growing at Goss's Bluff, Western Australia. A slow-growing, long-lived, fire-tolerant, pachycaul perennial up to 5 m tall with a long, terminal inflorescence emerging from a crown of strongly reflexed leaves

swaying of the hindquarters; the toxic principles present are not known (Jackson and Jacobs 1985; Mabberley 1987; Dowling and McKenzie 1993).

GYMNOSPERMS

EPHEDRACEAE

Ephedra alata (white shrubby horsetail): a shrub up to 1 m tall, with jointed, green stems, vestigial leaves and dioecious flowers. A native of North Africa, Arabia and the Middle East (ZB and ZE III, III–IV, IV), the plant contains about 1% of the useful alkaloid drug pseudo-ephadrine. The alkaloid content is highest in the autumn, falling to its lowest content during the winter rains. It is also grazed by livestock and used by nomads for fuel; the crushed and boiled plant is used locally in tanning to thicken skins and to remove their offensive smell (Townsend and Guest 1966b).

Ephedra equisetina, syn. *E. nebrodensis*: a 1 to 2-m-tall shrub with blue-green branches. It is widely distributed in the mountains of Turkestan in Central Asia to north-west China and the Altai Mountains of western Mongolia [ZB VII, VII(rIII)]. The green, unlignified twigs are an important source of the alkaloids ephedrine, N-methylephedrine and catine (total content 0.23–0.36%). The alkaloid contents drop during the rainy season or following a heavy rain. Used in traditional Chinese medicine, it is officially listed in the Chinese pharmacopoeia. Ephedrine is a sympathomimetic drug, imitating the action of the sympathetic part of the autonomic nervous system, raising blood pressure, increasing the heart rate, dilating the air passages and the pupil of the eye as well as constricting the mucous membranes of the nose and throat. The pharmaceutical action of ephedrine resembles that of adrenaline, and it is used as an anti-asthmatic, diaphoretic and diuretic to relieve catarrh, asthma and hayfever and the toxic effects of narcotics and soporifics. The shrub is also cultivated as an ornamental (Komarov 1960-1977, vol. 1; Townsend and Guest 1966b; Petrov 1972; Brown 1977; Krüssman 1985; Nechayeva 1985; Tang and Eisenbrand 1992).

Ephedra viridis (green ephedra or Mormon tea): an obligate, spring flowering, up to 2-m-=tall shr ub with thin, jointed and ridged, bright green, photosynthetic branches, and with the leaves reduced to non-photosynthesising, dry, awl-shaped, 3 to 6-mm-long, scale leaves, the bases of which persist. The roots may descend to 4–6 m or more. It is widely distributed from Utah and Wyoming to California, Arizona and New Mexico (ZB VIIa). The dried stems and flowers are used to prepare a palatable tonic or beverage known as Mormon or Brigham tea, containing both pseudoephedrin and tannin. Moderately palatable, it is a valuable winter grazing. The shrub is recommended for landscaping (Kearney and Peebles 1951; Havens et al. 1973; Krüssman 1985; USDA 1988, Dobrowolski et al. 1990).

PINACEAE

Pinus halepensis var. ***halepensis*** (Aleppo pine): an evergreen tree up to 25 m tall, the leaves in pairs, less than 10 cm long. Widely distributed through the Mediterranean region (ZB IV). Aleppo pine provides a coarse-grained and resinous timber much used in the Mediterranean countries for carpentry, boxes, pit-props, railway sleepers, telegraph poles and charcoal. It is highly regarded as a drought resistant species for hot, dry regions with a modest winter rainfall.

Pinus halepensis var ***brutia***, syn. *P. brutia* (Calabrian pine): differs from *P. halepensis* var. *halepensis* in the longer, 8 to 12-cm-long leaves, its distribution being the eastern Mediterranean region through Asiatic Turkey to Kurdistan (ZB and ZE IV, VII-IV). Calabrian pine may also be used for timber, fuel, resin and turpentine. The oleo-resins of both vars. *halepensis* and *brutia* yield about 20% turpentine, but whereas that of var. *halepensis* is dextrorotatory, that of var. *brutia* is laevorotatory. The var. *brutia* is also less drought tolerant than var. *halepensis*. The needles of the latter will begin to suffer some damage if their water content drops below 100% of their dry weight; the desiccation resistance of *P. nigra* is somewhat lower, and the needles begin to suffer after only 10–15% loss of their original weight, with death after a loss of 25% (Oppenheimer 1960).

References

Aboriginal Communities of the Northern Territory (1993) Traditional aborigine medicine in the Northern Territory of Australia. Conservation Commission of the Northern Territory of Australia, Darwin, 651 pp

Al-Ani HA, Strain B, Mooney HA (1972) The physiological ecology of diverse populations of the desert shrub *Simmondsia chinensis*. J Ecol 60:41–57

Al-Futuh AIM (1983) *Balanites aegyptiaca*, an underutilized raw material potential ready for agro-industrial utilization. UNIDO, Vienna, 100 pp

Al-Futuh AIM (1989) Study on the processing of *Balanites aegyptiaca* fruits for drug, food and feed. In: Wickens GE, Haq N, Day P (eds) New crops for food and industry. Chapman and Hall, London, pp 272–287

Al-Khazraji SM, Al-Shamaony LA, Twaij HAA (1993) Hypoglycaemic effect of *Artemisia herba-alba*. I Effect of different parts and influence of the solvent on hypoglycaemia activity. J Ethnopharmacol 40:163–166

Alados CA, Barroso FG, Aguirre A, Escós J (1996) Effects of early season defoliation on further herbivore attack of *Anthyllis cytisoides* (a Mediterranean browse species). J Arid Environ 34: 455–463

Ali SI (1986) Under-exploited plants of Pakistan. Econ Bot 11:17–25

Anderson DJ (1982) Environmentally adaptive traits in arid zone plants. In: Barker WR, Greenslade PJM (eds) Evolution of the flora of arid Australia. Peacock Publications, Frewville South Australia, in association with Australian Systematics Botany Society and ANZAAS, South Australia Division Inc, pp 133–139

Anderson DMW (1985) Gums and resins, and factors influencing their economic development. In: Wickens GE, Goodin JR, Field DV (eds) Plants for arid lands. Allen and Unwin, London, pp 342–356

Anderson, DMW (1993) Some factors influencing the demand for gum arabic (*Acacia senegal* (L.) Willd.) and other water-soluble tree exudates. Forest Ecol Managem 58: 1–18

Anderson DMW, Dea ICM (1968) Structural studies of some unusual forms of the gum from *Acacia senegal*. Carbohydr Res 6:104–110

Andrade-Lima D (1981) The caatingas dominium. Revi Bras Bot 4:149–153

Anonymous (l895) Some new ideas. The plants cultivated by aboriginal people and how used in primitive commerce. The [Daily] Evening Telegraph, Philadelphia vol 64, no 134, p 2, Thursday, 5 Dec 1895

Anonymous (1906) Guayule in Deutsch-Ostafrika. Tropenpflanzer 10:397

Anonymous (1959) L'Arbre du Ténéré. Bull Liaison Sahara 36:300–302

Archibold OW (1995) Ecology of world vegetation. Chapman and Hall, London, 510 pp

Arnold TH, Wells MJ, Wehmeyer AS (1985) Khoisan food plants: taxa with potential for future economic exploitation. In: Wickens GE, Goodin JR, Field DV (eds) Plants for arid lands. Allen and Unwin, London, pp 69–86

Arnon L (1992) Agriculture in dry lands: principles and practice. Developments in agricultural and managed-forest ecology, vol 26. Elsevier, Amsterdam 979 pp

Aronson J (1985) Economic halophytes – a global review. In: Wickens GE, Goodin JR, Field DV (eds) Plants for arid lands. Allen and Unwin, London, pp 177–188

Aronson JM (1989) Haloph. A data base of salt tolerant plants of the world. Office of Arid Lands Studies, University of Arizona, Tucson, 77 pp

Aronson J (1990) Preliminary observations on the cyanogenic properties of five populations of *Acacia caven* and four additional American species of subgenus *Acacia*. Bull Int Group Study Mimosoideae 18:87–9

Aronson J (1991) Desert plants of use and charm from northern Chile. Desert Plants 10:65–74, 79–86

Aronson J, Wickens GE, Birnbaum E (1990) An experimental technique for long-distance transport of evergreen or deciduous cuttings under tropical conditions. FAO/IBPGR Plant Genet Resour Newslett 81/82:47–48

Aronson J, Ovalle C, Aguilera L, Léon P (1994) Phenology of an 'imigrant' savanna tree (*Acacia caven*, Leguminosae) in the Mediterranean climate zone of Chile. J Econ Bot 27:55–70

Ashbel D (1942) Intensity of sun radiation-soil temperature. Hebrew University, Jerusalem (in Hebrew)

Askew K, Mitchell AS (1978) The fodder trees and shrubs of the Northern Territory. Division of Primary Industry, Extension Bull no 16. Government Printer of the Northern Territory, Darwin, 84 pp

Baillon MW (1894) Histoire naturelle des plantes, vol 5, atlas 3, pt 1, fasc 35. In: Grandidier A (ed) Histoire, physique, naturelle et politique de Madagascar, vol 35. Imprimerie Nationale, Paris

Balick MJ (1989) Native neotropical palms: a resource of global interest. In: Wickens GE, Haq N, Day P (eds) New crops for food and industry. Chapman and Hall, London, pp 323–332

Bamberg SA, Kleinkopf GE, Wallace A, Vollmer A (1975) Comparative photosynthetic production of Mojave Desert plants. Ecology 56:732–736

Barbour MG (1968) Germination requirements of the desert shrub *Larrea divaricata*. Ecology 49:915–923

Barr A, Chapman J, Smith N (1990) Traditional bush medicine. An aboriginal pharmacopoeia. Aboriginal Communities of Northern Territory of Australia, Darwin, 256 pp

Batanouny KH (1981) Ecology and flora of Qatar. Alden Press, Oxford for the University of Qatar, 245 pp.

Batanouny KH, Abdel Wahab MA (1973) VIII Root penetration of *Leptadenia pyrotechnica* (Forsk.) Decne, in relation to its water balance. Oecologia (Berl) 11:151–161

Baum BR (1978) The genus *Tamarix*. Israel Academy of Sciences and Humanities, Jerusalem, 193 pp

Baumer M (1983) Notes on trees and shrubs in arid and semi-arid regions. EMASAR phase II. FAO, Rome, 270 pp

Beadle NCW (1952) Studies in halophytes. I, The germination of seeds and establishment of the seedlings of five species of *Atriplex*. Ecology 33:49–62

Beard JS (1969) The natural regions of the deserts of Western Australia. J Ecol 57:677–711

Becker R (1983) Nutritional quality of the fruit from the chanar tree (*Geoffroes decorticans*). Ecol Food Nutr 13:91–97

Bell EA (1985) Foreword. In: Wickens GE, Goodin JR, Field DV (eds) Plants for arid lands. Allen and Unwin, London, pp vi-vii.

Benichou,A. (1962) Recherches critiques sur l'olivier de Laperrine. Institut de Recherches Sahariennes, Univerisité d'Alger, Mémoire 6:1–55

Berry J, Bretting PK, Nabham GP, Weber C (1981) Domesticated *Proboscidea parviflora*: a potential oilseed crop for arid lands. J Arid Environ 4:147–160

Bertiller MB, Beeskow AM, Coronto F (1991) Seasonal environmental variation and plant phenology in arid Patagonia, Argentina. J Arid Environ 21:1–11

Bewlag JD, Krochko JE (1982) Desiccation tolerance. In: Lange OL, Nobel PS, Osmond CB, Ziegler H (eds) Physiological plant ecology. II Water relations and carbon assimilation. Springer, Berlin Heidelberg New York, pp 325–378

Bhandari MM (1990) Flora of the Indian Desert. MPS Repros, Jodhpur, 435 pp

Billings WD (1957) Physiological ecology. Annu Rev Plant Physiol 8:375–392

Blackwell WH (1990) Poisonous and medicinal plants. Prentice Hall, Englewood Cliffs, NJ, 329 pp

Blau PA, Feeny P, Contardo L, Robson DS (1978) Allyglucosinolate and herbivorous caterpillars: a contrast in toxicity and tolerance. Science 200:1296–1298

Bloomfield F (1985) Miracle plants jojoba and yucca. Century Publishing, London, 64 pp

Böcher TW, Lyshede OB (1968) Anatomical studies in xerophytic apophyllous plants. I. *Monttea aphylla, Bulnesia retama* and *Bredemeyera colletioides*. K Dan Vidensk Selsk Biol Skr 16(3):1–44

Böcher TW, Lyshede OB (1972) Anatomical studies in xerophytic apophyllous plants. II, Additional species from South American shrub steppes. K Dan Vidensk Selsk Biol Skr 18(4):1–137

Booth FEM, Wickens GE (1988) Non-timber uses of selected arid zone trees and shrubs in Africa. FAO Conservation Guide 19. FAO, Rome, 176 pp

BOSTID (1990) Saline agriculture: salt tolerant plants for developing countries. National Academy Press, Washington, DC, 143 pp

Boulos L (1966) Flora of the Nile region in Egyptian Nuba. Feddes Repert 73:184–215

Boulos L (1983) Medicinal plants of North Africa. Reference Publications, Algonac, Michigan, 286 pp

Boulos L (1989) Egyptian desert plants with promising economic potential. Arab Gulf Sci Res 7:91–108

Boulos (1992) Notes on *Agathophora* (Fenzl) Bunge and *Cornula* Del. Studies in the Chenopodiaceae of Arabia, 5. Kew Bull 47:283–287

Bramwell D and Bramwell Z (1974) Wild flowers of the Canary Islands. Thornes, London , 261 pp

Brand JC, Cherikoff V (1985) The nutritional composition of Australian food plants of the desert regions. In: Wickens GE, Goodin JR, Field DV (eds) Plants for arid lands. Allen and Unwin, London, pp 53–68

Brand JC, Cherikoff V (1988) Australian aboriginal bushfoods: their nutritional value. In: Whitehead EE, Hutchinson CF, Timmermann BN, Varady RG (eds) Arid lands: today and tomorrow. Proc International Research and Development Conference, Tucson, Arizona, Oct 20–25, 1985. Westview Press, Boulder; Belhaven Press, London, pp 99–103

Brand JC, Rae C, McDonnell J, Lee A, Cherikoff V, Truswell AS (1983) The nutritional composition of Australian aboriginal foods. 1. Food Technol Aust 35:293–298

Brand JC, Cherikoff V, Truswell AS (1985) The nutritive composition of Australian aboriginal bushfoods. III. Seeds and nuts. Food Technol Aust 37:275–279

Brenan JPM (1959) Flora of tropical East Africa: Leguminosae subfamily Mimosoideae. Crown Agents, London, 173 pp

Briedé J-W (1990) The physiological ecology of *Gutierrezia sarothrae* (Pursh) Rusby in the Chihuahuan Desert of southern New Mexico. Thesis, New Mexico State University, Las Cruces

Briggs MK (1996) Riparian ecosystem recovery in arid lands. Strategies and references. University of Arizona Press, Tucson, 159 pp

British Pharmacopoeia Commission (1993) British Pharmacopoeia. vol 1. HMSO, London, 1281 pp.
Britton NL, Rose JN (1920) The Cactaceae. Descriptions and illustrations of plants of the cactus family. vol 2. Carnegie Publications of Washington, no 248. Washington DC, 248 pp
Brooks WH (1978) Jojoba – a North American desert shrub; its ecology, possible commercialization, and potential as an introduction into other arid areas. J Arid Environ 1:227–236
Brown DE, Lowe CH (1980) Biotic communities of the southwest. USDA Forest Service general technical report RM-78. Rocky Mountain Forest and Range Experiment Station, Fort Collins, Colorado
Brown DE (ed) (1982) Biotic communities of the southwest United States and Mexico. Desert Plants 4:3–741
Brown JAC (1977) Pears medical encyclopaedia. Sphere Books Ltd, London, 784 pp (revised by Hastin-Bennett AM)
Brown NE (1921) *Mesembryanthemum* and some new genera separated from it. Gard Chron 70:250
Brummitt RK (1992) Vascular plant families and genera. Royal Botanic Gardens, Kew, 804 pp
Brummitt RK, Powell CE (1992) Authors of plant names. Royal Botanic Gardens, Kew, 732 pp
Buchmann SL (1987) Floral biology of jojoba (*Simmondsia chinesis*), an anemophilous plant. Desert Plants 8:111–124
Cabrera AL (1955) Latin America. In: UNESCO (ed) Plant ecology, Reviews of Research, vol 6. UNESCO, Paris, pp 77–113
Cabrera AL (1961) Anatomy of some xerophilous plants of Patagonia. In: UNESCO (ed) Plant-water relationships in arid and semi-arid conditions. Proc Madrid Symposium. Arid Zone Research, vol 16. UNESCO, Paris, pp 235–239
Cabrera AL, Willink A (1973) Biogeografía de América Latina. General Secretariat of the Organization of American States, Washington, DC, 120 pp
Cajal J (1989) Uso de habitat por vicuñas y guanacos en la reserva San Guillermo, Argentina. Vida Silv Neotrop 2: 21–31
Changdu C, Liyun Z (1988) The extremely arid deserts of China. Chin J Arid Land Res 1:37–46
Charen AK, Sen DN (1985) The distribution of *Calligonum polygonoides* L. in western Rajasthan, India – a phytogeographical appraisal. J Arid Environ 8:121–131
Chatterton NJ, McKell CM, Strain BR (1970) Intraspecific differences in temperature-induced respiratory acclimation of desert salt bush. Ecology 51:545–547
Chew FS (1988) Searching for defensive chemistry in the Cruciferae, or, do glucosinolates always control interactions of Cruciferae and their potential herbivores and symbionts? No! In: Spencer KC (ed) Chemical mediation of coevolution. Academic Press, San Diego, pp 81–112
Chiej R (1984) The Macdonald encyclopedia of medicinal plants. Macdonald, London, 447 pp
Chippendale GM (1965) Desert plants and their use. Aust Nat Hist 15:115–118
Chippindall LKA (1955) A guide to the identification of grasses in South Africa. In: Meredith D (ed) The grasses and pastures of South Africa. Central News Agency, Cape Town, pp 1–527
Clavijero FJ (1789) Storia della California. Fenzo, Venice
Clayton WD, Renvoize SA (1986) *Genera Graminum*. Grasses of the world. Kew Bulletin additional series 13. HMSO, London, 389 pp
Clements FE (1907) Plant physiology and ecology. Constable, London, 315 pp
Cloudsley-Thompson JL (1977) Man and the biology of desert zones. Arnold, London, 182 pp
Cloudsley-Thompson JL, Chadwick M (1964) Life in deserts. Foulis, London, 218 pp

Cole MM (1960) Cerado, caatinga and pantanal: the distribution and origin of the savanna vegetation of Brazil. Geogr J 126:168–179

Cole MM (1986) The savannas. Biogeography and geobotany. Academic Press, London, 438 pp

Colomer JH S, Guevara JC, Marchi A, Estevez OR, Stasi CR (1991) Native grasses and nutrition of the breeding cow in the arid plain of Mendoza, Argentina. J Arid Environ 20: 113–118

Corner EJH (1966) The natural history of palms. Weidenfeld and Nicolson, London, 393 pp

Cotton CM (1996) Ethnobotany. Principles and applications. Wiley, Chichester, 424 pp

Coyle J, Roberts NC (1975) A field guide to the common and interesting plants of Baja California. Natural History Publishing Company, La Jolla, 206 pp

Crane E, Walker P, Day R (1984) Directory of important world honey sources. International Bee Research Association, London, 384 pp

Cribb AB, Cribb JW (1974) Wild food in Australia. Collins, Sydney, 240 pp

Crosswhite FS (1980) The annual suagaro harvest and crop cycle of the Papago, with reference to ecology and symbolism. Desert Plants 2:3–61

Crosswhite FS (1986) The bitter wild cucumber of the Gila River. Desert Plants 8:45–46

Curtis OF, Clark DG (1950) An introduction to plant physiology. McGraw-Hill, New York, 752 pp

Dahir AM (1985) Economic plants of Somalia. University of Somalia, Mogadishu

Danin A (1983) Desert vegetation of Israel and Sinai. Cana Publishing House, Jerusalem, 148 pp

Danin A (1996) Plants of desert dunes. Adaptations of desert organisms. Springer, Berlin Heidelberg New York, 169 pp

Danin A, Ganor E (1997) Trapping of airborne dust by Eig's meadowgrass (*Poa eigii*) in the Judean Desert, Israel. J Arid Environ 35: 77–86

Davis PH (1965) Introduction. In: Davis PH (ed) Flora of Turkey, vol 1. University of Edinburgh Press, Edinburgh, pp 1–26

Dayton WA (1948) Poisonous plants. In: Stefferud A (ed) Grass. The yearbook of agriculture. US Government Printing Office, Washington, DC, pp 729–734

De Martonne E (1962) Une nouvelle fonction climatologique: l'indice d'aridite. Meteorol Nouv 2:449–458

Devitt J (1992) Acacias: a traditional aboriginal food source in Central Australia. In: House APN, Harwood CE (eds) Australian dry-zone acacias for human food. Australia Tree Seed Centre, Canberra, pp 37–53

Diguet L (1895) Le jojoba (*Simmondsia californica* Nutt.). Rev Sci Nat Appl 1895:685–687

Distel RA, Fernández OA (1986) Productivity of *Stipa tenuis* Phil. and *Piptochaetium napostaense* (Speg.) Hack. in semi-arid Argentina. J Arid Environ 11:93–96

Distel RA, Fernández OA (1987) Leaf water potential trends in three grasses native to semiarid Argentina. J Range Managem 40:203–207

Dobrowolski JP, Caldwell MM, Richards JH (1990) Basin hydrology and plant root systems. In: Osmond CB, Pitelka LF, Hidy GM (eds) Plant biology of the basin and range. Ecological studies. Analysis and synthesis, vol 80. Springer, Berlin Heidelberg New York, pp 243–292

Doebley JF (1984) "Seeds" of wild grasses: a major food of southwestern indians. Econ Bot 38:52–64

Domingo F, Puigdefabregas J, Clark SC, Incoll LD, Lockwood JG (1991) Plant physiological behaviour in relation to water in a semiarid environment in southeast Spain. In: Kienitz G, Milly PCD, Van Genuchten MTH, Rosbjerg D, Shuttleworth WJ (eds) Hydrological interactions between atmosphere, soil and vegetation. Proceedings of a symposium held during the IUGG assembly, Vienna, August 1991. ASHS publication no 204, Wallingford, pp 335–343

Douglas M (1947) Utilization abstracts. Jojoba. Econ Bot 1:41

Dowling RM, McKenzie RA (1993) Poisonous plants: a field guide. Department of Primary Industries, Brisbane, 164 pp

Dransfield J (1986) Flora of tropical East Africa: Palmae. Balkema, Rotterdam, 55 pp

Duke JA, Ayensu ES (1985) Medicinal plants of China, 2 vols. Medical plants of the world, no 4. Reference Publications, Algonac Michigan, 705 pp

Eckardt FE (1965) Remarques préliminaires concernant la méthodologie de l'éco-physiologie végétale et l'organisation du colloque de Montpellier. In: Eckardt FE (ed) Methodology of plant eco-physiology. Proceedings of the Montpellier symposium. Arid zone research, vol 25. UNESCO, Paris, pp 1–10

Edwards G, Walker DA (1983) C_3, C_4 mechanisms, and cellular and environmental regulation of photosynthesis. Blackwell, Oxford, 542 pp

Ehrlich PR, Raven PH (1964) Butterflies and plants: a study in co-evolution. Evolution 18:586–608

Eiten G (1982) Brazilian "savannas". In: Huntley BJ, Walker BH (eds) Ecology of tropical savannas. Springer, Berlin Heidelberg New York, pp 25–47

Eiten G, Goodland R (1979) Ecology and management of semi-arid ecosystems in Brazil. In: Walker BH (ed) Management of semi-arid ecosystems. Developments in agricultural and managed-forest ecosystems, vol 7. Elsevier, Amsterdam, pp 277–300

El-Ghareeb R, Bidak LM (1992) Influence of protection from grazing on carbohydrates and nitrogen metabolites of three range species in the western Mediterranean coastal region of Egypt. J Arid Environ 23:271–278

El-Ghonemy AA, El-Din HK, El-Razzak HA (1974) *Thymelaea hirsuta* (L.)Endl. A possible cellulose raw material for paperpulp industry in Egypt. Proc Egypt Acad Sci 27:51–66

El-Hadidi MN (1971) Distribution of *Cyperus papyrus* and *Nymphaea lotus* in inland water of Egypt. Mitt Bot Staatssamml 10:470–475

El-Hadidi MN (1976) The riverine flora in Nubia. In: Rzóska J (ed) The Nile. History of an ancient river. Junk, The Hague, pp 87–91

El-Hadidi MN, Abdel Wahab, MA (1964) The morphology and anatomy of *Leptadenia pyrotechnica* (Forsk.) Decne. Bull Sci Techn (Assiut Univ) 7: 149–157

Elliger CA, Waiss AC, Laudin RE (1973) Simmondsin, an unusual 2-cyanomethylenecyclohexylglucoside from *Simmondsia californica*. J Chem Soc 1:2209–2212

Ellis RP (1974a) The significance of the occurrence of both Kranz and non-Kranz leaf anatomy in the grass species *Alloteropsis semialata*. S Afr J Sci 70:169–173

Ellis RP (1974b) Anomalous vascular bundle sheath structure in *Alloteropsis semialata* leaf blade. Bothalia 2:273–275

Emberger L (1955) Une classification biogéographique des climats. Rec Trav Lab Bot Zool Univ Montpellier 7:3–43

Evenari M (1985) Adaptations of plants and animals to the desert environment. In: Evenari M, Noy-Meir I, Goodall DW (eds) Hot deserts and arid shrublands. Ecosystems of the world, vol 12A. Elsevier, Amsterdam, pp 79–92

Everist SL (1981) Poisonous plants of Australia. Australian natural science library. Angus and Robertson, Sydney, 966 pp

Fahn A, Dembo N (1964) Structure and development of the epidermis on articulated Chenopodiaceae. Isr J Bot 13:177–192

FAO (1990) Specifications for identity and purity In: FAO (ed) Specifications for identity and purity of certain food additives. Emulsifiers, enzyme preparations, flavouring agents, food colours, thickening agents, miscellaneous food additives. FAO Food Nutrition paper 49, FAO, Rome, pp 23–25

Farines M, Soulier J, Charrouf M, Cave A (1984) Etude de l'huile des graines d'*Argania spinosa* L.; Sapotaceae II. Sterols, alcohols, triterpeniques et methylsterols de l'huile d'argan. Rev Fra des Corps Gras 31:443–448

Feeny P (1975) Biochemical co-evolution between plants and their insect herbivores. In: Gilbert LE, Raven PS (eds) Co-evolution of animals and plants. University of Texas Press, Austin, pp 3–19

Felger RS, Lowe CH (1967) Clinal variation in the surface-volume relationship of the columnar cactus *Lophocereus schottii* in northwestern Mexico. Ecology 48:530–536

Felger RS, Moser MB (1985) People of the desert and sea. Ethnobotany of the Seri indians. University of Arizona Press, Tucson, 435 pp

Felker P, Clark PR (1981) Nodulation and nitrogen fixation (acetylene reduction) in desert ironwood (*Olneya tesota*). Oecologia (Berl) 48:292–293

Ferri MG (1954) Water balance of the "caatinga", a semi-arid type of vegetation of northern Brazil. In: 8ième Congrès International de Botanique Paris. Rapports et communications. Parvenus avant le congrès, sect 7. André, Paris, pp 221–223

Ferri MG (1961) Problems of water relations of some Brasilian vegetation types, with special consideration of the concepts of xeromorphy and xerophytism. In: UNESCO (ed)Plant-water relationships in arid and semi-arid conditions. Proc Madrid Symposium. Arid zone research vol 16. UNESCO, Paris, pp 191–197

Ferri MG, Labouriau L (1952) Water balance from the "caatinga". I. Transpiration of some of the most frequent species of the "caatinga" of Paulo Afonso (Bahia) in the rainy season. Rev Bras Biol 12:301–312

Fitter AH, Hay RKM (1987) Environmental physiology of plants, 2nd edn. Academic Press, London, 423 pp

Ford RI (1978) Ethnobotany: historical diversity and synthesis. In: Ford RI (ed) The nature and status of ethnobotany. Anthropol Paper 67, Museum of Anthropology, Ann Arbor, Michigan, pp 33–49.

Friedman J, Orshan G, Ziger-Cfir Y (1977) Suppression of annuals by *Artemisia herba-alba* in the Negev desert of Israel. J Ecol 65:413–426

Gaff DF, Ellis RP (1974) Southern African grasses with foliage that recovers after dehydration. Bothalia 11:305–308

Gardner IC (1958) Nitrogen fixation in *Elaeagnus* root nodules. Nature 181:717–718

Gastó JM, Olivares EA, Braun WRH (1985) South America. In: Goodin JR, Northington DK (eds) Plant resources of arid and semi-arid lands. A global perspective. Academic Press, Orlando, pp 257–289

Gausman HW, Quisenberry JE (1990) Spectrophotometric detection of plant leaf stress. In: Katterman F (ed) Environmental injury to plants. Academic Press, San Diego, pp 257–280

Gentry HS (1957) Gum tragacanths in Iran. Econ Bot 11:40–63

Gentry HS (1958) The natural history of jojoba (*Simmondsia chinensis*) and its cultural aspects. Econ Bot 12:261–295

Gaussen H (1954) Théorie et classification des climats et des microclimats. 8ième Congres International de Botanique Paris, sect 7, 8. André, Paris, pp 125–130

Gentry HS (1972) The agave family in Sonora. Agriculture Handbook no.399. Agricultural Research Service, US Department of Agriculture, Washington, DC, 195 pp

Gibbs JG, Patten DT (1970) Plant temperatures and heat flux in a Sonoran Desert environment. Oecologia (Berl) 5:165–184

Gibson AC (1996) Structure-function relations of warm desert plants. Springer, Berlin Heidelberg New York, 215 pp

Gindel I (1970) The nocturnal behaviour of xerophytes grown under arid conditions. New Phytol 69: 399–404

Ginszburg C (1963) Some anatomical features of splitting of desert shrubs. Phytomorph 13:92–97

Glat D, Dobrenz AH, Palzkill DA (1981) Stomatal characteristics of jojoba, *Simmondsia chinensis* (Link) Schneider. Desert Plants 3:153–155

Göhl B (1981) Tropical feeds: feed information and nutritive values. FAO animal production and health series, no 12. FAO, Rome, 521 pp

Goldman EA (1916) Plant records of an expedition to Lower California. Contrib US Nat Herb 16:309–371

Gonzalez-Coloma A, Wisdom CS, Rundel PW (1988) Ozone impact on the antioxidant nordihydroguaiaretic acid content in the external leaf resin of *Larrea tridentata*. Biochem Syst Ecol 16:59–64

Goss RA, Benedict CR, Keithly JH, Nessler CL, Stipanovic RD (1984) *cis*-Polyisoprene synthesis in guayule plants (*Parthenium argentatum* Gray) exposed to low, non-freezing temperatures. Plant Physiol 74:534–537

Green PS, Wickens GE (1989) The *Olea europaea* complex. In: Tan K, Mill RR, Elias TS (eds) Plant taxonomy, phytogeography and related subjects. Davis and Hedge Festschrift. Edinburgh University Press, Edinburgh, pp 287–299

Greenfield MD, Shelley T (1987) Variation in host-plant quality: implications for territoriality in a desert grasshopper. Ecology 68:828–838

Greenwood EAN (1986) Water use by trees and shrubs for lowering saline groundwater. In: Barrett-Lennard EG, Malcolm CV, Stern WR, Wilkins SM (eds) Forage and fuel production from salt affected wasteland. Proceedings of a seminar at Cunderdin, Western Australia, 19–27 May, 1984. Elsevier, Amsterdam, pp 423–434 (reprinted from Reclam Reveg Res 5:423–434)

Grenot CJ (1974) Physical and vegetational aspects of the Sahara Desert. In: Brown GW (ed.) Desert biology. Specific topics on physical and biological aspects of arid regions, vol 2. Academic Press, New York, pp 103–164

Groeneveld DP (1988) Stem fluting in Nevada saltbush (*Atriplex torreyi*). In: Hall CA, Doyle-Jones V (eds) Plant biology of eastern California. University of California, White Mountain Research Station, pp 37–45

Guevara JC, Candia RJ, Braun WRH (1978) Range inventory of Mendoza (Argentina). In: Hyder DN (ed) Proceedings of the first international rangeland congress. Society for Range Management, Denver, pp 500–504

Gutale FS, Ahmed MA (1984) *Cordeauxia edulis* pigment, cordeauxiaquinone is deposited on bones and may stimulate hemopoiesis in rats. Riv Tossicol Sper Clin 14:57–62

Gutterman Y (1967) Seed dispersal and germination in *Blepharis persica* (Burm.) Kuntze. Isr J Bot 16:213–234

Gutterman Y (1972) Delayed seed dispersal and rapid germination as survival mechanism of the desert plant *Blepharis persica* (Burm.) Kuntz. Oecologia (Berl)10:145–149

Gutterman Y (1974) The influence of the photoperiodic regime and red-far red light treatments of *Portulaca oleracea* L. plants on the germinability of their seeds. Oecologia (Berl) 17:27–38

Gutterman Y (1982) Survival mechanisms of desert winter annual plants in the Negev highlands of Israel. In: Mann HS (ed) Scientific views on arid zone research, vol 1. Scientific Publishers, Jodhpur, pp 249–283

Gutterman Y (1983) Mass germination of plants under desert conditions. Effects of environmental factors during seed maturation, dispersal, germination and establishment of desert annual and perennial plants in the Negev highlands, Israel. In: Shuval HL (ed) Developments in ecology and environmental quality. Balaban ISS, Rehovat/Philadelphia, pp 1–10

Gutterman Y (1990) Seed dispersal by rain (ombrohydrochory) in some of the flowering desert plants in the deserts of Israel and the Sinai Peninsula. Mitt Inst Allg Bot Hamburg 23b:841–852

Gutterman Y (1993) Seed germination in desert plants. Adaptations of desert organisms. Springer, Berlin Heidelberg New York, 253 pp

Gutterman Y, Witztum A, Evenari M (1969) Physiological and morphological differences between populations of *Blepharis persica* (Burm.) Kuntze. Isr J Bot 18:89–95

Guy C (1990) Molecular mechanisms of cold climmation. In: Katterman F (ed) Environmental injury to plants. Academic Press, San Diego, pp 35–61

Hall JB, Walker DH (1991) *Balanites aegyptiaca*. A monograph. School of Agriculture and Forest Sciences, University of Wales, Bangor, 65 pp

Hall N, Boden RW, Christian CS, Condon RW, Dale FA, Hart AJ, Leigh JH, Marshall JK, McArthur AG, Russell V, Turnbull JW (1972) The use of trees and shrubs in the dry country of Australia. Australian Government Publishing Service, Canberra, 558 pp

Halligan JP (1975) Toxic terpenes from *Artemisia californica*. Ecology 56:999–1003

Harborne JB (1988) Introduction to ecological biochemistry, 3rd edn. Academic Press, London, 356 pp

Harche M, Mekhaldi A, Catesson A-M (1992) Structure et architecture pariéale des tissues foliares de l'*Aristida pungens* Desf. Bull Soc Bot Fr Lett Bot 2:127–134

Harder R, Schumacher W, Firbas F, von Denffer D (1965) Strasburger's textbook of botany, 28th edn. Longman Green, London, 846 pp (English translation by Bell P, Coombe D)

Harrison MN, Jackson JK (1958) Ecological classification of the vegetation of the Sudan. Forests Bulletin no 2 (New Series), Khartoum, 45 pp

Harshberger JW (l896a) Purposes of ethno-botany. Bot Gaz 21:146–154

Harshberger JW (l896b) Purposes of ethno-botany. Am Antiquarian 17(2):73–81

Hatch SL, Pluhar J (1993) Texas range plants. Texas A&M University Press, College Station, Texas, 326 pp

Hattersley PW, Watson L (1992) Diversification of photosynthesis. In: Chapman GP (ed) Grass evolution and domestication, Cambridge University Press, Cambridge, pp 38–116

Havens YH, Horton JS, James MS, Jay JE, Renney CW, Thatcher AP, Schmutz EM (1973) Landscaping with native Arizona plants. Natural Vegetation Committee, Arizona Chapter, Soil Conservation Society of America, University of Arizona Press, Tucson, 194 pp

Hayashi I (1981) Plant communities and their environments in the caatinga of northeast Brazil. Lat Am Stud 2:65–79

Heckathorn SA, DeLucia EH (1991) Effect of leaf rolling on gas exchange and leaf temperature of *Andropogon gerardii* and *Spartina pectinata*. Bot Gaz 152:263–268

Hedrick UP (ed) (1972) Sturtevent's edible plants of the world. Dover, New York, 686 pp.

Hegazy AK, Mansour KS, Abdel-Hady NF (1990) Alleopathic and autotoxic effects of *Anastacia hierachuntica* L. J Chem Ecol 16:2183–2193

Heiser CB (1985) Ethnobotany of the naranjilla (*Solanum quitoense*) and its relatives. Econ Bot 34:4–11

Hellmuth EO (1968) Eco-physiological studies on plants in arid and semi-arid regions in Western Australia. I. Autecology of *Rhagodia baccata* (Labill.) Moq. J Ecol 56:319–344

Hellmuth EO (1969) Eco-physiological studies on plants in arid and semi-arid regions in We stern Australia. II. Field physiology of *Acacia craspedocarpa* F. Muell. J Ecol 57:613–634

Hellmuth EO (1971) Eco-physiological studies on plants in arid and semi-arid regions of Western Australia. III, Comparative studies on photosynthesis, respiration and water relations of ten arid zone and two semi-arid zone plants under winter and late summer climatic conditions. J Ecol 59:225–259

Henderson TO, Farnsworth NR, Myers TC (1987) Biochemistry of recognized molluscicidal compounds of plant origin. In: Mott KE (ed) Plant molluscicides. Wiley, Chichester, pp 109–130

Henrickson J (1972) A taxonomic revision of the Fouquieriaceae. Aliso 7:439–537

Henrickson J (1977) Leaf production and flowering in ocotillos. Cact Succ J (US) 49:133–137

Hidy GM, Klieforth HE (1990) Atmospheric processes affecting the climate of the Great Basin. In: Osmond CB, Pitelka LF, Hidy GM (eds) Plant biology of the basin and range. Ecological studies. Analysis and synthesis, vol 80. Springer, Berlin Heidelberg New York), pp 17–45

Hill AF (1952) Economic botany. A textbook of useful plants and plant products. McGraw-Hill, New York, 560 pp

Hodgkinson KC, Griffin GF (1982) Adaption of shrub species to fires in the arid zone. In: Barker WR, Greenslade PJM (eds) Evolution of the flora of arid Australia. Peacock Publications, Frewville South Australia, in association with Australian Systematics Botany Society and ANZAAS, South Australia Division Inc, pp 145–152

Hoffmann JAE (1978) Flora silvestre de Chile. Zona central. Ediciones Fundacion Claudio Gay, Santiago, Chile, 255 pp

Hoffmann JAE (1989) Cactaceas. En la flora silvestre de Chile. Ediciones Fundacion Claudio Gay, Santiago, Chile, 272 pp

Holechek JL, Pieper RD, Herbel CH (1989) Range management principles and practices. Prentice Hall, Englewood Cliffs, 501 pp

Hoover MM, Hein MA, Dayton WA, Erlanson CO (1948) The main grasses for farm and home. In: Stefferud (ed) Grass. The yearbook of agriculture 1948. US Government Printing Office, Washington, DC, pp 639–700

Hou H-Y (1985) People's Republic of China. In: Goodin JR, Northington DK (eds) Plant resources of arid and semi-arid lands. A global perspective. Academic Press, Orlando, pp 233–255

Huang KC (1993) The pharmacology of Chinese herbs. CRC Press, Boca Raton, 388 pp

Huber W, Sankhla N, Ziegler H (1973) Eco-physiological studies on Indian arid zone plants. Oecologia (Berl) 13:65–71

Humphrey RR (1935) A study of *Idria columnaris* and *Fouquieria splendens*. Am J Bot 22:184–207

Humphrey RR (1974) The boojun and its home. *Idria columnaris* Kellogg and its ecological niche. University of Arizona Press, Tucson, 214 pp

Hunt D (1992) CITES Cactaceae checklist. Royal Botanic Gardens, Kew, 190 pp

Hunting Technical Services (1964) Report on the survey of geology, geo-morphology and soils, vegetation and present land use. UNSP land and water use survey in Kordofan province of the Republic of the Sudan, Doc Dox-Sud-A 26. FAO, Rome, 349 pp

Hunting Technical Services (1968) Land and water resources survey of the Jebel Marra area, Sudan Republic. Reconnaissance vegetation survey, 2 vols. FAO, Rome

Hutchinson J (1959) The families of flowering plants, vol 1, Dicotyledons, 2nd edn. Clarendon Press, Oxford, 510 pp

IBPGR/RBG (1984) Forage and browse plants for arid and semi-arid Africa. IBPGR, Rome, 293 pp

Ihlenfeldt H-D (1988) Pedaliaceae. In: Launert E (ed) Flora Zambesiaca, vol 8(3). Flora Zambesiaca Managing Committee, London, pp 86–113

Ireland AW (1948) The climate of the Sudan. In: Tothill JD (ed) Agriculture in the Sudan. Oxford University Press, London, pp 62–83

Irwin-Williams CC, Osmond CB, Dansie AJ, Pitelka LK (1990) Man and plants in the Great Basin. In: Osmond CB, Pitelka LF, Hidy GM (eds) Plant biology of the basin and range.

Ecological studies. Analysis and synthesis, vol 80, Springer, Berlin Heidelberg New York pp 1–15
Jacks GV, Tavernler R, Boalch DH (1960) Multilingual vocabulary of soil science. Land and Water Development Division FAO, Rome, 450 pp
Jackson DL, Jacobs SWL (1985) Australian agricultural botany. Sydney University Press, Sydney, 377 pp
Jacobson KM (1997) Moisture and substrate stability determine VA-mycorrhizal fungal community distribution and structure in arid grassland. J Arid Environ 35: 59–75
Jaffri JMH, El-Gad A (1980) Flora of Libya: 86 Fabaceae. Al-Faateh University, Tripoli, 304 pp
Jingtai W, Derbyshire E (1987) Climatic geomorphology of the north eastern part of the Qinghai-Xizang plateau, People's Republic of China. Geogr J 153:59–71
Johnson DV (1985) Present and potential economic usages of palms in arid and semi-arid areas. In: Wickens GE, Goodin JR, Field DV (eds) Plants for arid lands. Allen and Unwin, London, pp 189–202
Jones DA (1972) Cyanogenic glycosides and their function. In: Harborne JB (ed) Phytochemical ecology. Proceedings of the Phytochemical Association, Royal Holloway College, Englefield Green, Surrey, April 1971. Academic Press, New York, pp 103–124
Jones DA (1981) Cyanide and coevolution. In: Vennesland B, Conn EE, Knowles CJ (eds) Cyanide in biology. Academic Press, New York, pp 509–516
Joseleau JP, Ullmann G (1985) A relation between starch metabolism and the synthesis of gum arabic. Bull Int Group Study Mimosoideae 13:46–54
Kappen L, Lange OL, Schulze E-D, Evenari M, Buschbom U (1972) Extreme water stress and photosynthetic activity of the desert plant *Artemisia herba-alba* Asso. Oecologia (Berl) 10:177–182
Kassas M, Batanouny KH (1984) Plant ecology. In: Cloudsley-Thompson JL (ed) Sahara desert. Pergamon Press, Oxford, pp 77–90
Kaul RB (1986) Physical and floristic characteristics of the Great Plains. In: The Great Plains Association (ed) Flora of the Great Plains. University Press of Kansas, Kansas City, pp 7–10.
Kaul RN, Thalen DCP (1979) South-west Asia. In: Goodall DW, Perry RA (eds) Arid land ecosystems. International biological programme 16, 17, vol 1. Cambridge University Press, Cambridge, pp 213–271
Kearney TH, Peebles RH (1951) Arizona flora. University of California Press, Berkeley, 1032 pp
Kebin Z (1989) The growth of man-made forests of *Haloxylon ammodendron* and their soil water contents in the Minqin desert region, Gansu Province, China. J Arid Environ 17:109–115
Kebin Z, Kaiguo Z (1989) Afforestation for sand fixation in China. J Arid Environ 16:3–10
Kemp PR (1989) Seed banks and vegetation processes in deserts. In: Leck MA, Parker VT, Simpson RL (eds) Ecology of soil seed banks. Academic Press, San Diego, pp 257–281
Kemp PR, Cunningham GL (1981) Light, temperature and salinity effects on growth, leaf anatomy and photosynthesis of *Distichlis spicata*_(L.) Greene. Am J Bot 68(4):507–516
Kendrew WG (19961) The climates of continents. Clarendon Press, Oxford. 608 pp
Kiritsakis A, Markakis P (1991) Olive oil analysis. In: Linskene HF, Jackson JF (eds) Essential oils and waxes. Modern methods of plant analysis, new series vol 12. Springer, Berlin Heidelberg New York, pp 1–20
Kirmse RD, Pfister JA, de Queiroz JS (1983) Woody plants of the northern Ceará caatinga. Small Ruminant Collaborative Support Program, technical report series No 14. EMBRAPA/Centro Nacional de Pesquisa de Caprinos, Sobral, Ceará; Department of Range Science, Utah State University, Logan, 49 pp

Kloos H, McCullough FS (1987) Plants with recognized molluscicidal activity. In: Mott KE (ed) Plant molluscicides. Wiley, Chichester, pp 45–108

Kohler A (1967) Beiträge zur Kenntnis der ephemeren Vegetation an Südhang der Atacama-Wüste (Chile). Ber Dtsch Bot Ges 80:563–572

Komarov VL (ed) (1934–1964) Flora U.R.S.S., 30 vols. Izdatel'stvo Akademii Nauk SSSR, Moscow [English translation by Lanlau N, Lavoott P (1960–1997) Flora of the USSR, vols 1–27. Israel Program for Scientific Translations, Jerusalem]

Köppen W (1936) Das geographische System der Klimate. In: Köppen W, Geiger W (eds) Handbuch der Klimatologie, vol 1, part C. Borntraeger, Berlin

Krüssman G (1985) Manual of cultivated conifers, 2nd edn. Timber Press, Portland, 361 pp

Küchler AW (1964) Potential and natural vegetation of the conterminous United States. Am Geogr Soc Spec Publ 36, 116 pp

Kumar S, Shankar V (1982) Medicinal plants of the Indian desert: *Commiphora wightii* (Arnott)Bhand. J Arid Environ 5:1–11

Lange OL, Schulze E-D, Evenaria M, Koppen L, Buschbom U (1974) The temperature-related photosynthetic capacity of plants under desert conditions. I, Seasonal changes of the photosynthetic response to temperature. Oecologia (Berl) 17:97–110

Lange OL, Schultze E-D, Evenari M, Kappen L, Buschbom U (1975) The temperature-related photosynthetic capacity of plants under desert conditions. II. Possible controlling mechanisms of the seasonal changes of the photosynthetic response to temperature. Oecologia (Berl) 18:45–53

Launert E (1981) Edible and medicinal plants of Britain and northern Europe. Country Life Books, Twickenham, 288 pp

Lawless J (1992) The encyclopaedia of essential oils. Element Books, Shaftesbury, Dorset, 226 pp

Lazarides M (1992) Resurrection grasses (Poaceae) in Australia. In: Chapman GP (ed) Desertified grasslands: their biology and management. Academic Press, London, pp 213–234

Lazarides M, Hince B (eds) (1993) CSIRO handbook of economic plants of Australia. CSIRO Publications, East Melbourne, 330 pp

Le Houérou HN (1984) Rain use efficiency: a unifying concept in arid-land ecology. J Arid Environ 7:213–247

Le Houérou HN (1985) Forage and fuel plants in the arid zone of North Africa. In: Wickens GE, Goodin GR, Field DV (eds) Plants for arid lands. Allen and Unwin, London, pp 117–141

Le Houérou HN (1986) The desert and arid zones of northern Africa. In: Evenari M, Noy-Meir J, Goodall DW (eds) Hot deserts and arid shrublands. Ecosystems of the world, vol 12B. Elsevier, Amsterdam, pp 101–147

Le Houérou HN (1989) The grazing land ecosystems of the African Sahel. Ecological studies vol. 75. Springer, Berlin Heidelberg NewYork, 282 pp

Le Houérou HN (1995) Bioclimatologie et biogéographie des steppes arides du nord de l'Afrique. Diversité biologique, développement durable et désertisation. Options méditerranéennes, serie B: études et recherches no 10. CIHEAM, Montpellier; ACCT, Paris, 396 pp

Le Houérou HN (1996) Climate change, drought and desertification. J Arid Environ 34:133–185

Le Houérou HN, Popov GF, See L (1993) Agro-bioclimatic classification of Africa. Agrometeorology series working paper no. 6. FAO, Rome, 227 pp

Leakey RRB, Last FT (1980) Biology and potential of *Prosopis* species in arid environments, with particular reference to *P. cineraria*. J Arid Environ 3: 9–24

Lebrun J (1968) À propos du rytheme végétive de l'*Acacia albida* Del. Collect Bot 7, 33:625–636

Lekhak HD, Jha PK, Sen DN (1983) Ecological races: adaptation of *Aerva javanica* to the Indian desert. J Arid Environ 6:99–105

Léonard J (1981–1992) Contribution à l'etude de la végétation des deserts d'Iran. 10 parts. Jardin Botanique National de Belgique, Meise

Letty C (1959) *Bauhinia esculenta*. Flora Plant Afr 33: t1311

Levitt J, Sullivan CY, Krull E (1960) Some problems in drought resistance. Bull Res Council Isr Sect D 8: 173–180

Lewis G (1987) Legumes of Brazil. Royal Botanic Gardens, Kew, 369 pp

Li W-H (1979) The vegetation in Tibet of China and its economic significance. In: Goodin JR, Northington DK (eds) Arid land plant resources. ICASALS, Texas Tech University, Lubbock, pp 650–654

Ling YR (1991) The Old World *Seriphidium* (Bess.)Pooljaik. (Compositae). Bull Bot Res 11,4:1–40

Ling YR (1992) The Old World *Artemisia* Linn. (Compositae). Bull Bot Res 12,1:1–108

Lipu Y (1988) The ecological environment in the Tarim Basin and its protection. Chin J Arid Land Res l:53–56

Little EL (1983) Common fuelwood crops. A handbook for their identification. Communi-Tech Associates, Morgantown, West Virginia, 354 pp

Low T (1991) Wild herbs of Australia and New Zealand. Collins/Angus and Robertson, North Ryde, New South Wales, 160 pp

Lunt OR, Letey J, Clark SB (1973) Oxygen requirement for root growth in three species of desert shrubs. Ecology 54:1356–362

Mabberley DJ (1987) The plant book. A portable dictionary of the higher plants. Cambridge University Press, Cambridge, 706 pp

Mabry TJ, Difeo DR Jr, Sakakibara M, Bohnstedt CF, Siegler D (1977) Natural products chemistry of *Larrea*. In: Mabry TJ, Hunziker JH, Difeo DR (eds) Creosote bush. Biology and chemistry of *Larrea* in New World deserts. Dowden, Hutchinson and Ross, Stroudsberg, Pennsylvania, pp 115–134

MacMahon JA (1988) Warm deserts. In: Barbour MG, Billings WD (eds) North American terrestrial vegetation. Cambridge University Press, New York, pp 231–264

Maconochie JR (1982) Environmentally adaptive traits in arid zone plants. In: Barker WR, Greenslade PJM (eds) Evolution of the flora of arid Australia. Peacock Publications, Frewville South Australia, in association with Australian Systematics Botany Society and ANZAAS, South Australia Division Inc, pp 141–144

Maconochie JR (1985) Plants of the Australian arid zone – an undeveloped potential. In: Wickens GE, Goodin JR, Field DV (eds) Plants for arid lands. Allen and Unwin, London, pp 289–301

Mann HS, Shankarnaryan KA (1980) The role of *Prosopis cineraria* in an agropastoral system in western Rajasthan. In: Le Houérou HN (ed) Browse in Africa. International Livestock Centre for Africa, Addis Ababa, pp 437–447

Manousis T, Moore NF (1988) The olive tree. Biologist 35:7–12

Martín AJS (1983) Medicinal plants in central Chile. Econ Bot 37:216–227

Maslin BR (1988) Should *Acacia* be divided? Bull Int Group Study Mimosoid 16:54–76

Mason CT, Mason PB (1987) A handbook of Mexican roadside flora. University of Arizona Press, Tucson, 380 pp

McDaniel KC, Briedé J-W, Torell LA (1991) Classification, ecology, and economics of perennial snakeweed. In: James LF. Evans JO, Ralphs MH, Childs RD (Eds) Noxious range weeds. West View Press, Boulder, Colorado, pp 210–219.

McGinnies WG (1981) Discovering the desert. University of Arizona Press, Tucson, 276 pp

McGinnies WG, Goldman BJ, Paylore P (1968) Deserts of the world. An appraisal of research into their physical and biological environments. University of Arizona Press, Tucson, 788 pp

McLaughlin SP (1986) *Grindelia camporum*: potential cash crop for the arid southwest. Econ Bot 40:162–169

McLoud ES (1970) Waxes. In: Kroschwitz DI, Howe-Grant M (eds) Encyclopedia of chemical technology, vol 22, 2nd edn. Wiley, New York, pp 156–173

Meigs P (1953) World distribution of arid and semi-arid homoclimates. In: Kaul RN (ed) Reviews of research on arid zone hydrology. Arid zone research, vol 1. UNESCO, Paris pp 203–210

Meigs P (1966) Geography of coastal deserts. Arid zone research vol 28. UNESCO, Paris, 140pp

Meikle RD (1977, 1985) Flora of Cyprus, 2 vols. Royal Botanic Gardens, Kew, 1969 pp

Meinzer FC, Sharifi MR, Nilsen ET, Rundel PW (1988) Effects of manipulation of water and nitrogen regime on the water relations of the desert shrub *Larrea tridentata*. Oecologia (Berl) 77:480–486

Metcalfe CR (1983) Ecological anatomy and morphology, general survey. In: Metcalfe CR, Chalk L (eds) Anatomy of the dicotyledons, vol 2, 2nd edn. Clarendon Press, Oxford, pp 127–152

Migahid AM, Abdel Wahab MA, Batanouny KH (1972) Eco-physiological studies on desert plants. VII. Water relations of *Leptadenis pyrotechnica* (Forsk.) Decne. growing in the Egyptian Desert. Oecologia (Berl) 10:79–91

Miller AG, Morris M (1988) Plants of Dhofar. The southern region of Oman. Traditional, economic and medicinal uses. Office of the Adviser for Conservation of the Environment, Diwan of Royal Court, Muscat, 361 pp

Mirov NT (1952) *Simmondsia* or jojoba – a problem in economic botany. Econ Bot 6:41–47

Mitchell AA, Wilcox DG (1988) Arid shrubland plants of Western Australia. University of Western Australia Press, Nedlands and Western Australia Department of Agriculture, Perth, 84 pp

Mohammed S, Sen DN (1987) Proline accumulation in arid zone plants. J Arid Environ 13:231–236

Monod T (1973) Les déserts. Horizons de France, Paris, 247 pp

Mooney HA, Ehleringer J, Bjorkman O (1977) The energy balance of leaves of the evergreen desert shrub *Atriplex hymenelytra*. Oecologia (Berl) 29:301–310

Mooney HA, Gulmon SW, Rundel PW, Ehleringer J (1980) Further observations on the water relations of *Prosopis tamarugo* of the northern Atacama Desert. Oecologia (Berl) 44:577–580

Moquin-Tandon CHBA, Cosson E (1862) Description d'une nouvelle espèce d'*Anabasis*. Bull Soc Bot Fr 9:299–302

Morello J (1951a) Multiplicación de arbustos en las mesetas araucanas del Valle de Santa Maria (Tucumán). Bol Soc Argent Bot 3:207–217

Morello J (1951b) El bosque de algarrobo y la estepa de Jarilla en el Valle de Santa Maria(Provincia de Tucumán). Darwiniana 9:315–347

Morello J (1955) Estudios botanicos en las regiones aridas de la Argentina. ll. Transpiracion de los arbustos resinosos de follaje permanente del Monte. Rev Agron Noroeste Argent 1:385–524

Morello J (1958) Las provincia fitogeográfica del Monte. Opera Lilloana 2:1–155

Mors WB, Rizzini CT (1966) Useful plants of Brazil. Holden-Day, San Francisco, 166 pp

Mouret M (1985) Gummosis of acacias: current histological research. Bull Int Group Study Mimosoideae 13:38–45

Mozafar A, Goodin JR (1970) Vesiculated hairs: a mechanism for salt tolerance in *Atriplex halimus* L. Plant Physiol 45:62–65

Nabhan GP, Felger RS (1985) Wild desert relatives of crops: their direct uses as food. In: Wickens GE, Goodin JR, Field DV (eds) Plants for arid lands. Allen and Unwin, London, pp 19–33

NAS (1979) Tropical legumes: resources for the future. National Academy of Sciences, Washington, DC, 331 pp

NAS (1983) Firewood crops: shrub and tree species for energy production, vol 2. National Academy of Sciences, Washington, DC, 92 pp

Naval Staff Intelligence (1919) A handbook of Asia Minor, vol 1. Naval Staff Intelligence Department. HMSO, London, 365 pp

Nechayeva NT (1985) The Union of Soviet Socialist Republics. In: Goodin JR, Northington DK(eds) Plant resources of arid and semi-arid lands. A global perspective. Academic Press, Orlando, pp 291–317

Nobel PS (1977) Water relations and photosynthesis of a barrel cactus, *Ferocactus acanthodes* in the Colorado Desert. Oecologia (Berl) 27:117–133

Nobel PS (1980) Morphology, surface temperature and northern limits of columnar cacti in the Sonoran Desert. Ecology 61:1–7

Nobel PS (1991) Achievable productivities of certain CAM plants, basis for high values compared with C_3 and C_4 plants. New Phytol 119:183–205

Nobel PS (1996) Responses of some North American CAM plants to freezing temperatures and doubled CO_2 concentrations: implications of global climatic change for extending cultivation. J Arid Environ 34: 187–196

Noble JC (1982) The significance of fire in the biology and evolutionary ecology of mallee *Eucalyptus* populations. In: Barker WR, Greenslade PJM (eds) Evolution of the flora of arid Australia. Peacock Publications, Frewville South Australia, in association with Australian Systematics Botany Society and ANZAAS, South Australia Division Inc, pp 153–159

Noggle GR, Fritz GJ (1983) Introductory plant physiology, 2nd edn. Prentice-Hall, Englewood Cliffs, 627 pp

Nuttonson MY (1961) The physical environment and agriculture of Morocco, Algeria, and Tunisia with special reference to their regions containing areas climatically and altitudinally analogous to Israel. American Institute for Crop Ecology, Washington, DC, 600 pp

Obeid Mabarak M, Bari EA, Wickens GE, Williams MAJ (1982) The vegetation of central Sudan. In: Williams MAJ, Adamson DA (eds) A land between two Niles. Balkema, Rotterdam, pp 143–164

Oliveto EP (1972) Nordihydroguaiaretic acid. A naturally occurring antioxidant. Chem Ind 17:677–679

Oppenheimer HR (1960) Adaptation to drought: xerophytism. In: UNESCO (ed) Plant-water relationships in arid and semi-arid conditions. Reviews of research. Arid zone research. UNESCO, Paris, 15. pp 105–138

Orev Y (1992) Fodder, fuel, shade and shelter – some trees and shrubs for arid areas. Desertification Bull 21:55–58

Osborn DJ (1968) Notes on medical and other uses of plants in Egypt. Econ Bot 22:165–177

Osborn DJ, Krombein KV (1969) Habitat, flora, mammals and wasps of Gebel 'Uweinat, Libyan Desert. Smithsonian Contrib Zool no 11, 18 pp

Osmond CB, Björkman O, Anderson DJ (1980) Physiological processes in plant ecology. Toward a synthesis with *Atriplex*. Ecological studies, vol 36. Springer, Berlin Heidelberg New York, pp 468

Ozenda P (1977) Flora du Sahara, 2nd edn. CNRS, Paris, 622 pp

Parker J (1968) Drought-resistance mechanisms. In: Kozlowski TT (ed) Water deficits and plant growth, vol l. Academic Press, New York, pp 195–234

Parry M (1990) Climate change and world agriculture. Earthscan Publications, London, 157 pp

Pasternak D, Aronson JA, Ben-Dov J, Forti M, Mendlinger S, Nerd A, Sitton D (1986) Development of new arid zone crops for the Negev Desert of Israel. J Arid Environ 11:37–59

Pearcy LW, Harrison AT, Mooney HA, Björkman O (1974) Seasonal changes in net photosynthesis of *Atriplex hymenelytra* shrub growing in Death Valley, California. Oecologia (Berl) 17:111–121

Pedley L (1975) Review of the extra-Australian species of *Acacia* subgenus *Heterophyllum.* Contrib Qnld Herb 18, 24 pp.

Pedley L (1981) Classification of acacias. Bull Int Group Study Mimosoid 9:42–48

Petrov MP (1972) Asia. In: McKell CM, Blaisdell JP, Goodin JR (eds) Wildland shrubs – their biology and utilization. USDA Forest Service general technical report INT-1. Intermountain Forest and Range Experiment Station, Ogden, Utah, pp 37–50

Petrov MP (1976) Deserts of the world. Wiley, Toronto (Israel Program for Scientific Translations, Jerusalem; English translation of Pustyni zemnogo shara.Izdatel'stvo Nauka, Leningrad, 1973)

Petrov MP (1979) Land-use of semi-deserts in the U.S.S.R. In: Walker BH (ed) Management of semi-arid ecosystems. Developments in agricultural and managed-forest ecosystems vol 7. Elsevier, Amsterdam, pp 301–327

Pettit GR, Doubek DL, Herald DL (1991) Isolation and structure of cytostatic steroidal saponins from the African medicinal plant *Balanites aegyptiaca.* J Nat Prod 54:1491–1502

Pitot A (1956) Sur la structure et la morphologie interne de quelques plantes désertiques. Bull IFAN [A] 18:1054–1085

Pollak G, Waisel Y (1970) Salt secretion in *Aleuropus litoralis* (Willd.)Parl. Ann Bot 34:879–888

Porto ER, Silva AS (1988) Small-scale water management in farming systems in the Brazilian arid zones. In: Whitehead EE, Hutchinson CE, Timmermann BN, Varady RG (eds) Arid lands: today and tomorrow. Proc International Research and Development Conf, Tucson, Arizona, Oct 20–25, 1985. Westview Press, Boulder; Belhaven Press, London, pp 943–957

Potter-Basso G (1991) *Yucca baccata* from the beginning. Arid Lands Newslett 31:24–29

Price SR (1911) The roots of some North African grasses. New Phytol 10:328–339

Purseglove JW (l985) Tropical crops. Monocotyledons. Longman, London 607 pp

Quézel P (1965) La végétation du Sahara. Du Tchad à la Mauritanie. Fischer, Stuttgart, 333 pp

Quézel P (1969) Flore et végétation des plateaux du Darfur nord-occidental et du jebel Gourgeil (Rép. du Soudan). Centre Natl Recherche Coop Progr no 45, dossier 5. CNRS, Paris, 146 pp

Quézel P (1978) Analysis of the flora of the Mediterranean and Saharan Africa. Ann Missouri Bot Gard 65:479–534

Rabesandratana R (1984) Flora of the Malagasy southwest. In: Jolly A, Oberlé P, Albignac R (eds) Madagascar. Key environments. Pergamon Press, Oxford, pp 55–74

Ragonese AM (1987) La presencia de cristales en los pelos de varias especies de *Nassauvia* (Compositae). Darwinia 28:245–250

Ralph CP (1978) Observations on *Azorella compacta* (Umbelliferae), a tropical Andean cushion plant. Biotropica 10:62–67

Rauh W (1985) The Peruvian-Chilean deserts. In: Evenari M, Noy-Meir I, Goodall DW (eds) Hot deserts and arid shrublands. Ecosystems of the world, vol 12A. Elsevier, Amsterdam, pp 239–267

Raynal J (1973) Notes cypérologiques. Contribution no 19: la classification de la sous-famille des Cyperoideae. Adansonia [Sér 2] 13:145–171

Rea AM (1991) Gila River Pima dietary reconstruction. Arid Lands Newslett 31:3–10

Reichman OJ (1984) Spatal and temporal variation of seed distribution in Sonoran Desert soils. J Biogeogr 11:1–11

Renfrew JM (1973) Palaeoethnobotany. The prehistoric food plants of the Near East and Europe. Columbia University Press, New York, 248 pp

Rhoades DF (1977) Integrated antiherbivore, antidesiccant and ultraviolet screening properties of creosote bush resin. Biochem Syst Ecol 5:281–290

Rivett DE, Jones GP, Tucker DJ (1989) *Santalum acuminatum* fruit: a prospect for horticultural development. In: Wickens GE, Haq N, Day P (eds) New crops for food and industry. Chapman and Hall, London, pp 208–215

Robinson H (1977) Latin America, 4th edn. New certificate geography series. MacDonald and Evans, Plymouth, 523 pp

Rodin LE (1981) Primary productivity. In: Goodall DW, Perry RA (eds) Arid land ecosystems International biological programme 16, 17, vol 2. Cambridge University Press, Cambridge, pp 169–198

Rodriguez RR, Matthei O, Quezada MM (1983) Flora arbórea de Chile. Universidad de Concepción, Concepción, 408 pp

Rotherham ER, Blaxell DF, Briggs BG, Carolin RC (eds) (1975) Flowers and plants of New South Wales and southern Queensland. Reed, 191 pp

Roundy BA, Ruyle GB (1989) Effects of herbivory on twig dynamics of a Sonoran Desert shrub *Simmondsia chinensis* (Link) Sch. J Appl Ecol 26:701–710

Runciman HV (1986) Forage production from salt affected wasteland in Australia. In: Barrett-Lennard EG, Malcolm CV, Stern WR, Wilkins SM (eds) Forage and fuel production from salt affected wasteland. Proceedings of a symposium held at Cunderdin, Western Australia, 19–27 May 1984. Elsevier, Amsterdam, pp 17–29

Rundel PW (1982) Water uptake by organs other than roots. In: Lange OL, Nobel PS, Osmond CB, Ziegler H (eds) Physiological plant ecology. II Water relations and carbon assimilation. Springer, Berlin Heidelberg New York, pp 111–134

Rundel PW, Mahu M (1976) Community structure and diversity in a coastal fog desert in northern Chile. Flora 165:493–505

Sagredo R (1987) Flora de Almeria. Plantas vasculares de la Provincia. Instituto de Estudios Almerienses, Almeria, 552 pp

Sahni KC (1968) Important trees of the northern Sudan. UNDP and FAO, Khartoum, 138 pp

Sala OE, Golluscio RA, Lauenroth WK, Soriano A (1989) Resource partitioning between shrubs and grasses in the Patagonian steppe. Oecologia (Berl) 81:501–505

Salih OM, Nour AM, Harper DA (1991) Chemical and nutritional composition of two famine food sources used in the Sudan, mukheit (*Boscia senegalensis*) and maikah (*Dobera roxburghii*). J Sci Food Agric 57:367–377

Sammis TW, Gay LW (1979) Evapotranspiration from an arid zone plant community. J Econ Bot 2:313–321

Sánchez-Díaz MF, Mooney HA (1979) Resistance to water transfer in desert shrubs native to Death Valley, California. Physiol Plant 46:139–146

Sankary MN, Barbour MG (1972) Autecology of *Atriplex polycarpa* fron California. Ecology 52:1152–1162

Saunders CF (1976) Edible and useful wild plants of the United States and Canada. Dover Publications, New York, 275 pp (reprint of a work first published as: Useful wild plants of the United States and Canada. McBride, New York)

Savage DA, Costello DF (1948) Range management. In: Stefferud A (ed) Grass. The yearbook of agriculture 1948. US Government Printing Office, Washington DC, pp 522–537

Schimper AFW (1960) Plant-geography upon a physiological basis. Clarendon Press, Oxford, 839 pp. (Original 1903 in German)

Schmidt J, Stavisky N (1983) Uses of *Thymelaea hirsuta* (mitnan) with emphasis on hand papermaking. Econ Bot 37:310–321

Schnell R (1966) Problèmes phytogéographiques, écologiques et économiques de la Caatinga Brésilienne. J Agric Trop Bot Appl 13:59–90

Schultes RE, Hofmann A (1992) Plants of the gods, their sacred, healing and hallucinogenic powers. Healing Press, Rochester, Vermont, 192 pp

Schultze E-D, Lange OL, Buschbom U, Kappen L, Evenari M (1972) Stomatal responses to changes in humidity in plants growing in the desert. Planta 108:259–270

Schütz W, Milberg A (1997) Seed germination in *Launaea arborescens*: a continuously flowering semi-desert schrub. J Arid Environ 36:113–122

Sen DN, Mohammed s(1994) General aspects of salinity and biology of desert plants. In: Pessarakli M (ed) Handbook of plant and crop stress. Dekkor, New York, pp 125–145

Senock RS, Barrow JR, Gibbens RP, Herbel CH (1991) Ecophysiology of the polyploid shrub *Atriplex canescens* (Chenopodiaceae) growing in situ in the northern Chihuahuan Desert. J Arid Environ 21:45–57

Shachak M (1975) Some aspects of the structure and function of a desert ecosystem and its use in a teaching programme of a field studies centre. PhD thesis, Hebrew University of Jerusalem (in Hebrew with English summary)

Shaltout KH (1992a) Dimension analysis of *Thymelaea hirsuta* (L.) Endl. fibres. Feddes Rep 103:99–106

Shaltout KH (1992b) Nutrient status of *Thymelaea hirsuta* (L.) Endl. in Egypt. J Arid Environ 23:423–432

Sharp D (ed) (1983) Dictionary of chemistry. Penguin Books, London, 434 pp

Sherbrooke WC (1976) Differential acceptance of toxic jojoba seed (*Simmondsia chinensis*) by four Sonoran desert heteromyid rodents. Ecology 57:596–602.

Shreve F (1925) Ecological aspects of the deserts of California. Ecology 6:93–103

Shreve F (1933) Desert investigations. Carnegie Year Book 32:195–201

Shreve F (1939) Observations on the vegetation of Chihuahua. Mandroño 5:1–13

Shreve F (1942a) The desert vegetation of North America. Bot Rev 8:195–246

Shreve F (1942b) Grassland and related vegetation in northern Mexico. Madroño 6:190–198

Shreve F (1951) Vegetation of the Sonoran Desert. Carnegie Institution of Washington, Publ no 591, Washington, DC, 192 pp (reprinted in Shreve and Wiggins, vol 1, pp 1–186)

Shreve F, Wiggins IL (1964) Vegetation and flora of the Sonoran Desert, 2 vols. Stanford University Press, Stanford, 1740 pp

Simmons MH (1981) Acacias of Australia. Nelson, Melbourne, 323 pp

Sims PL (1988) Grasslands. In: Barbour MG, Billings WD (eds) North American terrestrial vegetation. Cambridge University Press, New York, pp 365–286

Sivakumar MVK (1990) Agroclimatic aspects of rainfed agriculture in the Sudano-Sahelian zone. In: ICRISAT (ed) Soil, crop, and water management in the Sudano-Sahelian zone. ICRISAT, Patancheru, India, pp 17–38

Skerman PJ, Riveros F (1990) Tropical grasses. FAO plant production series no 23. FAO, Rome, 832 pp

Slatyer RO (1960) Aspects of the tissue water relationships of an important arid zone species (*Acacia aneura* F.Muell.) in comparison with two mesophytes. Bull Res Counc Isr Sect D Bot 8:159–168

Slatyer RO (1961) Methodology of a water balance study on a desert woodland (*Acacia aneura* F. Muell.) community in central Australia. In: UNESCO (ed), Plant-water relationships in arid and semi-arid conditions. Proceedings of the Madrid symposium. Arid zone research, vol 16. UNESCO, Paris, pp 15–26

Smith LB (1945) The vegetation of Brasil. In: Verdoon F (ed) Plants and plant science in Latin America. Chronica Botanica, Waltham, Massachusetts, pp 297–302

Smith SD, Nowak RS (1990) Ecophysiology of plants in the intermountain lowlands. In: Osmond CB, Pitelka LF, Hidy GM (eds) Plant biology of the basin and range. Ecological Studies. Analysis and synthesis, vol 80. Springer, Berlin Heidelberg New York, pp 179–241

Smith SD, Monson RK, Anderson JE (1997) Physiological ecology of North American plants. Adaptations of desert organisms. Springer, Berlin Heidelberg, 286 pp

Song B (1990) Water studies of *Caragana microphylla* mosaic assembly in *Aneurolepidium chinense* steppe. In: Yang H (ed.) Proc Int Symp on Grassland Vegetation, Aug 15–20, 1987, Hohhot, People's Republic of China. Science Press, Beijing, pp 381–391

Soriano A (1961) Germination of *Stipa neaei* in relation to imbition and moisture level. UNESCO (ed) Plant-water relationships in arid and semi-arid conditions. Proc Madrid Symp. Arid zone research vol 16. UNESCO, Paris, pp 261–254

Soriano A (1972) South America. In: McKell CM, Blaisdell JP, Goodin JR (eds) Wildland shrubs – their biology and utilization. USDA Forest Service general technical report INT-1. Intermountain Forest and Range Experiment Station, Ogden, Utah, pp 51–59

Sprent JI (1985) Nitrogen fixation in arid environments. In: Wickens GE, Goodin JR, Field DV (eds) Plants for arid lands. Allen and Unwin, London, pp 215–229

Sprent JI, Sprent P (1990) Nitrogen fixing organisms. Chapman and Hall, London, 256 pp

Standley PC (1920–1926) Trees and shrubs of Mexico. Contrib US Nat Herb 23:1–170 (1920), 171–515–848 (1923), 849–1312 (1924), 1313–1721 (1926)

Stewart GR, Lee JA (1974) The role of proline accumulation in halophytes. Planta 120:279–289

Stienen H (1985) *Prosopis tamarugo* in the Chilean Atacama – ecophysiological and reforestation aspects. In: Wickens GE, Goodin JR, Field DV (eds) Plants for arid lands. Allen and Unwin, London, pp 103–116

Story R (1958) Some plants used by the bushmen in obtaining food and water. Mem Bot Surv S Afr 30. 115 pp

Street HE, Öpik H (1984) The physiology of flowering plants: their growth and development, 3rd edn. Arnold, London, 279 pp

Sudzuki HF (1969) Absorción foliar de humedad atmosférica en Tamarugo (*Prosopis tamarugo* Phil.). Minist Agric Bol Tecnol 30, 23 pp

Suliman AEGM, Jackson JK (1959) The heglig tree. Sudan Silva 9:1–4

Sundberg MD (1985) Trends in distribution of stomata in desert plants. Desert Plants 7:154–157 Szarek SR, Woodhouse RM (1976) Ecophysiological studies of Sonoran desert plants. I. Diurnal photosynthetic patterns of *Ambrosia deltoides* and *Olneya tesota*. Oecologia (Berl) 26:225–234

Szarek SR, Woodhouse RM (1977) Ecophysiological studies of Sonoran desert plants. II Seasonal photosynthesis patterns and primary production of *Ambrosia deltoides* and *Olneya tesosta*. Oecologia (Berl) 28:365–375

Tan K (1980) Studies in the Thymeliaceae II. A revision of the genus *Thymelaea*. Notes R Bot Gard Edinb 38:189–246

Tanaka S. (1976) Tanaka's Cyclopedia of edible plants of the world, 1st edn. Keigaku, Tokyo, 924 pp

Tang W, Eisenbrand G (1992) Chinese drugs of plant origin: chemistry, pharmacology, and use in traditional and modern medicine. Springer, Berlin Heidelberg New York, 1056 pp

Taylor FW (1985) The potential for the commercial utilization of indigenous plants of Botswana. In: Wickens GE, Goodin JR, Field DV (eds) Plants for arid lands. Allen and Unwin, London, pp 231–242

Thomson L (1992) Australia's subtropical dry-zone *Acacia* species with human food potential. In: House APN, Harwood CE (eds) Australian dry-zone acacias for human food. Australia Tree Seed Centre, Canberra, pp 3–36

Thornthwaite CW (1948) An approach towards a rational classification of climate. Geogr Rev 38(1):55–94

Thulin M, Warfa AM (1987) The frankincense trees (*Boswellia* spp., Burseraceae) of northern Somalia and southern Arabia. Kew Bull 42:487–500

Timmermann BN, Hoffmann JJ (1985) Resins from *Grindelia*: a model for renewable resources in arid environments. In: Wickens GE, Goodin JR, Field DV (eds) Plants for arid lands. Allen and Unwin, London, pp 357–368

Timmermann BN, Hoffmann JJ (1988) The potential for the commercial utilization of resins from *Grindelia camporum*. In: Whitehead EE, Hutchinson CE, Timmermann BN, Varady RG (eds) Arid lands: today and tomorrow. Proc International research and development Conf, Tucson, Arizona, Oct 20–25, 1985. Westview Press, Boulder; Belhaven Press, London, pp 1321–1339

Tootill E (ed) (1984) The Penguin dictionary of botany. Penguin Books, London, 291 pp

Townsend CC, Guest E (eds) (1966–1985) Flora of Iraq, Ministry of Agriculture, Baghdad (vol 1 1966a, vol 2 1966b, vol 9 1968, vol 3 1974, vol 4 1980, vol 8 1985)

Trione SO, Passera CB (1993) Growth and nitrogen status of *Atriplex lampa* seedlings under different water regimes. J Arid Environ 25:331–341

Turner JS, Picker MD (1993) Thermal ecology of an embedded dwarf succulent from southern Africa (*Lithops* spp.: Mesembryanthemaceae). J Arid Environ 24:361–385

Turner RM, Bowers JE, Burgess TL (1995) Sonoran Desert plants. An ecological atlas. University of Arizona Press, Tucson, 501 pp

Ullmann I (1985) Tagesgänge von Transpiration und stomatärer Leitfähigkeit Sahelischer und Saharischer Akazien in der Trockenzeit. Flora 176:383–409

UNESCO (1979) Map of the world distribution of arid regions. Explanatory note. MAB technical notes, no 7. UNESCO, Paris, 54 pp

USDA (1988) Range plant handbook. Dover, New York, 816 pp (reprint of a work first prepared by USDA Forest Service in 1937 and published by the Government Printer, Washington, DC)

Villagrán C, Arroyo MT, Armesto JJ (1982) La vegetacion de un transecto altitudinal en los Andes del norte de Chile (18–19°S). In: Veloso A, Bustos E (eds) El ambiente natural y las pobluciones humanas de los Andes de norte grande de Chile (Arica, lat 18°28'S), vol 1. Rostloc, Montevideo, pp 13–69

Volk OH (1964) Zur Kenntnis von *Harpagophyton procumbens* DC. Dtsche Apoth Z 104(18)373–376

Volkens G (1887) Die Flora der Aegyptisch-Arabischen Wüste auf Grundlage anatonisch-physiologischer Forschungen. Borsträger, Berlin, 156 pp

von Martius KFP (1824) Die Physiognomie des Pflanzenreiches in Brasilien. Lindauer, Munich, 36 pp

von Willert DJ, Eller BM, Werger MJA, Brinckmann E, Ihlenfeldt H-D (1992) Life strategies of succulents in deserts. With special reference to the Namib desert. Cambridge studies in ecology. Cambridge University Press, Cambridge, 340 pp

Waisel Y (1960a) Ecological studies on *Tamarix aphylla* (L.) Karst. I. Distribution and eproduction. Phyton (Argent) 15:7–17

Waisel Y (1960b) Ecological studies on *Tamarix aphylla* (L.) Karst. II. The water economy. Phyton (Argent) 15:17–27

Waisel Y (1961) Ecological studies on *Tamarix aphylla* (L.) Karst. III. The salt economy. Plant Soil 4:356–364

Waisel Y (1991) The glands of *Tamarix aphylla:* a system for salt secretion or for carbon concentration? Physiol Plant 83:506–510

Walter H (1963a) Climatic diagrams as a means to comprehend the variable climatic types for ecologcal and agricultural purposes. In: Ratter AJ, Whitehead FH (eds) Water relations of plants. Blackwell, Oxford, pp 3–9

Walter H (1963b) The water supply of desert plants. In: Ratter AJ, Whitehead FH (eds) Water relations of plants. Blackwell, Oxford, pp 199–205

Walter H, Breckle S-W (1985) Ecological systems of the geobiosphere. I Ecological principles in global perspective. Springer, Berlin Heidelberg New York, 242 pp

Walter H, Breckle S-W (1986) Ecological systems of the geobiosphere. II Tropical and subtropical zonobiomes. Springer, Berlin Heidelberg New York, 265 pp

Walter H, Breckle S-W (1989) Ecological systems of the geobiosphere. III Temperate and polar zonobiomes of northern Eurasia. Springer, Berlin Heidelberg New York, 581 pp

Walter H, Lieth H (1960–1967) Klimadiagramm-Weltatlas. Fischer, Jena

Wan C, Sosebee RE, McMichael BL (1993a) Broom snakeweed response to drought. I. Photosynthesis, conductance, and water use efficiency. J Range Managem 46:355–359

Wan C, Sosebee RE, McMichael BL (1993b) Broom snakeweed response to drought. II, Root growth, carbon allocation, and mortality. J Range Managem 46:360–363

Warwick NWM, Halloran GM (1991) Variation in salinity tolerance and ion uptake in accessions of brown beetle grass (*Diplachne fusca* (L.) Beauv.). New Phytol 119:161–168

Watt JM, Breyer-Brandwijk MG (1962) The medical and poisonous plants of southern and eastern Africa. 2nd edn. Livingstone, Edinburgh, 1457 pp

Weaver JE, Albertson FW (1956) Grasslands of the Great Plains; their nature and use. Johnsen, Lincoln, Nebraska, 395 pp

Weber DJ, Davis TD, McArthur ED, Sankhla N (1985) *Chrysothamnus nauseosus* (rubber rabbitbrush): multiple-use shrub of the desert. Desert Plants 7:172–180

Weiss RF (1988) Herbal medicine. Beaconsfield Publishers, Beaconsfield, 362 pp

Went FW (1971) Parallel evolution. Taxon 20:197–226

Went FW (1979) Germination and seedling behaviour of desert plants. In: Goodall DW, Perry RA (eds) Arid land ecosystems. International biological programme 16, 17, vol. 1. Cambridge University Press, Cambridge, pp 477–489

West NE (1988) Intermountain deserts, shrub steppes, and woodlands. In: Barbour MG, Billings WD (eds) North American terrestrial vegetation. Cambridge University Press, New York, pp 209–230

White F (1976) The underground forests of Africa: a preliminary review. Gard Bull (Singapore) 29:55–71

White F (1983) The vegetation of Africa. A descriptive memoir to accompany the UNESCO/AETFAT/UNSO vegetation map of Africa. Natural resources research 20. UNESCO, Paris, 356 pp

White F, Léonard J (1991) Phytogeographical links between Africa and south-west Asia. Flora Veg Mundi 9:229–246

Whitehead EE, Hutchinson CF, Timmermann BN, Varady RG (eds) (1988) Arid lands: today and tomorrow. Proceedings of an international research and development conference, Tucson, Arizona, Oct 20–25, 1985. Westview Press, Boulder; Belhaven Press, London, 1435 pp

WHO (1990a) Technological evaluation of certain food additives and contaminants. Food Addit Ser 26: 77–79

WHO (1990b) Evaluation of certain food additives and contaminants. WHO Tech Rep Ser 798: 24

Wickens GE (1969) A study of *Acacia albida* Del. Kew Bull 23:181–202

Wickens GE (1977a) Some of the phytogeographical problems associated with Egypt. Publ Cairo Univ Herb 7/8:223–230

Wickens GE (1977b) The flora of Jebel Marra (Sudan Republic) and its geographical affinities. Kew Bulletin additional series 5. HMSO, London, 368 pp
Wickens GE (1982a) Palaeobotanical speculations and Quaternary environments in the Sudan. In: Williams MAJ, Adamson DA (eds) A land between two Niles. Quaternary geology and biology of the central Sudan. Balkema, Rotterdam, pp 23–50
Wickens GE (1982b) The baobab – Africa's upside-down tree. Kew Bull 37:173–209
Wickens GE (1988) A preliminary ethnobotanical/agricultural survey of El Loa Province, Chile. Andean Project, London, 42 pp
Wickens GE (1990) What is economic botany? Econ Bot 44:12–28
Wickens GE (1992) Arid and semiarid ecosystems. In: Nierenberg WA (ed) Encyclopedia of earth system science, vol. 1. Academic Press, San Diego, pp 113–118
Wickens GE (1993) Vegetation and ethnobotany of the Atacama Desert and adjacent Andes in northern Chile. Opera Bot 121:291–307
Wickens GE (1995a) Llareta (*Azorella compacta*, Umbelliferae): a review. Econ Bot 49: 207–212
Wickens GE (1995b) Edible nuts. Non-wood forest products, vol 5. FAO, Rome, 198 pp
Wickens GE, Collier FW (1970) Some vegetation patterns in the Republic of the Sudan. Geoderma 6:43–59
Wickens GE, Storey INJ (1984) *Cordeauxia edulis*. SEPASAT dossier no 5. Royal Botanic Gardens, Kew
Williams JT, Farias RM (1972) Utilization and taxonomy of the desert grass *Panicum turgidum*. Econ Bot 26:13–20
Williams LO (1981) The useful plants of Central America. CEIBA 24:1–342
Williams OB (1982) The vegetation of present-day arid Australia: a biotic appraisal. In: Barker WR, Greenslade PJM (eds) Evolution of the flora of arid Australia. Peacock Publications, Frewville, South Australia, in association with Australian Systematics Botany Society and ANZAAS, South Australia Division Inc, pp 3–14
Willing T (1989) Tara nut (*Lemuropisum edule*, Caesalpinioideae) from south west Madagascar: a neglected potential crop for semi-arid areas. Royal Botanic Gardens, Kew, 29 pp
Wilson RC (1972) *Abronia*, I, distribution, ecology and habit of nine species of *Abronia* found in California. Aliso 7:421–437
Wilson RT (1978) The 'gizu'; winter grazing in the south Libyan Desert. J Arid Environ 1:327–344
Winkelman M (1991) Native Baja California plants: potential treatment for diabetes. Arid Land Plants 31:14–17
Winter K (1981) C_4 plants of high biomass in arid regions of Asia – occurrence of C_4 photosynthesis in Chenopodiaceae and Polygonaceae from the Middle East and USSR. Oecologia (Berl) 48:100–106
Witztum A, Gutterman Y, Evenari M (1969) Integumentary mucilage as an oxygen barrier during germination of *Blepharis persica* (Burm.) Kuntze. Bot Gaz 130(4):238–241
Wullstein LH (1991) Variation in N_2 fixation (C_2H_2 reduction) associated with rhizosheaths of Indian ricegrass (*Stipa hymenoides*). Am Midl Nat 126:76–81
Wullstein LH, Bruening ML, Bollen WB (1979) Nitrogen fixation associated with sandgrain root sheaths (rhizosheaths) of certain xeric grasses. Physiol Plant 46:1–4
Youngner VB, Naqvi HH, West J, Hashemi A (1986) *Parthenium* species of potential use in the improvement of guayule, *Parthenium argentatum*. J Arid Environ 11:97–102
Zin SR, Weiss RC (1980) La salud por medico de las plantas medicinales. Editorial Salesiana, Santiago, Chile, 387 pp
Zohary M (1940) On the "ghada"-tree of northern Arabia and the Syrian Desert. Palest J Bot Jerus Ser 1:413–416
Zohary M (1955) Geobotany. Sifriat Poalim, Tel Aviv, 591 pp (in Hebrew)

Zohary M (1962) Plant life of Palestine. Ronald Press, New York, 262 pp
Zohary M (1966, 1972) Flora Palaestina, 2 vols. The Israel Institute of Sciences and Humanities, Jerusalem, 364 + 489 pp
Zohary M (1973) Geobotanical foundations of the Middle East, 2 vols. Fischer, Stuttgart; Swets and Zeitlinger, Amsterdam, 739 pp

Taxonomic Indices

Families and genera of the vascular plants (Brummitt 1992) and authors of plant names (Brummitt and Powell 1992) are according to those recognised at the Royal Botanic Gardens, Kew. Accepted names are in **bold**, synonyms noted in the cited literature are in *italics*.

Botanical Index

Zoological Index

Subject Index

MIX
Papier aus verantwortungsvollen Quellen
Paper from responsible sources
FSC® C105338

If you have any concerns about our products,
you can contact us on
ProductSafety@springernature.com

In case Publisher is established outside the EU,
the EU authorized representative is:
Springer Nature Customer Service Center GmbH
Europaplatz 3, 69115 Heidelberg, Germany

Printed by Libri Plureos GmbH
in Hamburg, Germany